Palgrave Series in Indian Ocean World Studies

Series editor
Gwyn Campbell
McGill University
Montreal, Canada

This is the first scholarly series devoted to the study of the Indian Ocean world from early times to the present day. Encouraging interdisciplinarity, it incorporates and contributes to key debates in a number of areas including history, environmental studies, anthropology, sociology, political science, geography, economics, law, and labor and gender studies. Because it breaks from the restrictions imposed by country/regional studies and Eurocentric periodization, the series provides new frameworks through which to interpret past events, and new insights for present-day policymakers in key areas from labor relations and migration to diplomacy and trade.

More information about this series at
http://www.palgrave.com/gp/series/14661

Gwyn Campbell
Editor

Bondage and the Environment in the Indian Ocean World

palgrave
macmillan

Editor
Gwyn Campbell
Indian Ocean World Centre
McGill University
Montreal, Québec, Canada

Palgrave Series in Indian Ocean World Studies
ISBN 978-3-319-70027-4 ISBN 978-3-319-70028-1 (eBook)
https://doi.org/10.1007/978-3-319-70028-1

Library of Congress Control Number: 2017962757

Printed on acid-free paper

This Palgrave Macmillan imprint is published by Springer Nature
The registered company is Springer International Publishing AG
The registered company address is: Gewerbestrasse 11, 6330 Cham, Switzerland

Contents

Notes on Contributors

Subho Basu is currently an associate professor in the Department of History and Classics, McGill University. After completing his PhD at the University of Cambridge, UK, he worked as Smuts Hinduja Fellow at the Centre of South Asian Studies, University of Cambridge, and was elected a fellow of Wolfson College, Cambridge. In years following, he taught at the School of Oriental and African Studies, University of London, and Syracuse University. He has co-edited with Suranjan Das *Electoral Politics in South Asia*. Oxford University Press (SOAS Series) has published his monograph *Does Class Matter?* His other works include a co-edited volume with Crispin Bates, *Rethinking Indian Political Institutions*, recently published by Anthem Press, London. With Ali Riaz, Basu has co-authored *Paradise Lost: State Failure in Nepal*. Basu is currently working on two projects related to the construction of space, race, gender and nationalism and popular movements and the rise of Bengal Marxists. He has also authored numerous articles and book chapters on labour history.

Sravani Biswas completed her PhD degree at Syracuse University, New York, in 2016. She is currently a postdoctoral fellow at Nanjing Technology University, Singapore.

Gwyn Campbell is Canada Research Chair in Indian Ocean World History and founding Director of the Indian Ocean World Centre at McGill University. He is also General Editor of the Palgrave Series in Indian Ocean World Studies and Editor-in-Chief of the Journal of Indian Ocean World Studies (JIOWS). Born in Madagascar, he grew up in Wales, where he worked as a BBC radio producer in English and Welsh. He holds degrees in economic history from the Universities of Birmingham and Wales and has taught in India (Voluntary Service Overseas) as well as at universities in Madagascar, Britain, South Africa, Belgium, and France. He served as an academic consultant for the South African government in a series of inter-governmental meetings which led to the formation of an Indian Ocean regional association in 1997. He has published widely on Africa and the wider Indian Ocean world, including David Griffiths and the Missionary 'History of Madagascar' (Leiden: Brill, 2012), An Economic History of Imperial Madagascar, 1750–1895: The Rise and Fall of an Island Empire (Cambridge: Cambridge University Press, 2005), and (as editor) Africa and the Early Indian Ocean World Exchange to circa 1300 (London: Palgrave Macmillan, 2016) and The Structure of Slavery in Indian Ocean Africa and Asia (London: Routledge, 2004). In 2017 he received a Humboldt Award for his research and teaching in Indian Ocean world studies.

Ronald Kydd holds a PhD in Patristics from the University of St Andrews, Scotland and has taught at several colleges, seminaries and universities in Canada. He is currently Associate Professor of Church History at Tyndale Seminary in Toronto. He has also participated in conferences and consultations of the World Council of Churches in Spain, Syria, and Switzerland. His current research interests focus on early Christianity in Asia.

George Michael La Rue is Professor Emeritus of History at Clarion University of Pennsylvania. He did his doctoral field research in Dar Fur, Sudan, and currently works primarily on issues related to slavery in Sudan and Egypt. His recent publications include "Seeking Freedom in Multiple Contexts: An Enslaved Sudanese Woman's Life Trajectory, ca. 1800–1834," *Journal of Global Slavery* 2, no. 1–2 (2017): 11–43; "Land Documents in Dār Fūr Sultanate (Sudan, 1785–1875): Between Memory and Archives," *Afriques* (online), 2016, http://afriques.revues.org/1896; "Ismaÿl Urbain à la rencontre des Noirs d'Égypte: *liminalité* et empathie avec les esclaves dans la communauté saint-simonienne (1834–1836)," in *Les Saint-Simoniens dans l'Algérie du XIXe siècle: le combat du Français musulman Ismaÿl Urbain*, ed. Michel Levallois and Philippe Régnier (Paris: Riveneuve, 2016), 39–64; and "Treating Black Deaths in Egypt: Clot-Bey, African Slaves and the Plague Epidemic of 1834–35," in *Histories of Medicine and Healing in the Indian Ocean World*, ed. Anna Winterbottom and Facil Tesfaye (New York: Palgrave Macmillan, 2015).

Sue Peabody is Meyer Distinguished Professor of Liberal Arts and History at Washington State University, Vancouver, USA. Her book *Madeleine's Children: Family, Freedom, Secrets, and Lies in France's Indian Ocean Colonies* (Oxford University Press, 2017) explores the history of a family of slaves and free people in Bengal, France, Réunion, and Mauritius, 1750–1850.

Steven Serels is an associate research fellow at the Zentrum Moderner Orient, Berlin. He holds a master's degree (2007) and a PhD in history (2012), both from McGill University, and from 2012 until 2014 he was a postdoctoral fellow at Harvard University. His research focuses on the economic and environmental history of the southern Red Sea region. He is particularly interested in the history of poverty in the region. His first monograph is titled *Starvation and the State: Famine, Slavery and Power in Sudan 1883–1956* (2013).

Alessandro Stanziani is Professor in Global History at École des Hautes Études en Sciences Sociales (EHESS) and Research Director at *Centre national de la recherche scientifique* (CNRS). He is also the Director of the Institute of Global Studies, PSL University, Paris. He holds a PhD in economics from the University of Naples (1991) and a PhD in history from EHESS (1995). His more recent books include *Rules of Exchange: French Capitalism in Comparative Perspective, 18th–20th Centuries* (Cambridge: Cambridge University Press, 2012); *Bondage, Labor and Rights in Eurasia, 17th–20th Centuries* (New York: Berghahn, 2014); and *Sailors, Slaves, and Immigrants: Bondage in the Indian Ocean World, 1750–1914* (New York and London: Palgrave Macmillan, 2014).

Hideaki Suzuki is an associate professor at Nagasaki University, Japan. His interests include slavery and slave trade, the African diaspora and medieval Arab geography. He is the author of *Slave Trade Profiteers in the Western Indian Ocean* (Palgrave, 2017) and the editor of *Slavery as a Global Experience* (NUS Press, 2016). He has contributed to a number of journals and edited volumes.

James Francis Warren is Emeritus Professor of Southeast Asian Modern History at Murdoch University, Perth, Western Australia. He is an award-winning historian who has published numerous monographs, journal articles, and book chapters. His books include *Iranun and Balangingi: Globalization, Maritime Raiding and the Birth of Ethnicity* (2002); *The Sulu Zone 1768–1898: The Dynamics of External Trade, Slavery and Ethnicity in the Transformation of a Southeast Asian Maritime State*

(2007); and *Pirates, Prostitutes and Pullers: Explorations in the Ethno- and Social History of Southeast Asia* (2008).

Nigel Worden is Emeritus Professor of History at the University of Cape Town, South Africa. He has written extensively on the history of the early Cape Colony, focusing especially on slavery and the formation of social identities. His latest publications are *Cape Town between East and West: Social Identities in a Dutch Colonial Town* (2012) and *Honourable Intentions? Violence and Virtue in Australian and Cape Colonies, c 1750 to 1850* (2016), edited jointly with P. Russell.

List of Figures

List of Tables

CHAPTER 1

Introduction: Bondage and the Environment in the Indian Ocean World

Gwyn Campbell

The Context

The contributors to this volume examine the relationship between bondage and the environment in the Indian Ocean world (IOW). Although the term "Indian Ocean world" is becoming more widely used by scholars, it is sometimes used erroneously, so it is important to first define its usage here. The IOW refers to a vast macro-region, running from Africa (roughly east of a Cape to Cairo line) to the Far East. The rationale for considering this vast area as a "world" is the historical influence upon it of the monsoon system and related environmental factors. The monsoon system governs all IOW maritime spheres in the northern hemisphere and to about 12° south of the equator and much of their continental hinterlands. In the boreal summer, as the Asian continent warms, hot air rises and through a process of convection sucks in moist air from the seas to the south, creating the south-west monsoon. In winter, the reverse

I wish to acknowledge the assistance of Social Sciences and Humanities Research Council, Canada, and the Alexander von Humboldt Foundation in facilitating the research for, and writing of, this volume.

G. Campbell (✉)
McGill University, Montreal, QC, Canada

G. Campbell (ed.), *Bondage and the Environment in the Indian Ocean World*, Palgrave Series in Indian Ocean World Studies,
https://doi.org/10.1007/978-3-319-70028-1_1

process occurs, creating the south-east monsoon. Historical patterns of production and exchange have been heavily influenced, and to some degree still are, by the monsoons and associated systems of winds and currents that govern much of the region. Agricultural production, in which the vast bulk of the population was engaged, was largely shaped by the monsoon rains. The south-west monsoons bring rains that create the wet-crop (chiefly rice) cultivation zone of the littoral regions of the IOW, beyond which lie first the dry-crop (wheat, barley) cultivation zone, and beyond that generally arid pastoral regions. At the same time, the biannual changeover in monsoon winds and currents created the possibility for regular direct trans-oceanic exchange. Moreover, to the south of the monsoon system, south-east trade winds feed into the monsoon system. This permitted the development by the BCE/CE changeover of an IOW "global" economy—a sophisticated, regular system of long-distance exchange of commodities, peoples, ideas, and technologies across and beyond the IOW, albeit with considerable fluctuations, that has continued to the present day.[1]

It is also necessary to explain why the term "bondage" is used in the title of this volume. Systems of unfree labour existed throughout the IOW but varied greatly according to time and place. The term "bondage" is used here in preference to "slavery" because of the close association of the latter term to systems of servile labour in the ancient Mediterranean and early modern Atlantic worlds where "slavery" had a clear meaning. It there referred to an institution in which slaves formed a high proportion (20 per cent or more) of the total population, slave labour constituted the base of the economy, and slavery pervaded societal culture. Moreover, in such slave societies the term "slave" was also clear. It referred essentially to a chattel, owned by a master, which could be bought and sold like other commodities.

Such slave societies, defined by Moses Finley, were, according to Keith Hopkins, limited to only five locations in history: classical Athens, Roman Italy, and, in the age of the transatlantic slave trade, Brazil, the Caribbean, and the American South.[2] According to such definitions, there existed few

[1] See also Gwyn Campbell, "Africa and the Early Indian Ocean World Exchange System in the Context of Human-Environment Interaction," in *Africa and the Early Indian Ocean World Exchange to circa 1300*, ed. Gwyn Campbell (London: Palgrave Macmillan, 2016), 1–24.

[2] Moses Finley, *Ancient Slavery and Modern Ideology* (London: Chatto and Windus, 1980); Keith Hopkins, *Conquerors and Slaves: Sociological Studies in Roman History* (Cambridge: Cambridge University Press, 1978).

slave societies in the IOW outside plantation economies such as nineteenth-century Zanzibar and in the eighteenth and early nineteenth centuries on the Mascarene Islands of Mauritius and Réunion. Moreover, the identification of a slave as "chattel" in the terms defined by Finley and Hopkins, and followed by most historians of slavery in the Atlantic world, is a complex issue in non-European contexts. Chattel slaves could be found in European enclaves, notably the Mascarene Islands. However, chattel slavery represented one extreme of a wide spectrum of systems of unfree labour in the IOW, including slaves who occasionally, as in some agrestic societies in India, could be sold only with the land they lived on, and sometimes, as in the case of concubines in Muslim societies who had borne their master a child, could not be sold. In contrast to the Atlantic system, black Africans formed a minority of slaves in the IOW, constituting a majority only in Africa and, at certain periods, in other lands littoral to the western Indian Ocean. Slaves in the IOW comprised people of all ethnic and cultural backgrounds. In addition, the majority of those enslaved and trafficked were women and children, not men as in the transatlantic trade. Most of those subject to IOW bondage systems were accorded rights; some rose to positions of influence and wealth superior to that of nominally free peasants, a few became sovereigns and founded dynasties.[3]

A significant traffic in human beings had developed by at least 2000 BCE in the IOW. This trade experienced three major periods of demand-led expansion corresponding to sustained bursts of IOW-wide economic growth: from about 300 BCE to 300 CE, between the ninth and thirteenth centuries CE, and again from around 1830. In the intervening periods, which were generally characterized by economic stagnation or decline, human trafficking was dominated by supply factors.[4]

It is impossible in a brief volume to evaluate the historical relationship between bondage and the environment in all IOW societies and from early times to the present. The contributions here concentrate on

[3] Gwyn Campbell, "Introduction: Slavery and other forms of Unfree Labour in the Indian Ocean World," in *The Structure of Slavery in Indian Ocean Africa and Asia*, ed. Gwyn Campbell (London: Frank Cass, 2004), vii–xxxii.

[4] Jack Goody, "Slavery in Time and Space," in *Asian and African Systems of Slavery*, ed. James L. Watson (Berkeley: University of California Press, 1980), 18; André Wink, *Al-Hind: The Making of the Indo-Islamic World* (Leiden: Brill, 1996), 30–31; Angela Schottenhammer, "Slaves and Forms of Slavery in Late Imperial China (Seventeenth to Early Twentieth Centuries)," in *Structure of Slavery*, 143–54.

the period from the mid-eighteenth century. However, they clearly demonstrate that from early times the environment played a central, sometimes determining role in the processes by which people entered forms of bondage and human trafficking, although those forms varied according to time and place.

Environment and the IOW Global Economy

Environment critically influenced the rise and development of the IOW global economy. The term "environment" covers a wide range of factors. It can mean the natural environment at any given place and time. Thus Nigel Worden (in Chap. 5 of this volume) identifies for Cape Colony the creation of a slave "environment." He notes that Dutch East India Company (VOC) and white settler control of the Cape was effectively limited to the port, garrison, farmsteads and a few Company outposts. In a littoral colony backed by a vast hinterland, this opened up possibilities for evasion that were frequently taken by slaves. However, survival in the arid interior was difficult without help from the indigenous Khoi and San—who, initially at least, often proved hostile. Thus, slaves survived largely by learning how to exist in the interstices in the colonial landscape. It was there that they created an alternative slave environment, taking advantage of natural topography, notably Table Mountain, and raiding settler farms. Again, Hideaki Suzuki (in Chap. 9 of this volume) notes that one of the main reasons for the success of the East African slave trade, despite growing British anti-slave trade measures, was the knowledge that indigenous IOW slavers possessed of the regional climatic and marine environment. Suzuki's study focuses notably upon the area around the Lamu Archipelago, which was strategically placed for tapping slave sources on the East African coast to the south and running slaves to Arabia and the Persian Gulf to the north.

In more general terms, the environment can also refer to climatic change over the *longue durée*. In this sense, patterns emerge in the relationship between changes in the climate, the IOW global economy, and IOW structures of bondage. The foundations of the IOW global economy were laid by two main environmental factors. The first was global warming, which characterized the end of the last great Ice Age, some 10,000 years ago, and which facilitated a growth in human population and exploitation of regions of the world hitherto covered by glaciers. The second was aridification from around 6000 BCE. This promoted the domestication of plants and animals, techniques of water conservation and control, and the rise of sedentary agricultural communities whose productivity encouraged the production of agricultural surpluses, craft specialization, and long-distance exchange. This laid the basic structure for the development of the IOW global economy. However, it was the strengthening of the monsoon system from around 300 BCE, associated with global warming, that provided the catalyst for the emergence of a sophisticated and durable system of trans-IOW exchange.[5] Regular monsoon rains fundamentally shaped the "wet" (e.g. rice) and "dry" (e.g. wheat) cultivation regions of Asia. In addition, the monsoon winds, which blow in a regular alternating fashion from the south-west during the northern hemisphere summer and from the north-east during its winter, facilitated the rise of direct trans-oceanic sail in the IOW. This largely complemented terrestrial long-distance exchange, which also followed seasonal patterns and was largely restricted to the drier winter season.

The first major upsurge in the IOW global economy lasted until about 300 CE, following which it experienced a marked downturn until around 850 CE. This was due in part to continued global warming, which was in turn influenced by human activity: the boom from 300 BCE to 300 CE resulted from enhanced production, which emanated from, among other factors, significantly increased destruction of forest and burning of wood—which led to falling precipitation and, through rising gashouse emissions, global warming. Concentrations of human population in close proximity to restricted water sources and animals also facilitated the rise and spread of diseases. The net result was economic and demographic stagnation, accentuated by sulphur-rich volcanism and the first plague pandemic in the mid-first millennium, that only began to be corrected by about 800 CE. Between the ninth and thirteenth centuries, the IOW global economy experienced its second major boom, assisted by the onset of a warmer, wetter climate—the so-called "medieval warm period." However, it subsequently experienced another long period of stagnation that was critically affected by climatic factors—the "Little Ice Age" (LIA)—which peaked in the seven-

[5] John Gribben and H. H. Lamb, "Climatic Change in Historical Times," in *Climatic Change*, ed. John Gribben (Cambridge: Cambridge University Press, 1978), 70; Reid A. Bryson and Christine Padoch, "On the Climates of History," in *Climate and History*, ed. Robert I. Rotberg and Theodore K. Rabb (Princeton: Princeton University Press, 1981), 10–12.

teenth century and which forms the context for the argument I advance in Chap. 3 of this volume that, contrary to the currently dominant view, communities in highland Madagascar sought to retain rather than export servile labour and that because of environmental constraints, highland Madagascar could not have supplied from 1500 to 1750 thousands of slaves for the Swahili maritime slave trade on a regular annual basis.[6]

Most papers in this volume focus on the period between the mid-eighteenth and early twentieth centuries. The adverse climatic factors associated with the LIA continued to significantly influence human activity into the early nineteenth century. For example, the indications are that the western Himalayas and the region to their south experienced lower temperatures and possibly lower monsoon rainfall from 1778 to 1796, 1815 to 1817, and 1832 to 1835 as well as higher temperatures and precipitation from 1730 to 1777.[7] Drought and famine frequently followed periods of low monsoon rainfall, which were in turn often associated with El Niños from 1702, 1707 to 1709, 1727, and 1788 to 1796.[8] Following the drought of 1770 in Bengal and Bihar, the harvest failed, famine ensued, a smallpox epidemic erupted, and some 10 million people died, including one-third of the population of Bengal. Cultivation was abandoned, and brigandry became rife.[9] In the district of Purnia, in Bihar, half of the population perished in the famine of 1770, and many of the survivors fled to Nepal.[10] In 1783, famine affected almost all of India, and possibly 11 million died.[11] From 1788 to 1794, possibly 11 million people died in India, and many

[6] See Gwyn Campbell, "Environment and Enslavement in Highland Madagascar, 1500–1750: The Case for the Swahili Slave Export Trade Reassessed," Chap. 3 of this volume.

[7] H. P. Borgaonkar et al., "Assessment of Tree-Ring Analysis of High-Elevation *Cedrus deodara* D. Don from Western Himalaya (India) in Relation to Climate and Glacier Fluctuations," *Dendrochronologia* 27, no. 1 (2009): 64, 66.

[8] Richard Grove, "El Niño Chronology and the History of Socio-Economic and Agrarian Crisis in South and Southeast Asia 1250–1900," in *Land Use—Historical Perspectives: Focus on Indo-Gangetic Plains*, ed. Y. P. Abrol, Satpal Sangwan, and Mithilesh K. Tiwari (New Delhi: Allied Publishers, 2005), 139, 143–46, 148, 151, 157; Richard Grove, "The Great El Niño of 1789–93 and its Global Consequences: Reconstructing an Extreme Climate Event in World Environmental History," *Medieval History Journal* 10, no. 1–2 (2007): 76–78, 81.

[9] Vinita Damodaran, Rob Allen, and James Hamilton, "Climate Signals, Environment and Livelihoods in C17th India in a Comparative Context," http://foodsecurity.exeter.ac.uk/wp-content/uploads/2015/09/17th-century.pdf (accessed 30 August 2017).

[10] Grove, "The Great El Niño of 1789–93," 78.

[11] Grove, "The Great El Niño of 1789–93," 80.

villages were deserted,[12] while in 1792 alone, up to half of the population of the Northern Circars, in Madras Presidency, perished.[13] Cholera epidemics in 1781 (Madras), 1783 (Uttar Pradesh), and 1790 (Madras), as well as the first major cholera pandemic of 1817 (starting in Bengal), erupted during rains that followed extensive El Niño-induced drought.[14] Again, in the Middle East, environmental conditions worsened from about 1780,[15] provoking continued population decline until the early 1900s.

Indian Ocean Africa, including Madagascar, experienced increased aridity between the mid- to late seventeenth century and early eighteenth century that resulted in food shortages, periodic famine, a decline of concentrated settlements, and population dispersal. For instance, Egypt experienced long arid intervals from *ca.* 1625 to 1799,[16] most of which correlated with El Niño or Southern Oscillation (ENSO) events.[17] Droughts, harvest shortfalls, and major famines occurred in 1791–1792 and 1794–1796.[18] Long arid periods were punctuated by years of excessive rainfall and floods in 1743, 1745, 1801, 1818, 1822, and 1829, always causing damage and sometimes, as in 1817, considerable mortality among humans and animals.[19] Years of drought (and drought followed by excessive rainfall) also precipitated diseases that caused high mortality among humans and animals. The drought of 1718 was closely

[12] Grove, "The Great El Niño of 1789–93," 83.

[13] Grove, "The Great El Niño of 1789–93," 82.

[14] Grove, "El Niño Chronology," 167.

[15] Sam White, "Rethinking Disease in Ottoman History," *International Journal of Middle East Studies* 42, no. 4 (2010): 558–59; J. M. Russell and T. C. Johnson, "Little Ice Age Drought in Equatorial Africa: Intertropical Convergence Zone Migrations and El Niño–Southern Oscillation Variability," *Geology* 35, no. 1 (2007): 23.

[16] Grove, "The Great El Niño of 1789–93," 86; Alan Mikhail, "Unleashing the Beast: Animals, Energy, and the Economy of Labor in Ottoman Egypt," *American Historical Review* 118, no. 2 (2013): 332–34; Haim Tsoar, "Desertification in Northern Sinai in the Eighteenth Century," *Climatic Change* 29 (1995): 433–35.

[17] C. A. Spinage, *African Ecology: Benchmarks and Historical Perspectives* (Berlin: Springer, 2012), 82.

[18] Michael M. Santoro et al., "An Aggregated Climate Teleconnection Index Linked to Historical Egyptian Famines of the Last Thousand Years," *The Holocene* 25, no. 5 (2015): 873; Daniel A. Schiffman, "Rabbinical Perspectives on Money in Seventeenth-Century Ottoman Egypt," *European Journal of the History of Economic Thought* 17, no. 2 (2010): 182; Grove, "The Great El Niño of 1789–93," 86.

[19] Alan Mikhail, "An Irrigated Empire: The View from Ottoman Fayyum," *International Journal of Middle East Studies* 42, no. 4 (2010): 581; Spinage, *African Ecology*, 85–86.

followed by plague,[20] which had a major economic fallout. Alexandria alone experienced 59 plague epidemics between 1701 and 1844. The worst outbreaks were in 1701, 1736, 1759, 1785–1787—which decimated livestock in southern Egypt[21]—1790–1792, when an epizootic and plague epidemic killed large numbers of cattle and people, and 1799–1800.[22] These affected the entire Red Sea and eastern Mediterranean region.[23] Again, in 1801 and 1802, Jirja, in central Egypt, was hit by a plague epidemic and cattle epizootic that killed many people and animals.[24] Animals were also especially vulnerable to hail storms and bouts of extreme cold. Lightning in November 1804 and hail in November 1808 and February 1809 caused considerable damage and fatalities, particularly to cattle, while many humans and animals perished in the severe winter of 1813.[25] In addition, Egypt was afflicted by residual diseases. A number of lethal parasitic diseases plagued the Nile, while gastrointestinal and pulmonary infections appear to have been responsible for most deaths, notably of infants, in Cairo and Alexandria.[26] Peasants were also commonly afflicted with inflated legs—probably elephantiasis (*Lymphatic filariasis*), caused by parasitic worms spread by mosquitoes, and Guinea worm infestation.[27]

In 1786, in the middle of a major environmental crisis, Ottoman forces invaded in an attempt to suppress the rebel Mamluks (Beys) but succeeded only in pushing them into Upper Egypt and exciting popular revolt. Most of the Ottoman troops evacuated Egypt in 1787.[28] A prolonged period of drought from 1794 also provoked popular uprisings[29] and a collapse of

[20] Schiffman, "Rabbinical Perspectives on Money," 182; Spinage, *African Ecology*, 85–86.

[21] Grove, "The Great El Niño of 1789–93," 86; Mikhail, "Unleashing the Beast," 331–32.

[22] Michael J. Reimer, "Ottoman Alexandria: The Paradox of Decline and the Reconfiguration of Power in Eighteenth-Century Arab Provinces," *Journal of the Economic and Social History of the Orient* 37, no. 2 (1994): 113; Mikhail, "Unleashing the Beast," 332–34; Grove, "The Great El Niño of 1789–93," 86.

[23] Mikhail, "Unleashing the Beast," 333; Reimer, "Ottoman Alexandria," 127–28.

[24] Mikhail, "Unleashing the Beast," 335.

[25] Mikhail, "Unleashing the Beast," 335.

[26] White, "Rethinking Disease in Ottoman History," 557.

[27] Gary Leiser and Michael Dols, "Evliyā Chelebi's Description of Medicine in Seventeenth-century Egypt: Part I: Introduction," *Sudhoffs Archiv* 71, no. 2 (1987): 204–05; Richard Pankhurst, "The Medical Activities in Eighteenth Century Ethiopia of James Bruce the Explorer," *Medizinhistorisches Journal* 17, no. 3 (1982): 274–75.

[28] Reimer, "Ottoman Alexandria," 137.

[29] Grove, "The Great El Niño of 1789–93," 86.

central authority and general security, causing theft and brigandry to become rampant.[30] This opened the way for the French invasion and occupation from 1798 to 1801, which accentuated the general crisis, and during which, from 1799 to 1800, another famine and severe plague epidemics occurred.[31]

Ethiopia experienced dry conditions from about 1675, with droughts probably in 1694, 1715, and 1716, followed by periods of good rainfall, notably from 1737 to 1749 and 1755 to 1759, interrupted by a drought in 1750.[32] A further long arid period started, possibly in 1781, certainly by 1790,[33] that continued to about 1840,[34] with major droughts in 1800 and 1826–1827,[35] interrupted by years of exceptionally high rainfall in 1801, 1817, 1818, 1822, and 1829.[36] From around 1700, Ethiopia experienced intensified deforestation and soil erosion.[37] In 1769, a smallpox epidemic, which had already killed over 1,000 people in the neighbouring mainland port of Arkiko, hit Massawa, an island port,[38] which like all lowland Ethiopia was subject to endemic malaria.[39] From 1768 to 1769, the epidemic also hit highland Ethiopia,[40] where elephantiasis and dysentry were

[30] Mikhail, "Unleashing the Beast," 333–34.

[31] Reimer, "Ottoman Alexandria," 137–39; Mikhail, "Unleashing the Beast," 334.

[32] Iain Darbyshire, Henry Lamb, and Mohammed Umer, "Forest Clearance and Regrowth in Northern Ethiopia during the Last 3000 years," *The Holocene* 13, no. 4 (2003): 537–46; Henry F. Lamb et al., "Oxygen and Carbon Isotope Composition of Authigenic Carbonate from an Ethiopian Lake: A Climate Record of the Last 2000 Years," *The Holocene* 17, no. 4 (2007): 517–26; Hubert H. Lamb, *Climate: Present, Past and Future*, vol. 2 (London: Routledge, 2011), 470, footnote 2; Hubert H. Lamb, *Climate, History and the Modern World* (London: Routledge, 1995), 236; Spinage, *African Ecology*, 82; E. Aynekulu et al., "Dieback Affects Forest Structure in a Dry Afromontane Forest in Northern *Ethiopia*," *Journal of Arid Environments* 75, no. 5 (2011): 499; Spinage, *African Ecology*, 88.

[33] Dagnachew Legesse et al., "Environmental Changes in a Tropical Lake (Lake Abiyata, Ethiopia) during Recent Centuries," *Palaeography, Palaeoclimatology, Palaeoecology* 187, no. 3–4 (2002): 233; Spinage, *African Ecology*, 82, 85–86.

[34] Grove, "The Great El Niño of 1789–93," 87; Spinage, *African Ecology*, 82.

[35] Legesse et al., "Environmental Changes in a Tropical Lake," 254; David M. Anderson, "The Beginning of Time? Evidence for Catastrophic Drought in Baringo in the Early Nineteenth Century," *Journal of Eastern African Studies* 10, no. 1 (2016): 49.

[36] Spinage, *African Ecology*, 81, 85–86.

[37] Darbyshire, Lamb, and Umer, "Forest Clearance and Regrowth in Northern Ethiopia," 537–46.

[38] James Bruce, quoted in Pankhurst, "Medical Activities," 258.

[39] James Bruce, quoted in Pankhurst, "Medical Activities," 259–60, 269–71.

[40] Pankhurst, "Medical Activities," 261–62, 265–66.

also present.[41] Ethiopia was in 1800 and 1826–1827 again afflicted by droughts that were followed by famine.[42]

The southern sector of Indian Ocean Africa experienced a long period of significant rainfall until about 1750, although there were intervals of drought, probably from 1725 to 1729.[43] From the mid-eighteenth century, the climate abruptly changed and the entire equatorial region was until 1840 affected by major drought, notably from 1749 to 1755 (probably), from 1761 to 1769, from *ca.* 1785 to 1792, and especially from 1808 to 1826 (the worst for seventy years).[44] This region of the IOW was also characterized by famine, cattle epidemics, high human and animal mortality, mass migration, and depopulation.[45] Moreover, intervening regions and those to the north of the equatorial zone continued to experience a long cool and arid period that lasted into the nineteenth century.[46] Lake Malawi became progressively drier up to 1780. Lake Tanganyika levels were generally low between 1550 and 1800, while Lake Bosumtwi experienced dramatically low levels until 1850.[47] In south-eastern Africa, Mozambique also continued to suffer dry arid conditions, with severe droughts from *ca.* 1580 until the 1810s, notably in *ca.* 1700 and in the 1790s.[48] Natal experienced good rainfall from 1786 to 1796 and 1812 to

[41] Pankhurst, "Medical Activities," 272–73.

[42] Legesse et al., "Environmental Changes in a Tropical Lake," 254; Anderson, "Beginning of Time?," 49.

[43] Anderson, "Beginning of Time?," 47; Russell and Johnson, "Little Ice Age Drought," 21–24; Spinage, *African Ecology*, 88.

[44] Anderson, "Beginning of Time?," 47–48, 51; Massimiliano Ghinassi et al., "Shoreline Fluctuations of Lake Hayk (Northern Ethiopia) during the Last 3500 Years: Geomorphological, Sedimentary, and Isotope Records," *Palaeogeography, Palaeoclimatology, Palaeoecology* 365–366 (2012): 209–10; Spinage, *African Ecology*, 88.

[45] Anderson, "Beginning of Time?," 51, 58; Gayle McGlynn et al., "Palaeolimnological Evidence of Environmental Change over the Last 400 Years in the Rwenzori Mountains of Uganda," *Hydrobiolgia* 648, no. 1 (2010): 120.

[46] Sharon E. Nicholson et al., "Temperature Variability over Africa during the Last 2000 Years," *The Holocene* 23, no. 8 (2013): 1085–94; Richard Seager, "Blueprints for Medieval Hydroclimate," 9, http://www.ldeo.columbia.edu/res/div/ocp/pub/seager/Seager_etal_QSR.pdf (accessed 28 August 2017).

[47] Russell and Johnson, "Little Ice Age Drought," 22.

[48] Anneli Ekblom, "Forest-Savanna Dynamics in the Coastal Lowland of Southern Mozambique since c. AD 1400," *The Holocene* 18, no. 8 (2008): 1247–57; Isla S. Castañeda, Josef P. Werne, and Thomas C. Johnson, "Wet and Arid Phases in the Southeast African Tropics since the Last Glacial Maximum," *Geology* 35, no. 9 (2007): 825; Dirk Verschuren, "Decadal and Century-Scale Climate Variability in Tropical Africa during the Past

1819, but arid conditions from 1675 to 1770 and after 1810, with prolonged drought in 1789 and from 1799—when there was severe famine in Natal and Zululand—to the 1810s.[49] Southern Africa generally experienced a cold and dry climate from 1690 to 1740, a wetter period until the 1790s, and then another cold arid interval until about 1830, with extreme droughts in *ca.* 1700.[50]

The coral luminescence record confirms that from 1500 to 1800, north-east Madagascar experienced a significantly wetter climate than today,[51] while studies for eastern Madagascar generally indicate periods of relatively good rainfall from 1617 to 1724, 1746 to 1762, and after 1771, broken by arid episodes from 1724 to 1746 and 1785 to 1799.[52] North-west Madagascar was affected by notable wet periods in 1725 and 1790 and dry intervals centred at around 1700, 1709 (probably),[53] and 1770. From 1700 to 1750, it was affected by possibly the highest frequency of El Niño events since 1550.[54] The central plateau experienced intense drought in the first two decades of the eighteenth century and from the 1790s to 1810s.[55] There are references in oral traditions from

2000 Years," in *Past Climate Variability through Europe and Africa*, ed. R. W. Battarbee et al. (Dordrecht: Kluwer, 2004), 153.

[49] Grove, "The Great El Niño of 1789–93," 87; Spinage, *African Ecology*, 186–88; Verschuren, "Decadal and Century-Scale Climate Variability," 153.

[50] Verschuren, "Decadal and Century-Scale Climate Variability," 153; P. D. Tyson, W. Karien, K. Holmgre,n and G. A. Heiss, "The Little Ice Age and Medieval Warming in South Africa," *South African Journal of Science* 96, no. 3 (2000): 121; Sharon E. Nicholson, "The Historical Climatology of Africa," in *Climate and History: Studies in Past Climates and their Impact on Man*, ed. T. M. L. Wigley et al. (Cambridge: Cambridge University Press, 1981), 257.

[51] Craig Alexander Grove, "Madagascar's Climate History Unlocked by Giant Corals" (PhD diss., Vrije Universiteit, Amsterdam, 2012), 12–13.

[52] Glen M. MacDonald and Roslyn A. Case, "Variations in the Pacific Decadal Oscillation over the Past Millennium," *Geophysical Research Letters* 32, no. 8 (2005): L08703; Rosanne d'Arrigo and Rob Wilson, "On the Asian Expression of the PDO," *International Journal of Climatology* 26, no. 12 (2006): 1607–17.

[53] Jacques Galas, J.-P. Locci, René Grosso, and Sylvestre Clap, *Histoire de Vaucluse* (Avignon: A. Barthélemy, 1993), 10.

[54] George A. Brook et al., "A High-Resolution Proxy Record of Rainfall and ENSO since AD 1550 from Layering in Stalagmites from Anjohibe Cave, Madagascar," *The Holocene* 9, no. 6 (1999): 695–705.

[55] F. Gasse and E. van Campo, "A 40,000-Yr Pollen and Diatom Record from Lake Tritrivakely, Madagascar, in the Southern Tropics," *Quaternary Research* 49, no. 3 (1998): 307; Verschuren, "Decadal and Century-Scale Climate Variability," 153.

the area of present-day Imerina to three severe famines. The first, the *Tsimiofy* (lit. "do not peel"), which occurred during the reign of Andriamasinavalona (r. 1675–1710), in which many perished, was allegedly caused by violent winds that destroyed the rice crop. It is possible that the *Tsimiofy* famine occurred from *ca.* 1696 to *ca.* 1703,[56] the *Mavovava* ("yellow mouth") in *ca.* 1747, and another, the "younger famine" (which caused less deaths than the *Mavovava*), possibly from 1755 to 1756.[57] Smallpox also affected the highlands following increased exposure to foreign trade from the late eighteenth century, the first major epidemic occurring some time between 1791 and 1794, with others erupting in 1817 and 1833–1835.[58] South-west and central west coastal Madagascar probably experienced generally dry arid conditions, with severe droughts, from *ca.* 1580 to the 1810s.[59]

In the early to mid-nineteenth century, the IOW economy experienced dramatic change with the rise of the international economy marked by the Industrial Revolution, modern capitalist expansion, and a burgeoning demand for tropical and semi-tropical products. This process was assisted from the mid-nineteenth century by more favourable (i.e. warmer and wetter) climatic conditions. However, increased warfare, displacement of peoples, and dramatic improvements in transport—notably the advent of steamships—all contributed in the IOW to disease and high mortality that persisted well into the twentieth century.[60]

[56] Galas, *Histoire*, 10.

[57] [François] Callet, *Histoire des rois: Tantaran ny Andriana*, trans. George S. Chapus and Emmanuel Ratsimba (Tananarive: Éditions de la Librairie de Madagascar, 1974), 404; see also H. C. V. Leibbrandt, *Precis of the Archives of the Cape of Good Hope: Journal, 1699–1732* (Cape Town: W. A. Richardson, 1896), 302; Spinage, *African Ecology*, 188, table 4.1; David Ernest Hutchins, *Cycles of Drought and Good Seasons in South Africa* (Wynberg: "Times" Office, 1889), 69.

[58] Callet, *Histoire des rois*, 505, 522; Charles Theodore Hilsenberg and Wenceslaus Bojer, "A Sketch of the Province of Emerina, in the Island of Madagascar, and of the Huwa, its Inhabitants; Written during a Year's Residence," in *Botanical Miscellany*, vol. 3, ed. William Jackson Hooker (London: John Murray, 1833), 261; Freeman to Ellis, Tamatave, 3 October 1833 and Vohitsara, 11 October 1833, School of Oriental & African Studies, London, London Missionary Society Archives, Madagascar Incoming Letters, B4.F4.JC.

[59] Ekblom, "Forest-Savanna Dynamics," 1247–57; Castañeda, Werne, and Johnson, "Wet and Arid Phases," 825; Verschuren, "Decadal and Century-Scale Climate Variability," 153.

[60] Greg Bankoff and Joseph Christensen, eds., *Natural Hazards and Peoples in the Indian Ocean World* (New York: Palgrave Macmillan, 2016).

Environment, the IOW Global Economy, and Bondage

The long-term cycles in the IOW global economy, in which environmental factors and human activity were intricately intermeshed, also critically influenced structures of servile labour. Human bondage first developed as a significant social institution during the Neolithic Revolution in the centralized hierarchical polities of the Middle East, South Asia, and the Far East. This was first apparent in ancient Mesopotamia, where aridification prompted the application of methods designed to collect, store, and distribute water that was essential to crop cultivation. Such methods necessitated heavy and continuous investments of labour, which the elite secured through geographically concentrating their subject population and directing their labour. Such bonded subjects enabled the production of agricultural surpluses sufficient to maintain the ruling elite and, eventually, specialist artisan and merchant classes. Agricultural surpluses and the imposition of forced labour also permitted the formation of standing armies used by the elite to expand territorially. When successful, these armies generally chose to kill rather than enslave male opponents because they were likely to resist and revolt, thus requiring heavy surveillance and, whenever necessary, physical restraint. By contrast, victorious generals enslaved captive women and children, who were easier to control and more likely than adult men to adopt, and adapt to, the language, culture, and mores of the victor society.[61] Slave-raiding quickly became a major motive of such campaigns, and a class of specialist slave-raiders emerged as a result.[62]

Women and child captives were taken back to the imperial centre, where they were distributed among court members and military officers or sold to other members of the elite. This is the reputed origin of bondage as a significant institution in Mesopotamia in the third millennium BCE.[63] As bonded peasants already provided most menial productive labour, the majority of the enslaved were absorbed into court or elite

[61] Goody, "Slavery in Time and Space," 32–34; Gwyn Campbell, "The State and Pre-Colonial Demographic History: The Case of Nineteenth Century Madagascar," *Journal of African History*, 31, no. 3 (1991): 415–45.

[62] For example, see Bok-rae Kim, "Debt Slaves in Old Korea," in *Bonded Labour and Debt in the Indian Ocean World*, ed. Gwyn Campbell and Alessandro Stanziani (London: Routledge, 2013), 165–72.

[63] Goody, "Slavery in Time and Space," 18.

households.[64] Probably some two-thirds of these bondspeople were female.[65] As their numbers grew, the bulk of the enslaved in elite households were considered as objects of conspicuous consumption, displayed on private occasions to guests and visitors and on public occasions to reflect the wealth and power of the owner.

A second early form of entry into bondage occurred when conquering armies overran settled peasant communities from which they met no resistance. As such communities were agriculturally productive, they were generally maintained in a form of bondage that tied them to the land they occupied and worked. The conquered land was distributed by the court to leading supporters, who thus also governed and controlled the bondspeople on it and the product of their labour. This is most clearly seen in agrestic forms of servitude that developed in south India. However, it is also reflected in forms of bondage, such as that which developed after the Persian conquest of Babylonia in 539 BCE, when Persians forced servile Babylonians to help construct the palace of Darius I (r. 521–486 BCE) in Susa,[66] and in Imerina (Madagascar) in the late eighteenth century, where warlords tied newly subjected peoples to the land and organized them into between 100 and 1,000 strong territorially based units that they used in unremunerated forced labour (*fanompoana*) for public works.[67]

Over the *longue durée*, probably the largest number of people entered bondage as a result of debt. Indebtedness is a universal phenomenon that directly imposes obligations upon the debtor to the creditor that, when formally witnessed and agreed to, have the sanction of customary law even where this differs from statutory law.[68] This could and did occur in times of both economic expansion and recession, although the phenomenon of bondage as a result of debt was probably greater during periods of economic downturn. Traditionally, enslavement was legally enforced for debtors and their relatives in many IOW regions. In Imperial Madagascar, for

[64] For example, see Muhammad A. Dandamaev, *Slavery in Babylonia* (DeKalb, Illinois: Northern Illinois University Press, 2009), 560; C. Martin Wilbur, *Slavery in China during the Former Han Dynasty 206 BC–AD 25* (New York: Russell & Russell, 1943), 115.

[65] Gerda Lerner, "Women and Slavery," *Slavery & Abolition* 4, no. 3 (1983): 179.

[66] Dandamaev, *Slavery in Babylonia*, 560.

[67] Gwyn Campbell, "Introduction: Slavery and other forms of Unfree Labour in the Indian Ocean World," in *The Structure of Slavery in Indian Ocean Africa and Asia*, ed. Gwyn Campbell (London: Frank Cass, 2004), vii–xxxii.

[68] See William Gervase Clarence-Smith, "Debt and Coercion of Labour in Islamic Legal Tradition," in *Bonded Labour and Debt*, 21–30; Gwyn Campbell, "Debt and Slavery in Imperial Madagascar, 1790–1861," in *Bonded Labour and Debt*, 45–55.

instance, creditors could, through the application of law, enslave a debtor, his wife, and his children.[69] This was also the customary practice in Thailand and Malaya.[70]

High legal interest rates—50 per cent or more in late Chosun Korea (1392–1910)—caused many debtors to default on money or goods borrowed.[71] Indeed, because interest rates were so exorbitant, those entering debt relations often knew that the likely outcome of their action would be enslavement—to the degree that for some it was tantamount to voluntary entry into bondage.[72] Enslavement removed debtors from the proportion of the population subject to taxation by the state. In that sense, the deliberate use by individuals of extortionate interest rates and other mechanisms to induce debt, and thus enslavement of others, could be viewed as an economic challenge to state authorities. In this context, it is interesting that certain states indigenous to the IOW introduced measures forbidding the enslavement of free subjects for debt. This was the case in all Muslim societies governed by the Sharia, but also occurred elsewhere, as in Korea under the Chosun dynasty in 1392.[73]

Although those falling into debt were overwhelmingly adult males, the majority of those enslaved as a result of male indebtedness were, generally by choice of the male debtor, women and girls.[74] This was because they were often both the most vulnerable members of a household, extended family, or community and the most in demand as slaves. In particularly dire times, sons were also sold, although in China and Korea every attempt was made to spare the eldest son in order that he might perform the ancestral rites and offer a hope for the family line to continue.[75] Indebtedness

[69] Gwyn Campbell, *An Economic History of Imperial Madagascar, 1750–1895: The Rise and Fall of an Island Empire* (Cambridge: Cambridge University Press, 2005), 295–96.

[70] Bruno Lasker, *Human Bondage in Southeast Asia* (Chapel Hill: University of North Carolina Press, 1950), 147, 150.

[71] See Kim, "Debt Slaves in Old Korea."

[72] Kim, "Debt Slaves in Old Korea."

[73] Kim, "Debt Slaves in Old Korea"; see also Paul E. Lovejoy, "Pawnship and Seizure for Debt in the Process of Enslavement in West Africa," in *Debt and Slavery in the Mediterranean and Atlantic Worlds*, ed. Gwyn Campbell and Alessando Stanziani (London: Pickering & Chatto, 2013), 63–75.

[74] Kim, "Debt Slaves in Old Korea"; Campbell, "Debt and Slavery in Imperial Madagascar, 1790–1861."

[75] James Watson, "Transactions in People: The Chinese Market in Slaves, Servants, and Heirs," in idem. ed. *Asian and African Systems of Slavery*, 227–30; Kim, "Debt Slaves in Old Korea."

was normally expressed in monetary terms, although it was often incurred in non-cash forms such as food or tools. If the debt was paid off, an enslaved debtor could regain non-slave status.[76] In some regions, such as Korea, enslavement of local people for debt ensured for them permanent and inherited slave status, whereas in others, such as Imerina (Madagascar), enslaved Merina subjects (but not non-Merina) could be redeemed or redeem themselves.[77]

Here, slavery needs to be distinguished from debt-bondage (although the two could certainly overlap). Enslavement for indebtedness was involuntary and mostly independent of familial relations, whereas in most cases people were pushed into debt-bondage by their relatives, chiefly the elder males, as a credit-securing strategy. Mortgaging a child or wife to raise a loan was common practice in the IOW from early times; by the nineteenth century debt-bondage embraced a vast range of people, and it continued into the twentieth century.[78] Certainly indebtedness and debt-bondage are frequently linked to environmental factors. Harvests were often used as collateral on loans used to finance weddings and funerals, so that in times of lean harvests that resulted from intemperate weather, male peasants often incurred debt, which in turn frequently resulted in the enslavement of a family member, usually a girl. During catastrophes, people often entered debt-bondage or slavery in return for subsistence as a survival strategy, either voluntarily, as was the case of many *dvija* caste members in India, or propelled by their kin group.[79]

[76] Anthony Reid, "Introduction: Slavery and Bondage in Southeast Asian History," in idem, ed., *Slavery, Bondage and Dependency in Southeast Asia* (St Lucia: University of Queensland Press, 1983), 10; Peter Boomgaard, "Human Capital, Slavery and Low Rates of Economic and Population Growth in Indonesia, 1600–1910," in *Structure of Slavery*, 83–96.

[77] Kim, "Debt Slaves in Old Korea"; Campbell, "Debt and Slavery in Imperial Madagascar, 1790–1861."

[78] Edward M. Harris, "Did Solon Abolish Debt-Bondage?" *Classical Quarterly* 52, no. 2 (2002): 418; Boomgaard, "Human Capital"; Karine Delaye, "Slavery and Colonial Representations in Indochina from the Second Half of the Nineteenth to the Early Twentieth Centuries," in *Structure of Slavery*, 129–42; see also Watson, "Transactions in People," 228–36.

[79] Martin A. Klein, "Introduction: Modern European Expansion and Traditional Servitude in Africa and Asia," in idem, ed. *Breaking the Chains: Slavery, Bondage and Emancipation in Modern Africa and Asia* (Madison, Wisconsin: University of Wisconsin Press, 1993), 11; Utsa Patnaik and Manjari Dingwaney, eds., *Chains of Servitude: Bondage and Slavery in India* (Hyderabad: Sangam Books, 1985), 25–26.

Those subject to debt-bondage could outnumber slaves. For example, they were possibly the most numerous social category in Majapahit, in Java, while in central Thailand in the eighteenth and nineteenth centuries they formed up to 50 per cent of the total population. Again, debt-bondspeople could sometimes be exchanged, as could other servile people as part of a marriage dowry or a monastery donation.[80] The servitude to which those in debt-bondage were subject was generally taken as paying off interest on the loan they had contracted, to which was added the cost of lodging, feeding, and clothing the debtor. Consequently, the debt in most cases increased and servitude could become permanent, even hereditary, at which point there was little to distinguish debt-bondage from slavery.[81] Indeed, in nineteenth-century Thailand, where up to 50 per cent of the population of the central provinces were trapped in debt-bondage, if the interest owed by a debt-bondsman grew to a level exceeding the original loan, the master deemed him a bad investment and customarily sold him, albeit on unfavourable terms.[82]

A similar relationship existed between environmental factors and forms of debt and pawnship that were widespread in the IOW. In ancient Mesopotamia, indebted men could either sell their dependents (preferably female slaves, concubines, daughters, or wives) as slaves, or pawn them, in order to pay off the debt. In the case of pawns, should the debt not be repaid within a stipulated time, they generally became the slaves of their new master.[83] In medieval Japan, creditors were legally permitted to seize the children of debtors until such time as the debt had been paid off; and should the debt not be paid, the creditors had recourse to local authorities who could transform the pawn into a permanent slave.[84] The general term appears to have been ten years.[85] Pawnship was also common in eastern Africa, where a girl was frequently given as security for a loan. There, the creditor did not have the right to sell the pawn, whose labour he could use until the moment the family paid off the debt. However, in cases where an

[80] Boomgaard, "Human Capital."

[81] Matthew S. Hopper, "Debt and Slavery among Arabian Gulf Pearl Divers," in *Bonded Labour and Debt*, 103–18; Bok-rae Kim, "*Nobi*: A Korean System of Slavery," in *Structure of Slavery*, 155–68; Reid, "Introduction," 12.

[82] Lasker, *Human Bondage in Southeast Asia*, 151; Reid, "Introduction," 12.

[83] Lerner, "Women and Slavery," 184–85.

[84] Thomas Nelson, "Slavery in Medieval Japan," *Monumenta Nipponica* 59, no. 4 (2004), 475–76.

[85] Nelson, "Slavery in Medieval Japan," 485.

inordinate amount of time passed with little or no sign or repayment, the creditor's ownership of the pawn became established.[86] As Edward Alpers has noted, from the late eighteenth century, a rapidly growing slave export trade substantially altered the mechanisms of pawnship in East Africa, ultimately transforming it into an institution for the creation of slaves.[87] Thus the links between debt, pawnship, and enslavement were strongest along the littoral and the main trade routes into the interior. However, supply factors, notably famine, also played a significant role in inducing household heads to "sell" kin, chiefly girls and women, into pawnship.[88] This marked East Africa apart from sub-Saharan West Africa, where pawns could not be reduced to slave status—although if a female pawn gave birth to a child by her "master," that child was considered to be his slave—while on the Muslim–non-Muslim commercial frontier, cultural "misunderstandings" often led to pawns being carried as slaves to Muslim markets.[89]

Here, however, a distinction needs to be drawn between Muslim and non-Muslim societies. This is of major significance in the IOW, where, between the seventh and thirteenth centuries, Islam spread from its Middle Eastern heartland (*Dar al-Islam*), southwards down the East African littoral, and eastwards through South to South-East Asia. Initially, a *hadith* (tradition) relating the sale by the Prophet of a debtor was used to justify enslaving insolvent debtors, but given its uncertain legitimacy, the *hadith* was declared invalid from *ca.* 800 CE. As, according to the Sharia, free-born people could legally neither be pawned nor enslaved for debt, this generally also held true in practice in the Islamic heartland of the Middle East.[90] However, the distinction between indebtedness, debt-bondage, and slavery was often blurred, a notable example being in the Gulf pearl diving industry of the late nineteenth and early twentieth century.[91] Under the Sharia, enslavement for debt was also forbidden for *dhimmi* (non-Muslim subjects in an Islamic state) Jews, Christians, and Zoroastrians, who, as "people of the book" (possessing scriptural traditions similar to

[86] Jan-George Deutsch, "Notes on the Rise of Slavery and Social Change in Unyamwezi, c. 1860–1900," in *Slavery in the Great Lake Region of East Africa*, ed. Henri Médard and Shane Doyle (Oxford: James Currey, 2007), 89.

[87] Edward A. Alpers, "Pawnship and Slavery in Nineteenth-Century East Africa," in *Bonded Labour and Debt*, 31–43.

[88] Alpers, "Pawnship and Slavery," 31–43.

[89] Lovejoy, "Pawnship and Seizure for Debt."

[90] Clarence-Smith, "Debt and Coercion of Labour in Islamic Legal Tradition."

[91] Hopper, "Debt and Slavery among Arabian Gulf Pearl Divers."

the Quran), were granted protected status under Muslim rulers. However, Muslim societies were highly commercial and inevitably produced debtors who, in many instances because of gaps in the law, or in Muslim societies outside the Middle East where elements of pre-Islamic customary law were upheld, were often ensnared in forms of debt-bondage and slavery. Also, there was no firm legal ruling on the purchase of people enslaved for debt outside Muslim and *dhimmi* communities.[92]

Environment and Bondage in the Context of the Rise of the International Economy

In the eighteenth and early nineteenth centuries, the grip of the LIA slowly began to diminish. However, the process was uneven, being characterized by considerable climatic fluctuations that continued to critically influence living standards and life expectations, and thus impoverishment, indebtedness, and bondage. In the northern hemisphere, temperatures began to fall beginning in 1770, with the 1805–1820 period being the coldest on record in Europe, North America, and Japan. It is in this context that James Warren (in Chap. 4 of this volume) analyses the effects of the eruption of Mount Macaturin in 1765 on the slaving structure in the Philippine region. The eruption caused high mortality and destroyed the economy of south-west Mindanao, forcing thousands of desperate Iranun to flee to other regions in the Indonesian archipelago. However, their entrepreneurial spirit became manifest as, by the 1780s, they emerged as paramount slave traders—an activity that they pursued with vigour into the early nineteenth century as regional maritime trade intensified in the context of a rapidly expanding international economy. Again, Sue Peabody (in Chap. 6 of this volume) examines the impact in the late eighteenth and early nineteenth centuries of a particularly severe series of natural disasters, notably cyclones, drought, and a volcanic eruption, on the French slave colony of Bourbon (present-day Réunion). Through a painstaking investigation of the history of members of an enslaved family on the Routier estate, she reveals in a highly nuanced manner the impact of such disasters not only on the economy of the island, which fuelled demand for slaves well into the era when the slave trade was forbidden, but also on the fortunes of the Routiers and, most significantly, on individual slave members of the Routier household.

[92] Clarence-Smith, "Debt and Coercion of Labour in Islamic Legal Tradition."

From 1825, temperatures across the IOW started to increase, but the Far East was affected by low temperatures from 1870 to 1900, with exceptionally cold weather hitting south China from 1876 to 1895.[93] Moreover, between 1830 and 1900 the southern hemisphere was in the grip of a colder climate. Longer-term climatic change was complicated by shorter-term factors. Strong ENSO effects, associated with severe droughts followed in consecutive years by unusually heavy rain, were experienced in 1844–1846, 1876–1878, and 1899–1900. Again, high volcanic dust veil marked the periods 1783–1788, 1832–1838, and 1884. Both the ENSO effect and high dust veil indexes correlate with crop failures, food shortages, and disease.[94]

Those whose livelihoods were undermined by the ravages of disease, drought, or other natural catastrophes, warfare, and famine were often pushed into slavery or irredeemable debt, the ultimate consequence of which was bondage. Throughout the region, impoverishment forced families into debt, which was normally expressed in monetary terms, although it was often incurred in non-cash forms such as food or tools. Commonly, household heads attempted to meet debt payments through the sale or mortgage of a household member, notably a girl, but if in dire straits also boys and wives. This was a pattern across the IOW, from East Africa to the Far East, where in Korea alone the percentage of the slave population who entered slavery voluntarily as a consequence of destitution rose from under 1 per cent to almost 50 per cent during the nineteenth century.[95]

[93] Helmut E. Landsberg, "Past Climates from Unexploited Written Sources," in *Climate and History*, ed. Robert I. Rotberg and Theodore K. Rabb (Princeton: Princeton University Press, 1981), 61–62.

[94] William S. Atwell, "Volcanism and Short-Term Climatic Change in East Asian and World History, c.1200–1699," *Journal of World History* 12, no. 1 (2001): 31–40; Candace Gudmundson, "El Niño and Climate Prediction," in *El Niño: Overview and Bibliography*, ed. A. M. Babkina (New York: Nova Science, 2003), 5–28.

[95] Kim, "Debt Slaves in Old Korea"; Ei Murakami, "Two Bonded Labour Emigration Patterns in Mid-Nineteenth-Century Southern China: The Coolie Trade and Emigration to Southeast Asia," in *Bonded Labour and Debt*, 153–64; Alpers, "Pawnship and Slavery in Nineteenth-Century East Africa"; see also Lucie C. Hirata, "Free, Indentured, Enslaved: Chinese Prostitutes in Nineteenth-Century America," *Signs* 5, no. 1 (1979): 4; Jack L. Dull, ed., *Han Social Structure* (Seattle: University of Washington Press, 1972), 110; Fred Morton, "Small Change: Children in the Nineteenth-Century East African Slave Trade," in *Children in Slavery Through the Ages*, eds. Gwyn Campbell, Suzanne Miers, and Joseph C. Miller (Athens: Ohio University Press, 2009), 59.

The incidence of disease also increased, at least in some regions of the IOW, often reducing entire communities to poverty. For instance, two decades of drought and diminished harvests experienced by most of eastern Africa from *ca.* 1820 to 1840, and a generally significantly warmer and wetter period from *ca.* 1850 to 1880, created an environment more conducive to disease vectors. Similarly, the expansion of irrigation and construction projects in the more disease-friendly tropics created pools of stagnant water (ideal breeding grounds for disease vectors of, for example, bilharzia and malaria). Large troop, population, and cattle movements along with hugely increased transport facilities resulted in increased diffusion across the IOW of diseases, including those like cholera (of Ganges delta origin) that had before been regionally contained. Among the major diseases to spread widely were malaria, which proved a greater killer than any other disease or famine, cholera (an epidemic in 1834–1835 killed an estimated one-third of the Egyptian population), the plague, and smallpox. Human diseases took a particularly heavy toll on forced labour and malnourished and displaced peoples, all of whom possessed lowered physiological resistance. By the same token, cattle diseases, such as the 1863–1864 rinderpest outbreak in Egypt that killed an estimated 700,000 animals and the great rinderpest plague of 1889–1897 that had an 85 per cent mortality rate among unprotected cattle in eastern and southern Africa, inflicted significant socio-economic damage.

Moreover, increased trade and migration and improvements in transport and communications in the nineteenth century helped diseases to break out of old disease-tolerant environments and spread to hitherto protected environments with devastating consequences. For example, from 1817, a cholera endemic in Bengal was carried by British troops to Nepal and Afghanistan and then by British ships from 1820 to 1822 to Sri Lanka, South-East Asia, China, and Japan, as well as to Muscat, from where it travelled to the Persian Gulf and East Africa. By 1831, cholera was established in Mecca: between 1831 and 1912 it broke out forty times during the *Hadj*, with pilgrims carrying it to every part of the Muslim world.[96] A similar pattern developed with other killer diseases such as smallpox and malaria. Venereal disease, which was rarely as dramatic but which did severely affect health and fertility, also spread rapidly throughout the

[96] William H. McNeill, *Plagues and Peoples* (New York: Anchor Books, 1976), 231–34.

IOW. Only from the mid-twentieth century did modern medicine and effective treatment begin to counter this pattern in some IOW regions.[97]

The spread of virulent diseases had considerable demographic impact in the IOW. Cholera killed approximately 13 per cent of Cairo's population in the 1831 outbreak and almost 30 million Indians in epidemics that swept India in the second half of the century. In China, a combination of epidemics, experienced in 36 of the 61 years from 1811 to 1872, and almost constant warfare from 1850 to 1878 left vast tracts of the centre and north depopulated; in the T'ai-p'ing and Nien uprisings alone, some 25 million people died. In the famine of 1877–1879, a further 10 million perished in northern China. By 1900, the Chinese population stood at only 450 million.[98] Natural factors frequently complicated the situation. Sravani Biswas and Subho Basu (in Chap. 10 of this volume) examine the conjuncture of environmental forces—the 1876 cyclone and subsequent cholera epidemic in eastern Bengal—and human forces—British allegiance to free market forces and rapacious Indian landlords and moneylenders—in impoverishing local peasants and forcing them to join already plentiful pools of Indian migrant labour vulnerable to different forms of bonded work.

All this occurred in the context of the rise and development of the international economy, which resulted in acute manpower shortages in the IOW. Pre-industrial economies were heavily labour-intensive. With the rise of the international economy in the nineteenth century, there was greater recourse to capital-intensive means of production, notably in the northern United States and Germany. However, in other industrial regions, and in non-industrial zones, production continued to be labour-intensive. In the IOW, as elsewhere, the enhanced commercial opportunities offered by the creation of an international economy greatly increased demand for labour that was required, for instance, for the clearing of vegetation and cultivation of cash crops (e.g. cloves, coffee, cotton, cocoa, tea, sugar); the collection of forest (e.g. wax, rubber, exotic hardwoods), animal (e.g. ivory, rhino horn, oil, skins, feathers) and marine (e.g. pearls, turtle shell, whale oil, ambergris) products; the exploitation of minerals (e.g. gold, tin); porterage, storage, loading, and offloading of vessels; and the manning of ships.

[97] For example, see Campbell, "The State and Pre-Colonial Demographic History," 415–45.

[98] Gwyn Campbell, "Servitude and the Changing Face of Demand for Labor in the Indian Ocean World, c.1800–1900," in *Slavery and the Slave Trades in the Indian Ocean World: Global Connections and Disconnections*, eds. Bob Harms, Bernard Freamon, and David W. Blight (New Haven: Yale University Press, 2013), 23–44.

Simultaneously, indigenous and European powers engaged in state or empire-building projects that involved large numbers of soldiers, administrators, and workers. Some of these included indigenous industrial experiments, such as those which occurred in nineteenth-century Egypt, Madagascar, and China.[99] However, this placed great structural strain on the labour market as growth on the demand side failed to result in an expansion in the supply of labour for two main reasons: uneven demographic growth and the prevalence of systems of bonded labour. Before the eighteenth century, Asia possessed a greater population than any other continent, but while Europe's population more than doubled in the nineteenth century from 190 million to 423 million (that of Britain, Germany, and the United States increased almost fivefold in the 100 years prior to 1914), it took almost 200 years for the population of Asia to double from about 415 million to 970 million between 1700 and 1900. Thus, whereas in 1750, Asia possessed 64 per cent of the global population, Europe 21 per cent, and Africa 13 per cent, by 1900, Asia's share of the global population had fallen to 57 per cent and Africa's to 8 per cent, while Europe had increased its share to 25 per cent and the Americas to 10 per cent.[100]

Population growth rates in Africa expanded significantly only from the 1920s.[101] The conventional view is that demographic stagnation in Africa was a result of instability, petty warfare, and especially the slave trade.[102] Slave exports, which by mid-century may have bled sub-Saharan Africa of half of its potential population of 100 million, were compounded by the impact of human and animal diseases and droughts. As George Michael La Rue points

[99] For example, see Jean Batou, "Muhammad-'Ali's Egypt: A Command Economy in the 19th Century?," in *Between Development and Underdevelopment: The Precocious Attempts at Industrialization of the Periphery (1800–1870)*, ed. Jean Batou (Geneva: Droz, 1991), 181–218; Campbell, *Economic History*; Albert Feuerwerker, *China's Early Industrialization: Sheng Hsuan-huai (1844–1916) and Mandarin Enterprise* (Cambridge, Mass.: Harvard University Press, 1958).

[100] Sumit Guha, "The Population History of South Asia from the Seventeenth to the Twentieth Centuries: An Exploration," in *Asian Population History*, eds. Cuirong Liu et al. (Oxford: Oxford University Press, 2001), 74; Campbell, "Servitude and the Changing Face of Demand for Labor."

[101] Dennis D. Cordell and Joel W. Gregory, eds., *African Population and Capitalism: Historical Perspectives* (Boulder: Westview Press, 1987). For Madagascar, see Campbell, "The State and Pre-Colonial Demographic History."

[102] Patrick Manning, "The Slave Trade: The Formal Demography of a Global System," *Social Science History* 14, no. 2 (1990): 255–79; Patrick Manning, *Slavery and African Life: Occidental, Oriental and African Slave Trades* (Cambridge: Cambridge University Press, 1990); Martin Klein, "Simulating the African Slave Trade," *Canadian Journal of African Studies* 28, no. 2 (1994): 296–99.

out (in Chap. 8 of this volume), peaks in Egyptian imports of Sudanese slaves in 1835–1840 and 1861–1865 were associated with regional drought, famine, and disease—both human, as in the period from 1835–1836 when cholera and plague epidemics struck, and animal as when from 1861–1865 there occurred cattle murrain and rinderpest outbreaks, which reduced Egyptian cattle stocks and sharply increased the demand for human labour.

European colonization in the late nineteenth century also had an impact. However, it is now clear that demographic stagnation in Madagascar—conventionally attributed to French colonial policies from 1895—was evident from the early 1830s and resulted from causes both man-made (warfare, slave-raiding, economic mismanagement, labour exploitation) and natural (climatic variations, disease, cyclones). Many of these factors also affected the neighbouring African continent, which in addition suffered acutely from rinderpest and other cattle diseases, notably from the 1880s. It is therefore likely that the same mixture of forces that were active in moulding demographic trends in Madagascar were present in the rest of eastern and southern Africa from the early nineteenth century.[103]

In all, the commercialization associated with the rise of a truly international economy, climate change, disease, and an increase in the power and reach of the state resulted in increased demand for labour. Further, indigenous states and authorities claimed the labour of all "free" subjects, which was a problem for Europeans. Indeed, in the eighteenth and nineteenth centuries, some IOW states, such as Korea and Imerina, purposefully applied selective measures of abolition, without compensation to the former masters, in order to "free" slaves formerly tied up in the private sector, which then increased the population subject to taxation (in the form of money and labour). As Ronald Kydd points out in his study of the Tang emperor Wu Zong (r. 840–846), such conflicts between the state and non-state owners of slaves (in this case Buddhist monasteries in China) occurred before the modern era. Moreover, adverse climatic conditions had always played a significant role in determining imperial actions. However, with the growth of the international economy from the early nineteenth century, a major confrontation between state and non-state actors occurred over the issue of servile labour. Thus, in Korea, the state abolished the public *nobi* system of hereditary enslavement in 1801—a measure that undermined slavery (it was finally abolished in

[103] Campbell, "Servitude and the Changing Face of Demand for Labor."

1894)[104] —and in Thailand, successive state decrees ensured that slavery had largely disappeared by the end of the nineteenth century.[105] Thereby, in both countries, state access to forced labour increased. Again, in Madagascar in June 1877, the Merina court "emancipated" all African slaves imported since 1865. No compensation was paid to the former owners, and the private sector was deprived at a stroke of an estimated 150,000 slaves, a considerable portion of its labour reserve, whom the imperial court immediately classified as *Tsierondahy*, a category of the *Mainty* (traditional hereditary slave) caste. A few *Masombika* (slaves imported from Mozambique) in and around Antananarivo were genuinely liberated, often finding employment in the construction sector or porterage. However, most were grouped into distinct forced labour units—hence constituting "simply the appropriation on the part of the Government of so much private though stolen property."[106]

There are also striking incidences in the nineteenth century of indigenous states imposing such heavy impositions on the "free" population that it drove many into destitution, indebtedness, and forms of bondage or even slavery. Egypt under Muhammed Ali and Imperial Madagascar under Radama I and Ranavalona I are major examples.[107] Such a process was also promoted by the imposition by indigenous states of high interest rates—up to 100 per cent in nineteenth-century Madagascar. Consequently, defaulting on debts incurred was common.[108] Indeed, the nineteenth century witnessed the apogee of indigenous IOW slave trades. Because of the conventional focus on black slaves, the only IOW region with concentrated estimates for the slave export trade is eastern Africa. There, estimates

[104] Kim, "*Nobi.*"

[105] Martin Klein, "The Emancipation of Slaves in the Indian Ocean," in *Abolition and its Aftermath in Indian Ocean Africa and Asia*, ed. Gwyn Campbell (London: Routledge, 2005), 206.

[106] Robinson to Hunter, Tamatave, 28 June 1877, cited in Gwyn Campbell, "Unfree Labour and the Significance of Abolition in Madagascar, c.1825–97," in *Abolition and its Aftermath in Indian Ocean Africa and Asia*, ed. Gwyn Campbell (London: Routledge, 2005), 66–82. see also Campbell, "Unfree Labour and the Significance of Abolition in Madagascar, c.1825–97," 65–82.

[107] Campbell, "Debt and Slavery in Imperial Madagascar, 1790–1861"; Campbell, *Economic History*; Khaled Fahmy, *All the Pasha's Men: Mehmed Ali, his Army and the Making of Modern Egypt* (Cambridge: Cambridge University Press, 1997); Afaf Lufti Al-Sayyid Marsot, *Egypt in the Reign of Muhammad Ali* (Cambridge: Cambridge University Press, 1990).

[108] For example, see Kim, "Debt Slaves in Old Korea"; Campbell, "Debt and Slavery in Imperial Madagascar, 1790–1861."

for the number of slaves exported by ship in the long nineteenth century range from 800,000 to over two million, and from about 72,000 to double that figure in Madagascar.[109]

Moreover, until the abolition of slavery in European territory, in British colonies in 1833, in French colonies in 1848, and in the Dutch East Indies in 1860, European-held territories in the IOW initially depended largely on slaves to meet their labour demands. For instance, Robert Farquhar, the first British governor of Mauritius, delayed anti-slave import measures in acknowledgement of the cheap labour requirements of local sugar planters.[110] In India, where government by the English East India Company formed the pretext for the lack of rigorous application of abolition in 1833, there were an estimated eight to nine million indigenous slaves in 1841[111]—double the number of black slaves in the United States in 1865. Again, European powers often declared newly conquered territories to be "protectorates" and thus avoided enforcing some abolitionist measures compulsory in colonies. Complete bans on slavery in European-controlled territories occurred fitfully well into the twentieth century. In Africa, the internal slave traffic remained buoyant for some fifty years after the banning of the external slave trade.

Steven Serels (in Chap. 11 of this volume) analyses the impact on slavery and slave trading of the famine that struck the Sudan, Ethiopia, and Eritrea in the late 1880s and 1890s. The famine and the waves of epidemics that followed killed between one-third and two-thirds of the population. The region was simultaneously hit by rinderpest, which decimated livestock. As a result, vast swathes of formerly cultivated land were deserted. In such conditions, many people sold themselves into slavery rather than

[109] Ralph A. Austen, "The 19th Century Islamic Slave Trade from East Africa (Swahili and Red Sea Coasts): A Tentative Census," in *Economics of the Indian Ocean Slave Trade*, ed. William Gervase Clarence-Smith (London: Cass, 1989), 29, 31, 33; Abdul Sheriff, *Slaves, Spices & Ivory in Zanzibar* (London: James Currey, 1987), 226; Campbell, *Economic History*, 55–56, 238; Edward A. Alpers, *Ivory and Slaves in East Central Africa: Changing Patterns of International Trade to the Later 19th Century* (Berkeley: University of California Press, 1975), 151, 185–87.

[110] Clare Anderson, "The Bel Ombre Rebellion: Indian Convicts in Mauritius, 1815–53," in *Abolition and its Aftermath*, 50–65; see also Samuel Pasfield Oliver, "Sir Robert Townsend Farquhar and the Malagasy Slave Trade," *Antananarivo Annual and Madagascar Magazine* 15 (1891): 319–21.

[111] Edward Balfour, *The Cyclopædia of India and of Eastern and Southern Asia: Commercial, Industrial and Scientific, Products of the Mineral, Vegetable, and Animal Kingdoms, Useful Arts and Manufactures*, vol. 3 (London: Bernard Quaritch, 1885), 674.

starve. Serels notes that some existing slaves seized the opportunity presented by a dramatic rise in human and animal mortality, and thus a dramatic decrease in both the workforce and stock of draft animals, to renegotiate the terms of their relationship with, or to flee from, their masters. However, famine conditions also led indigenous elites to launch heightened armed raids on surviving agricultural communities, the members of which were often seized and sold as slaves, while in the late 1890s, Anglo-Egyptian officials deliberately implemented policies that encouraged the reconstruction of a slave-based agricultural system in the Sudan.

Similarly, in Somalia, the colonial regime initially permitted European settler farmers access to slave labour and even returned fugitive slaves to their owners,[112] and in the Sudan, effective measures to curtail slavery were taken only in the late 1920s.[113] On the eastern side of the IOW, the French first seriously applied anti-slavery measures in Indochina in 1897, while the British abolished slavery in Hulsawng valley in eastern Burma only in 1926. Slavery was outlawed in the Netherlands Indies in 1860, but the Dutch then possessed only one-quarter of the Indonesian territory that was to pass under their control by 1910—in much of which they tolerated slavery. Slavery endured in remoter regions of French Indochina and the Dutch Indies into the 1940s. In the Middle East, drawn into the British informal empire after the First World War, abolitionist pressure remained muted until after 1945. Indeed, in the IOW generally, there were few swift and effective abolition measures by European powers.[114]

Also, while using the suppression of slavery as a pretext for the imposition of colonial rule, European authorities in the region needed to retain the goodwill of local slave-owning elites. Thus, while increasingly hindering the maritime slave trade in Africans in the western IOW, colonial authorities neglected both the maritime slave trade in non-Africans and most human trafficking overland, notably of females and children.

[112] Omar A. Eno, "The Abolition of Slavery and the Aftermath Stigma: The Case of the Bantu/Jareer People on the Benadir Coast of Southern Somalia," in *Abolition and its Aftermath*, 83–89.

[113] Suzanne Miers, "Slavery and the Slave Trade in Saudi Arabia and the Arab States on the Persian Gulf, 1921–63," in *Abolition and its Aftermath*, 120–36; see also Igor Kopytoff and Suzanne Miers, "African 'Slavery' as an Institution of Marginality," in idem eds., *Slavery in Africa: Historical and Anthropological Perspectives* (Madison: University of Wisconsin Press, 1977), 72; Suzanne Miers and Martin A. Klein, eds., "Introduction," in *Slavery and Colonial Rule in Africa* (London: Frank Cass, 1999), 1–2, 4–5.

[114] Campbell, "Servitude and the Changing Face of Demand for Labor."

They also proved reluctant to attack indigenous institutions of slavery lest it spark revolt, as occurred in the Muslim province of the southern Philippines after abolition was enforced in 1904.[115]

Through imposing monetary taxes, promoting commercialization, and enforcing credit contracts, European colonial regimes facilitated a growth in indebtedness. At the same time, colonial authorities both maintained tight budgetary regimes that avoided funding public welfare programmes and distinguished debt-bondspeople from "true" slaves, whose condition they attributed solely to violent capture. As a result, debt-bondage and enslavement through debt expanded considerably across the IOW, affecting a wide range of people including farmers mortgaging future harvests, potential grooms borrowing a bride price, small traders living off credit from larger merchants, the ubiquitous rural gamblers of South-East and East Asia, opium addicts in nineteenth-century China, and Gulf pearl divers.[116] India is a prime example of these impacts. In a century characterized by rising taxation and years of famine, "freedom" for members of the former slave outcastes, who had deliberately been kept destitute and debarred from land ownership, translated into the liberty to starve. Some adopted sharecropping, but with two-thirds of the crop paid to the landlord, the risk of failure was high. In order to survive, many entered debt-bondage, which was from 1859 reinforced by the Breach of Contract Act. In some areas of India, members of the most depressed castes formed the overwhelming bulk of those in debt-bondage. The situation closely resembled slavery in that bondage could be inherited, and the vast majority of bonded people had their geographical mobility restricted.[117]

Indentured Labour and the Environment

Indentured labour constituted the major form of non-slave labour in the wake of abolition. Although conventionally this has been portrayed as a phenomenon that arose in the nineteenth century, indentured labour was

[115] William Gervase Clarence-Smith, "Islam and the Abolition of the Slave Trade in the Indian Ocean," in *Abolition and its Aftermath*, 137–49; Michael Salman, "The Meaning of Slavery: The Genealogy of 'an Insult to the American Government and to the Filipino People'," in *Abolition and its Aftermath*, 180–97.

[116] Hopper, "Debt and Slavery among Arabian Gulf Pearl Divers"; Boomgaard, "Human Capital"; Delaye, "Slavery and Colonial Representations"; Schottenhammer, "Slaves and Forms of Slavery."

[117] Patnaik and Dingwaney, *Chains of Servitude*, 29–31.

an old institution in the IOW, where there were clear and often overlapping relations with European forms of indentureship and the concept of "servant." For instance, indentureship was a feature of eighteenth-century Cape society, where it was either formalized by contracts or imposed ad hoc, as with Khoi and San boys captured by Dutch farmers in the interior. The captives were forced to work until the age of 25, by which time they were often married with sons who were subject to similar obligations. Many parents refused to abandon their children and so remained tied to the farm for life. Following abolition of the slave trade in 1807, the Caledon Code of 1809 formalized indenture in the Cape through a "pass" system, which restricted San and Khoi to farms in a system of "virtual slavery." Thus, Worden argues, the first effective formal abolition in the Cape was not official Abolition in 1834 but the 1828 repeal of the Caledon Code.[118] Upon abolition in the Cape and Mauritius, ex-slaves, who received no financial assistance, were declared "apprentices" and obliged to continue working for a fixed period for their old masters. In the Cape the system ended in 1838, but on Mauritius it continued until mid-century. Many apprentices fell into debt-bondage to their old employer or chose to continue working for him in order to remain with their children.[119]

A modified indenture system arose through the failure of the 1834 (British) and 1848 (French) abolitionist measures to transform ex-slaves into pliant wage labourers. Generally involving five-year contracts, it channelled manpower resources to enterprises both within and external to the IOW, such as the sugar plantations of the Fiji Islands. Recruits comprised essentially poverty-stricken Indians, Chinese, and, for the French islands, Africans. By the end of the nineteenth century, approximately one million Indian indentured labourers were employed in India; two million were also shipped to overseas plantations between 1834 and 1920. Contemporary observers and subsequent historians have underlined that the recruitment, transport, and living and working conditions of indentured labourers were often similar to those of slaves.[120] Alessandro Stanziani (in Chap. 7 of this volume), in analysing the French colony of Réunion, emphasizes, like Peabody, the vulnerability of a small island to periodically devastating

[118] Nigel Worden, "Indian Ocean Slavery and its Demise in the Cape Colony," in *Abolition and its Aftermath*, 29–49.

[119] Worden, "Indian Ocean Slavery and its Demise"; see also Marina Carter, *Servants, Sidars and Settlers: Indians in Mauritius, 1834–1874* (Delhi: Oxford University Press, 1995).

[120] Campbell, "Servitude and the Changing Face of Demand for Labor."

cyclones and notes the destructive impact they had not so much on human life and cash crops (sugar proved remarkably resilient to violent storms) but on food crops. This caused the price of provisions to rise, with a consequent pressure on small planters to compensate by continuing to clandestinely import slave labour (albeit under the guise of the *engagé* system) and recruit cheap Chinese and Indian indentured labour. However, the 1860s and 1870s witnessed intense cyclonic activity, while the price of sugar, the staple cash crop, declined dramatically from the 1870s. In a context of a rapidly rising population, petty planters chose to reduce worker food rations or even retain their wages, thus forcing labourers into a cycle of sometimes insurmountable debt. Moreover, the French authorities largely permitted the planters a free hand in labour recruitment.

Attention in the IOW has focused on indentured labour in European plantations and mines[121]—including Australia from the discovery of gold there in 1851—but recent research has also revealed forms of indigenous indentureship. European sources indicate that by the mid-1840s, an estimated 15,000 Chinese workers were being carried in Chinese junks to Bangkok every year.[122] However, Chinese sources suggest that the number of Chinese migrants involved in forms of indentured labour schemes was far greater than previous estimates based on European language sources and that some 7.7 million people emigrated from south China in the period 1851 to 1901—some 87 per cent to South-East Asia. Moreover, most of this migration serviced indigenous IOW rather than European demands for labour. Adam McKeown considers that only about 4 per cent of such emigrants formed "indentured" labour on European estates and that even these "were still often bound to other Chinese through a variety of debt and contract schemes with widely varying levels of obligation."[123]

[121] For more on Indian indentured labour, see, for example, Sudhansu Bimal Mookherrji, *The Indenture System in Mauritius, 1837–1915* (Calcutta: K. L. Mukhopadhyay, 1962); Hugh Tinker, *A New System of Slavery: The Export of Indian Labour Overseas, 1830–1920* (London: Oxford University Press, 1974); Carter, *Servants, Sirdars and Settlers*; Hubert Gerbeau, "Engagés and Coolies on Réunion Island, Slavery's Masks and Freedom's Constraints," in *Colonialism and Migration: Indentured Labour before and after Slavery*, ed. Piet C. Emmer (Dordrecht, The Netherlands: Martinus Nijhoff, 1986), 209–36; Keya Dasgupta, "Plantation Labour in the Brahmaputra Valley: Regional Enclaves in a Colonial Context," in *Abolition and its Aftermath*, 169–79.

[122] A. D. Blue, "Chinese Emigration and the Deck Passenger Trade," *Journal of the Hong Kong Branch of the Royal Asiatic Society* 10 (1970): 80, 83.

[123] Adam McKeown, "Global Chinese Migration, 1850–1940," paper presented at 5th Conference of the International Society for the Study of Chinese Overseas (ISSCO V) Helsignor,

Throughout, as with enslavement, environmental factors, including climate change, floods, and droughts, as well as associated events such as poor or failed harvests, famine, and disease were major factors driving impoverished people into indentureship. Debt was a central feature of the indenture recruitment system, as advances were made from the chief broker to lesser brokers, who in turn often made advances to the heads of households of enlisted workers or local agents through whom they were recruited. Recruits were then obliged to remit sums to their home regions so that such advances could be repaid. Others were already in debt and signed up to indenture schemes in order to earn the money to repay debts. Workers were also obliged to repay the capital and interest on the cost of transport to their place of employment.[124]

The same factors also led to the rise of an unprecedented trans-IOW traffic in impoverished females to serve as sex slaves——a traffic that, like forms of debt-bonded labour, is still vibrant today. The trade in females for sexual purposes was an age-old phenomenon. However, in the pre-1800 era it was chiefly confined to elite markets, and the women involved were generally highly valued and enjoyed a lifestyle superior to that of peasant women. This was particularly the case with concubines, courtesans, and singing girls but may also have been true of the majority of slave prostitutes. The sex trafficking that arose in the 1800s was of a different order. Nineteenth-century demand emanated chiefly from the emergence of permanent and concentrated masses of poor contract workers and soldiers. For example, the migration of millions of Chinese male labourers to IOW centres such as Singapore led to a huge demand for females for sexual purposes that was met by a traffic in mostly involuntary prostitutes, notably from impoverished rural communities in Japan and China. Large concentrations of soldiers led to a similar demand. While the intermediaries were often women—frequently ex-prostitutes—who generally hoodwinked parents and girls into thinking that they would be offered a legitimate job in a distant city, European authorities, eager to minimize tension among males on work sites and in army camps, often colluded in the trade. Thus British authorities in Hong Kong facilitated the flow of Chinese and Japanese girls to become prostitutes in Singapore, and British army officials did likewise for girls from the Himalayas to service the sexual needs

Denmark, May 2004. For the traditional view, see, for example, Blue, "Chinese Emigration"; Perisa Campbell, *Chinese Coolie Emigration to Countries within the British Empire: To Countries within the British Empire* (London: Routledge, 1971).

[124] Murakami, "Two Bonded Labour Emigration Patterns."

of soldiers in India. In many cases, "advances" were given to the girl's parents, the capital and interest on which, along with the cost of travel, lodging, clothing, and food, were deducted from the girl's earnings from male clients, often to the extent that she was permanently in debt.[125]

In sum, in the IOW there has always been a close relationship between adverse environmental events, human agency, and bondage. These sometimes aligned more or less directly, as in the early centralized polities that were at the heart of the Neolithic Revolution, where growing aridity led ruling elites to geographically restrict their subjects and forcibly direct their labour towards water conservation and irrigation schemes. At other times, the relationship was more blurred, as when natural disasters pushed ordinary people into indebtedness, which subsequently resulted in bondage for the debtor, or when a combination of adverse environmental factors, disease, and human (from landlord to colonialist), state, or market forces propelled peasants to migrate in search of work that was often bonded. However, it is apparent that such factors characterized systems of IOW bondage from early times, and—in conjunction with the forces of globalization—are at the heart of modern-day forms of bondage.

[125] Maria Jaschok and Suzanne Miers, "Women in the Chinese Patriarchal System: Submission, Servitude, Escape and Collusion," in *Women and Chinese Patriarchy: Submission, Servitude and Escape*, eds. Maria Jaschok and Suzanne Miers (London: Zed Books, 1994), 19–20; James Francis Warren, "Chinese Prostitution in Singapore: Recruitment and Brothel Organisation," in *Women and Chinese Patriarchy*, 77–105.

CHAPTER 2

Abolition in the Midst of Turmoil: The Case of the Tang Emperor Wu Zong (814–846 CE)

Ronald Kydd

The Environmental, Socio-Political, and Economic Context

Wu Zong came to the throne during an era of turbulent environmental conditions. China comprises four major climatic zones: the Mongolian Plateau and north, central, and south China. Long-term weather conditions could differ dramatically from region to region, those operative in one region supporting stability and growth at times when contrasting

At the outset I wish to acknowledge my indebtedness to Ms. Frances Ren. Ms. Ren was born and raised in Mainland China. She is a graduate of Tyndale Seminary where I was able to retain her as a Research Assistant through a SSHRCC grant. She holds masters degrees in several disciplines, but what made her essential to this project is her knowledge of Classical Chinese, modern Chinese languages, and Chinese history. There were some Chinese documents which I wanted her to see, but in addition to them, she located many others which turned out to be vital primary materials. The importance of her work will become obvious as the paper unfolds.

R. Kydd (✉)
Tyndale University College and Seminary, Toronto, ON, Canada

G. Campbell (ed.), *Bondage and the Environment in the Indian Ocean World*, Palgrave Series in Indian Ocean World Studies,
https://doi.org/10.1007/978-3-319-70028-1_2

conditions in another region led to deprivation and social unrest.[1] In their work, palaeoclimatologists have used a wide range of "tools" extending from archaeology and historical records to "proxy data" gathered from stalagmites and tree rings.[2] The general consensus that can be extracted from their work is:

> Archaeological evidence … reveals that cycles of desertification and decreasing biological productivity occurred frequently and in almost all cases, dynasties flourished when desertification was reversing and biological activity was increasing, and collapsed when desertification expanded in their core regions.[3]

Prolonged climate change bringing drought and cooler temperatures compromised food production for people and animals. In extreme circumstances, it could result in famine and socio-political destabilization. As Xunming Wang and collaborators argue, "a dynasty cannot survive for long if adverse climate prevents its people from producing sufficient food to maintain a strong economy and a strong army capable of defending the dynasty or even expanding its borders."[4]

Ge QuanSheng and collaborators found that the period from 541 to 810, which covers all of the Sui dynasty's reign and two-thirds of the Tang, was one of the warmest in Chinese history.[5] However, Xunming and collaborators state that desertification began to occur in north China in the 790s and spread into central China by the 820s, resulting in food scarcity. He and his associates carried their assessment further, stating, "Unrest recurred in central China due to extreme droughts after 840 AD."[6] Other studies concur with their analysis.[7]

[1] Ka-wai Fan, "Climatic Change and Dynastic Cycles in Chinese History: A Review Essay," *Climate Change* 101 (2010): 565–73; Xunming Wang et al., "Climate Desertification, and the Rise and Collapse of China's Historical Dynasties," *Human Ecology* 38 (2010): 157–72; Zhibin Zhang et al., "Periodic Climate Cooling Enhanced Natural Disasters and Wars in China during AD 10–1900," *Proceedings of the Royal Society B* 277 (2010): 3745–53; Ge QuanSheng et al., "General Characteristics of Climate Changes during the Past 2000 Years in China," *Science China Earth Sciences* 56 (2013): 321–29.

[2] Xunming, "Climate Desertification," 159; Zhibin, "Periodic Climate Cooling Enhanced Natural Disasters and Wars," 3745; Ge et al., "General Characteristics of Climate Changes," 328.

[3] Xunming, "Climate Desertification," 157; Zhibin, "Periodic Climate Cooling Enhanced Natural Disasters and Wars," 3748.

[4] Xunming, "Climate Desertification," 170.

[5] Ge, "General Characteristics of Climate Changes," 322.

[6] Xunming, "Climate Desertification," 164.

[7] See Zhibin, "Periodic Climate Cooling Enhanced Natural Disasters and Wars," 3747; Jianxin Cui and Hong Chang, "The Possible Climate Impact on the Collapse of an Ancient Urban City in Mu Us Desert, China," *Regional Environmental Change* 13 (2013): 358.

The impact of three ethnic groups from Central Asia created a further complication. During the time of Tang's maximum expansion (mid-seventh to mid-eighth centuries CE), a number of these groups had come under imperial Chinese rule. One group, the Sogdians, first appear in Persian inscriptions in the sixth century BCE. They came to occupy an area covering approximately modern Uzbekistan and Tajikistan,[8] and they subsequently spread into China, where many eventually took up residence in the two capitals of the Tang, Chang'an, and Luoyang.[9] Another group, the Turks, were first recorded in Mongolia in the sixth century CE.[10] The third was the Uygurs from Mongolia, who by the early eighth century CE had forged strong relations with the Chinese.[11]

These groups significantly influenced the political instability that followed the death of Emperor Xuan Zong (r. 721–756), notably the An Lushan Rebellion (755–763), which, according to Colin Mackerras, was "by far the largest and most important rebellion in medieval China. Although the Tang court eventually defeated the rebellion, it was dealt a blow so heavy that it never fully recovered."[12] An Lushan was probably the son of a Sogdian father and a Turkish mother, both of whom came from prominent families.[13] The reasons for his rebellion were diverse.[14] In 755 he declared himself emperor and launched his assault on the Tang, taking Luoyang, one of their capitals, the next year. As his army approached Luoyang, the imperial court fled, leaving the capital to fall to the usurper in 756.[15] An Lushan was assassinated in 757, but the rebellion raged until 763 when, with the help of the Uygurs, the Tang dynasty finally crushed it.[16]

[8] V. Hansen, "New Work on the Sogdians, the Most Important Traders on the Silk Road, A.D. 500–1000," *T'oung Pao, Second Series* 89, no. 1 (2003): 149.

[9] Hansen, "New Work on the Sogdians," 154.

[10] M. R. Drompp, "Imperial State Formation in Inner Asia: The Early Tuikic Empires (6th to 9th Centuries)," *Acta Orientalia Academiae Scientiarum Hung* 58, no. 1 (2005): 105. Like the Sogdians, they would eventually play an important role in China. In the transition from the Sui dynasty to the Tang in 617/618, they supported first one side and then the other.

[11] C. Mackerras, "Uygur-Tang Relations, 744–840," *Central Asian Survey* 19 (2000): 223.

[12] Mackerras, "Uygur-Tang Relations, 744–840," 224.

[13] H. S. Levy, "Review: *The Background of the Rebellion of An Lu-shan* by Edwin G. Pulleyblank," *Journal of the American Oriental Society* 75, no. 3 (July–September 1955): 188–92; Edwin Pulleyblank, *The Background of the Rebellion of An Lu-shan*, vol. 4, London Oriental Series (London: Oxford University Press, 1955). Levy agrees with the view of Pulleyblank.

[14] Levy, "Review," 190.

[15] Mackerras, "Uygur-Tang Relations, 744–840," 224.

[16] Mackerras, "Uygur-Tang Relations, 744–840," 224.

However, the victory was far from secure, and the catastrophic consequences were felt throughout the remainder of the Tang dynasty's rule. The capital cities changed hands several times during the rebellion, forcing the government to vacate and exposing the population to socio-economic chaos and rampant political insecurity. Food production plunged and transport networks collapsed; government assistance was impossible, leading to starvation and disease; the monetary system was non-functional[17]; and there was enormous loss of life during and after the rebellion, a period which Niall Ferguson considers to have witnessed one of the greatest mass killings in human history.[18] Steven Pinker estimated the number of deaths at 36 million.[19]

The rebellion and its aftermath undermined central power and influence for the rest of the Tang dynasty, and other factors accentuated the crisis. For example, a Tang general who had played an important role in defeating An Lushan's army rebelled against the imperial throne with assistance from Tibetan forces that occupied Chang'an in 763 and the Uygurs. The rebellion failed,[20] but by 787 the Tibetans had seized much of the territory in the north-west that was previously held by the Chinese.[21] Again, in 821, Emperor Mu Zong (795–824) was forced to suppress a rebel governor on one of his frontiers.[22]

Religion

Even more critical an issue for Wu Zong was the influence of Buddhism. The first written reference to Buddhism in China dates from 65 CE, with Buddhism appearing as an established community of monks and laity,

[17] See Jacques Gernet's comments in his review of Denis C. Twitchett's *Financial Administration under the T'ang Dynasty* (Cambridge: Cambridge University Press, 1963) in *Journal of the Economic and Social History of the Orient* 7 (1964): 325; Drompp, 106.

[18] N. Ferguson, "The Next War of the World," *Foreign Affairs* 85, no. 5 (2006): 61.

[19] Steven Pinker, *The Better Angels of our Nature: Why Violence has Declined* (London: Penguin, 2011), 195, 707, footnote 13; J. D. Durand, "The Population Statistics of China, A.D. 1953," *Population Studies* 13, no. 3 (1960): 224. Writing in 1960, John Durand dismissed that figure as incredible, and many continue to agree with him. He said, "Even if such a huge loss were conceivable, it would be naive to suppose that an accurate account of survivors could have been carried out in the midst of the ensuing chaos." In fact, Durand doubts that any census taken from *ca.* 760 to the end of the Tang reign in 907 could be viewed as having achieved complete coverage. He was questioning the ability to arrive at an accurate count rather than denying massive loss of life.

[20] Mackerras, "Uygur-Tang Relations, 744–840," 226.

[21] Hansen, "New Work on the Sogdians," 156.

[22] Mackerras, "Uygur-Tang Relations, 744–840," 226.

indicating that it had arrived some time earlier, but it became popular early in the third century CE, when it quickly attracted a large following across all social strata.[23] Buddhism expanded further in subsequent centuries, with the number of monks and nuns growing from some 24,000 in the early fifth century to well over ten times that figure by the early ninth century.[24]

Much of the drawing power of Buddhism resided in the monks' abilities to perform extraordinary works of wonder. For the most part, these were monks—Mark Lewis has called them exorcists, magicians, and mediums—who wandered freely outside the discipline of monasteries and who could become entranced and burn parts of their bodies or drive nails through their flesh without pain.[25] For the public, these self-mutilations and self-immolations attested to the depth of devotion of the monks, who as a result gained followers among the poor and the rich—the latter often donating vast sums of money to the monasteries, sometimes even liquidating their own fortunes.[26]

Through non-literary evidence, Robert H. Sharf has pointed to a stream of Buddhism, the characteristics of which included veneration of relics, stupas, images, and sacred texts; filial piety; offering of material goods to the *samgha* (community); believing that merit moving one along the path to nirvana could be transferred from one person to another; appeasing local spirits; monks controlling personal property; tremendous wealth for some monasteries; invocation of deities through shamanism; ancestor worship; cult veneration of aboriginal gods and holy men; appeasement of spirits and ghosts; and ritual possession.[27] Gernet insists that this did not mean that orderly, structured religion had been lost in Buddhism, which continued to offer a moral doctrine and a liturgy: "It

[23] Mark Lewis, *China between Empires: The Northern and Southern Dynasties*, History of Imperial China, ed. Timothy Brook, vol. 2 (Cambridge, MA, and London: Belknap Press of Harvard University Press, 2009), 204–08; Jacques Gernet, *Buddhism in Chinese Society: An Economic History from the Fifth to the Tenth Centuries*, trans. Franciscus Verellen (New York: Columbia University Press, 1995), 279; *Jiu Tang Shu*, [*Old Book of Tang*], vol. 18, ed. Liu Xu et al. (Shanghai: Zhonghua Book Company, 1963), 605.

[24] Gernet, *Buddhism in Chinese Society*; *Jiu Tang Shu*, [Old Book of Tang], vol. 18, p. 606; see also *Tang Hui Yao* [*Institutional History of Tang*], ed. Wang Pu (Taipei: World Book, 1963), vol. 84, p. 1553.

[25] Lewis, *China between Empires*, 211.

[26] Gernet, *Buddhism in Chinese Society*, 233–47.

[27] R. H. Sharf, *Coming to Terms with Chinese Buddhism: A Reading of the Treasure Store Treatise*, vol. 14, Studies in East Asian Buddhism (Honolulu: Kuroda Institute, University of Hawai'i Press, 2002), 14.

may appear that there exists a fundamental difference between these religious aspects of Buddhism and its magical practices. Yet the opposition is only apparent."[28] However, later he added that

> it was by that path [the exuberant religion Sharf described], through the pursuit of holiness for its own sake, that certain rudiments of Buddhist doctrine were able to spread imperceptibly, among the people. Indeed, it is difficult to imagine how they could have done so otherwise.[29]

Charity, too, played an important role in the growth of Chinese Buddhism. Compassion was required of both clergy and laity for the indigent, frail, and less fortunate in society.[30] As Gernet notes: "Many monks practiced charity on their own initiative, providing care for the sick, making gifts of cures, distributing food to the hungry."[31] These pious acts certainly garnered attention and evoked deep appreciation, especially from the poor.

However, as Sharf notes, Buddist monks and laity also adopted some practices that were explicitly proscribed in Buddhist scripture.[32] This came in large part from the extraordinary and growing wealth of monasteries due to donations from the wealthy.[33] In principle, communities of monks could own worldly goods, but they were to have no contact with the way those goods were managed or used. For example, one of the codes governing monastic behaviour forbade "gardening, planting, grain and silk, possession of servants, raising domestic poultry and animals, money and treasure, bedding and cooking pots, gold, jewellery, and all heavy things."[34]

[28] Gernet, *Buddhism in Chinese Society*, 253.

[29] Gernet, *Buddhism in Chinese Society*, 254.

[30] C. Benn, *Daily Life in Traditional China: The Tang Dynasty* (Westport, Conn: Greenwood, 2002), 28.

[31] Gernet, *Buddhism in Chinese Society*, 219.

[32] Sharf, *Coming to Terms with Chinese Buddhism*, 14, 288, footnote 18. Sharf also cites eight studies among many others that give support to his claim.

[33] Lewis, *China between Empires*, 207.

[34] Zhipan, "Fozu Tongji" ["A Chronicle of Buddhas and the Patriarchs"], *Dazheng Xinxiu Dazang Jing (Taisho Tripitaka)* vol. 49, no. 2035 (Chinese Buddhist Electronic Text Association), http://www.cbeta.org/result/normal/T49/2035_004.htm. Monks' property was classified as "light" and "heavy," with the heavy being possessions being of greater value. See He Ziquan, "Fojiao Jinglv Guanyu Sengni Siyou Caichan De Guiding" ["Stipulations Regarding Private Properties of Monks and Nuns in Buddhist Scriptures"] (1982), 158–81, in *Wushi Nian Lai Han Tang Fojiao Siyuan Jingji Yanjiu* [*Studies for the Past Fifty Years on Buddhist Monastery Economy in Han and Tang Dynasties*], ed. He Ziquan (Beijing: Beijing Normal University Press, 1986), 167–70.

However, regulations such as these were brushed aside. As Gernet notes: "One of the most striking features of the life of the Buddhist communities in China is the great freedom left to the monks in the management of their property. Virtually all of them devoted themselves to activities of a profane nature, and principally to the practice of usury."[35] Buddhism brought to China auctions, compound interest, equitable mortgages, non-kinship association (monasteries), the accumulation of merit through the belief in karma, self-giving (with the ideal being the Bodhisattvas—people able to reach nirvana, but delaying so doing out of compassion for suffering beings), and a distinction between risk capital and repayable loans.[36] Also, Buddhist monks managed their land assets by introducing a new understanding of property: they held exclusive rights to land that had been given to them or that they had purchased.[37] Consequently, some massive, opulent monasteries began to appear alongside thousands of smaller centres and retreats.[38] Insofar as individual monks were not permitted to hold possessions privately, revenue from donations, land rents, and other sources was placed in what was called the "Inexhaustible Treasury" of the respective Buddhist communities.[39] As its name implies, this treasury could never be emptied. It served as a source of assistance to the poor and suffering, as a pool of capital to finance the building of grain mills and oil presses, and as a source for extending credit. The mills and presses were rented to peasants,[40] and short- and long-term loans of seed, cloth, and money were made to both peasants and the elites at interest rates varying between 50 per cent (to peasants) and 120 per cent (to the elite). In law, the usually stipulated interest rate on loans was 25 per cent.[41] In this way, Gernet argues, "commerce became fully integrated into the devout

[35] Gernet, *Buddhism in Chinese Society*, 92.

[36] S. A. M. Adshead, *T'ang China: The Rise of the East in World History* (Basingstoke: Palgrave Macmillan, 2004), 85.

[37] Gernet, *Buddhism in Chinese Society*, 93.

[38] He Ziquan, "Zhonggu Shidai Zhi Zhongguo Fojiao Siyuan" ["Buddhist Monasteries in Medieval China"] (1934), 1–54, in *Wushi Nian Lai Han Tang Fojiao Siyuan Jingii Yanjiu* [*Studies for the Past Fifty Years on Buddhist Monastery Economy in Han and Tang Dynasties*], ed. He Ziquan (Beijing: Beijing Normal University Press, 1986), 43.

[39] Lewis, *China between Empires*, 214.

[40] Gernet, *Buddhism in Chinese Society*, 95.

[41] Benn, *Daily Life in Traditional China*, 31–32; Gernet, *Buddhism in Chinese Society*, 171–77.

activities of the *samgha* (Buddhist community)"[42] which effectively introduced "a form of modern capitalism into China."[43]

Buddhism and Slavery

Further, Buddhist institutions became major slaveholders. Edwin G. Pulleyblank argues that "chattel slavery, the free buying and selling of human beings like other pieces of merchandise," began to develop in the Warring States Period (480–221 BCE),[44] and that from the Han era (202 BCE–220 CE) "slaves (nu-pi), human beings who in some respects, if not all, were equated with property and could be bought and sold like domestic animals or inanimate objects, appear as a group having a distinct customary and legal status in society."[45] The process of enslavement varied. Some were hereditary slaves or enslaved as a punishment for revolt, others were war captives, and yet others comprised tribute from other states.[46]

Pulleyblank emphasizes the fundamentally penal origin of enslavement in China. Conviction of a crime resulted in a "good" (*liang*) person coming to be seen as "base," or "ignoble" (*chien*). He argues that *chien* was the general social category to which all slaves were seen to belong.[47] In clarification, he adds, "The condition of a slave did not stand in Chinese law in contrast to that of a 'free man,' an expression which would have been scarcely possible in Chinese prior to the introduction of western ideas in the nineteenth century, since no conception of a positive freedom could exist in any organic society where every man's position was defined by his social relationships."[48]

Under Tang legislation, slaves had dual status. They were people, but they were classified along with property.[49] Gernet notes that they,

[42] Gernet, *Buddhism in Chinese Society*, 153.

[43] Gernet, *Buddhism in Chinese Society*, 228.

[44] E. G. Pulleyblank, "The Origins and Nature of Chattel Slavery in China," *Journal of the Economic and Social History of the Orient* 1 (1958): 193.

[45] Pulleyblank, "The Origins and Nature of Chattel Slavery in China," 186.

[46] *Tang Hui Yao*, vol. 86, ed. Wang Pu, p. 1569.

[47] Pulleyblank, "The Origins and Nature of Chattel Slavery in China," 204.

[48] Pulleyblank, "The Origins and Nature of Chattel Slavery in China," 204.

[49] Yang Jiping, "Tangdai de Nubi, Buqu Yu Tongpu, Jiaren, Jingren" ["Slaves, Private Retainers, and Servants, Family Persons, Pure Persons"], *Zhongguo Shi Yanjiu* [*Chinese History Studies*] 3 (1996): 53–54.

"constituted commercial goods that could be used for any purpose."[50] Pulleyblank acknowledged that "In the commentary to the T'ang code slaves are several times defined as 'the same as property' or 'the same as domestic animals or property' or 'comparable to property.' They could be freely bought or sold,"[51] but he stated that he knew of no pre-Tang text equating slaves and property.[52] Yang Jiping asserts directly that "slaves were people," debased (*jian*) though they were, and that there were special regulations in the Tang Code that applied directly to them.[53] There were laws to protect slaves. Pulleyblank states that what the Tang Code showed was "that it was *full* human rights and not *all* human rights that were denied to a slave."[54]

A blurring of status often occurred for slaves. Some were treated very favourably; however, whether they were favoured or not, there was a Tang Code, and before the law, slaves were slaves. They could be marketed, and they had no freedom of movement. A runaway was subject to 60 blows with the heavy bamboo rod for the first day away and more if he or she was gone for longer.[55] Punishment for a slave who injured a free person was much heavier than for a free person who injured a slave.[56]

Over the centuries, Buddhist institutions became major slave owners.[57] This was problematic given that their scriptures stated: "The ones who hold the pure commandments should not sell nor trade, nor build houses, nor possess or raise people, slaves or domestic animals; they need to be far away from all planting and all treasures as if they were avoiding a fire pit; they should not cut grass or trees, nor cultivate or dig the land."[58] The apparent inconsistency was dealt with in two ways. First, slaves were

[50] Gernet, *Buddhism in Chinese Society*, 126.

[51] Pulleyblank, "The Origins and Nature of Chattel Slavery in China," 212.

[52] Pulleyblank, "The Origins and Nature of Chattel Slavery in China," 219.

[53] Yang Jiping, "Tangdai de Nubi, Buqu Yu Tongpu, Jiaren, Jingren," 53–54.

[54] Pulleyblank, "The Origins and Nature of Chattel Slavery in China," 214.

[55] Benn, *Daily Life in Traditional China*, 40.

[56] Pulleyblank, "The Origins and Nature of Chattel Slavery in China," 214–15.

[57] He Ziquan, "Zhonggu Dazu Siyuan Linghu Yanjiu" ["Studies on the 'Owned Household Persons' of the Nobles and Monasteries in Medieval Times"], (1936), 66–99, in *Wushi Nian Lai Han Tang Fojiao Siyuan Jingii Yanjiu* [*Studies for the Past Fifty Years on Buddhist Monastery Economy in Han and Tang Dynasties*] (Beijing: Beijing Normal University Press, 1986), 79–80; Yang Jiping, "Tangdai de Nubi, Buqu Yu Tongpu, Jiaren, Jingren," 3, 57.

[58] Kumarajiva, "Foyi Jiaojing Lunshu Jieyao" ["Extracted Commentary of Sutra on the Buddha's Bequeathed Teaching"], *Dazheng Xinxiu Dazang Jing* [*Taisho Tripitaka*] 40, no.

permitted to be taken into a monastery because their labour made it possible for monks to avoid things that would have made them impure. The monks were thus freed to "devote themselves entirely to pious activities."[59] He Ziquan views this as a rationalization that would lead to serious consequences.[60] It meant that few things had to be avoided by monasteries as long as there was someone other than the monks to handle them. Second, new slaves were handed over to the community. This meant that monasteries could take advantage of slave labour without any of the monks actually owning slaves.[61] However, slaves, *nubi*, could be bought and sold.[62] In addition to slaves, Buddhist monasteries had several other categories of people working for them, including hired workers and those who could rent, own land, and marry but could not leave their owner.[63] They were *ling hu*: "owned households."[64]

The Purge of Buddhism

Wu Zong's reign was characterized by a major purge of Buddhism. This has been explained in part by religious enmity. The emperor, a committed Taoist, was deeply disturbed by the ubiquity and the influence of Buddhism, once asking derisively, "Why do we need a small western religion to compete with me!"[65] He summoned to court Zhao Guizhen, a

1820, Chinese Buddhist Electronic Text Association, http://www.cbeta.org/result/normal/T40/1820_001.htm.

[59] Gernet, *Buddhism in Chinese Society*, 70. See Daocheng, "Shishi Yaolan" ["Manual of Buddhist Practices"], *Dazheng Xinxiu Dazang Jing* [*Taisho Tripitaka*] 54, no. 2127, Chinese Buddhist Electronic Text Association, http://www.cbeta.org/result/normT54/2127.003.htm

[60] He Ziquan, "Fojiao Jinglv Guanyu Sengni Siyou Caichan De Guiding," 162.

[61] He Ziquan, "Fojiao Jinglv Guanyu Sengni Siyou Caichan De Guiding," 165.

[62] Jiang Boquin, "Tang Xizhou Si Jiaren Nubi De Fangliang" ["Release of the Family Persons and Slaves in Xizhou Temple of Tang"], (1982), 202–19, in *Wushi Nian Lai Han Tang Fojiao Siyuan Jingji Yanjiu* [*Studies for the Past Fifty Years on Buddhist Monastery Economy in Han and Tang Dynasties*], ed. He Ziquan (Beijing: Beijing Normal University Press 1986), 211.

[63] For hired workers see Yang Jiping, "Tangdai de Nubi, Buqu Yu Tongpu, Jiaren, Jingren," 57.

[64] He Ziquan, "Zhonggu Dazu Siyuan Linghu Yanjiu," 66, 87.

[65] *Jiu Tang Shu*, [Old Book of Tang], *Old Book of Tang*, vol. 18, p. 606; see also Ennin, *Ru Tang Qiu Fa Xunli Xingji Jiaozhu* [*Ennin's Diary: The Record of a Pilgrimage to China in Search of the Law*], annotated by Ono Katsutoshi (Shijiazhuang: Huashan Literature and Art Publishing House, 1992), 440.

prominent Taoist monk, who not only guided the emperor in his personal quest to learn how to become immortal but also promoted a lavish programme of erecting Taoist buildings.[66]

However, even more significant was the economic challenge that Buddhism posed to the Chinese state. This stemmed from two major factors. The first was that the metal required to produce religious statues and ornaments in monasteries and the homes of the wealthy took a large amount of copper and precious metals out of circulation, which resulted in scarcities for the imperial treasury of gold, silver, and especially copper.[67] The second major economic problem posed by Buddhism was that it removed increasing numbers of people from the "free" categories that the state could tax and command to perform labour duties.[68] Many became religious personnel, while others laboured as enslaved peasants—many of them being indebted to Buddhist creditors.[69] In fact, during late Tang, taxes had risen dramatically, prompting many peasants to seek refuge in Buddhist monasteries. Some became monks for predominantly spiritual motives, but many others purchased ordination certificates as tax shelters and never entered temples. This drove up the number of peasants with religious status, all of whom enjoyed tax exemption. At the same time, others without adequate financial resources voluntarily became monastic slaves.[70] All this posed major difficulties for an emperor whose authority was under challenge and whose tax revenue had shrunk as a result of falling crop yields and livestock production resulting from worsening climatic conditions and political unrest.

[66] See, *Jiu Tang Shu,* [Old Book of Tang] vol. 18, pp. 585–87, 600; *Tang Hui Yao*, ed. Wang Pu, vol. 30, p. 563; Ennin, *Ru Tang Qiu Fa Xunli Xingji Jiaozhu*, 451. Ironically, Wu Zong is thought to have died of alchemical poisoning from a potion that may have come from Zhao Guizhen's cooking pots. See *Jiu Tang Shu*, *Old Book of Tang*, vol. 18, p. 610; T. H. Barrett, *Taoism under the Tang: Religion and Empire during the Golden Age of Chinese History* (London: Wellsweep, 1996), 87.

[67] He Ziquan, "Zhonggu Shidai Zhi Zhongguo Fojiao Siyuan," 46. See also Gernet, *Buddhism in Chinese Society*, 15, 17; Wolfram Eberhard, review of *Les aspects économiques du bouddhisme dan la société chinoise du Ve au Xe siècle* by Jacques Gernet, *Pacific Affairs* 30 (1957): 272.

[68] He Ziquan, "Zhonggu Shidai Zhi Zhongguo Fojiao Siyuan," 44.

[69] Gernet, *Buddhism in Chinese Society*, 14.

[70] Zhang Gong, "Tangdai Siyuan Nubi Jieung Lueshuo" ["Brief Survey of the Monastic Slave Class in the Tang Dynasty"], *Shehui Kexue Zhanxian* [*Social Science Front*] 3 (1986): 182.

Wu Zong first moved against the Manicheans, who were ethnic Uygurs. Ennin reported, "In mid-April of the 3rd Year of Huichang (843), an edict was issued, ordering the Manichaean priests across the country to be killed. Their heads are to be shaved and they are to be dressed in Buddhist robes and to be killed looking like Buddhist sramana (monks). Manichaean priests were the ones respected by the Uygurs."[71] A subsequent edict, issued on 9 October 843, identified Buddhist monks and nuns who were to be forced back into ordinary life,[72] focusing on those perceived to be the least worthy. These included practitioners of alchemy and sorcery; those who used incantations; those who bore the marks of whippings, tattoos, or other signs of forced labour; draft dodgers; and those judged morally reprehensible, such as monks who were married or who disregarded Buddhist discipline. In all, 3491 monks and nuns were laicized as a result of this edict, and their money, grain, lands, and estates were confiscated by the government.[73]

In March of 845, Wu Zong finally turned to his attention to Buddhism, his primary target. First, he ordered that a census should be taken of all Buddhist slaves in China. They were grouped into three categories. Those with military training would be sent to the military; younger unskilled ones would be sold; and the old and weak would be taken by the palaces.[74] In total, 150,000 male and female slaves were freed.[75] In April 845, Wu Zong also ordered a census of Buddhist temples (monasteries), monks, and nuns, and subsequently found that there were 4600 large, government-funded temples, 40,000 privately funded shrines, and 260,500 monks and nuns.[76] In July 845, following the completion of the census, Wu Zong launched an all-out assault on Buddhism.[77] The Department of the Secretariat reported to him stating:

[71] Ennin, *Ru Tang Qiu Fa Xunli Xingji Jiaozhu*, 416. The Uygurs became Manicheans in 763.

[72] Zhang, "Tangdai Siyuan Nubi Jieung Lueshuo," 182.

[73] Ennin, *Ru Tang Qiu Fa Xunli Xingji Jiaozhu*, 409; Gernet, *Buddhism in Chinese Society*, 375, footnote 11.

[74] Ennin, *Ru Tang Qiu Fa Xunli Xingji Jiaozhu*, 458–59.

[75] Jiu Tang Shu, *Old Book of Tang*, vol. 18, p. 606; Pu, *Tang Hui Yao*, ed. Wang Pu, vol. 84, p. 1553; Zhang, "Tangdai Siyuan Nubi Jieung Lueshuo," 182; Gernet, *Buddhism in Chinese Society*, 139.

[76] Jiu Tang Shu, *Old Book of Tang*, vol. 18, p. 604. See also Zhang, "Tangdai Siyuan Nubi Jieung Lueshuo," 182.

[77] *Jiu Tang Shu*, *Old Book of Tang*, vol. 18, p. 604.

> The temples were destroyed, and the bronze statues and bells were given to the salt and iron officer to make coins; the iron statues were handed to each state to make farming tools; the gold, silver, and bronze statues were melted and given to the Ministry of finance. The nobles and regular families were instructed to return gold, silver, and bronze statues to the government within a month after the edict was issued; if they disobey, the salt and iron officer will punish them according to law. The mud, wood, and stone statues should be kept in the temples. The Secretariat reported again, the monks and nuns should not belong to the Board of Worship. Please classify them under the Ministry of Foreign affairs. Please return foreigners to their original countries.[78]

Then came the grand imperial edict from Wu Zong in August 845:

> I heard Buddhism was never mentioned in the ancient three dynasties; only after Han and Wei, did Buddhism become popular. From then on the foreign customs were spread so widely that it ruined the national custom without notice. It tempted the people, and crowds became addicted. Across the country the monks and the temples increase daily. They exhausted labour in construction, robbed others' profits in gold and silver, abandoned the emperor and parents for teachers, betrayed spouses because of the tenants. Nothing has exceeded it in damage. In addition, if one man does not farm, there will be someone in hunger because of him; if one woman does not breed silk worms, there will be someone in coldness because of her. Today there are countless monks and nuns, and all of them depend on the peasants for food, the silkworm breeders for clothes. There are countless temples all with extravagant decorations.[79]

The emperor added: "Across the nation, 4600 temples have been destroyed, 260,500 monks and nuns laicized to be tax-payers, more than 40,000 shrines destroyed, thousands of acres of fertile land confiscated, and 150,000 male and female slaves received to be taxpayers. The monks and nuns now belong to the Ministry of Foreign Affairs, which show that it is a foreign religion."[80] Then, almost as an afterthought, Wu Zong appended, "I also ordered over 3000 Nestorian and Zoroastrian monks to be laicized, so that they will not pollute the Chinese custom," and

[78] *Jiu Tang Shu*, *Old Book of Tang*, vol. 18, p. 605.

[79] Jiu Tang Shu, *Old Book of Tang*, vol. 18, p. 605.

[80] Jiu Tang Shu, *Old Book of Tang*, vol. 18, p. 606; see also Pu, *Tang Hui Yao*, vol. 84, p. 1553.

concluded, "From now on, we should teach people about simplicity, learn 'Wu Wei' [no action][81] in Taoism, simplify the politics and make a success."[82]

On 22 June 845, Ennin spoke of visiting a large and famous monastery and finding that its residents had left and that its money, property, and slaves had been confiscated. The buildings were silent, awaiting demolition.[83] Ennin also recorded that by November 845, the decrees had all been carried out.[84] We read, "For the monks and nuns in the monasteries they have demolished, if their deeds violate the Buddhist commandments, it is proper to laicize them ... The other monks and nuns will be ordered to move to the big monasteries to live in their rooms."[85] Thus, not all Buddhist monks were laicized, and some 49 monasteries with 800 monks had been preserved by imperial dictate.[86] Those who survived the purge were judged to be faithful to Buddhist doctrine and practices.[87]

Conclusion

The liberation of 150,000 slaves by means of one edict and over one summer was truly astounding. What happened to these freed slaves? Du Mu, a poet who lived during the events (803–852), said that each one received 6.67 hectares of land,[88] a number confirmed by Gernet.[89] Zhang stated that the freed slaves were registered as farmers to be taxpayers.[90] One wonders about the degree to which these plans were implemented. If they

[81] Or "letting things take their own course."

[82] Jiu Tang Shu, *Old Book of Tang*, vol. 18, p. 606.

[83] Ennin, *Ru Tang Qiu Fa Xunli Xingji Jiaozhu*, 476.

[84] Ennin, *Ru Tang Qiu Fa Xunli Xingji Jiaozhu*, 458. See also Jiang Boquin, "Tang Xizhou Si Jiaren Nubi De Fangliang," 210.

[85] Li Fang, ed., *Wen Yuan Ying Hua* [*Finest Blossoms in the Garden of Literature*], vol. 429 (Beijing: Zhongguo Book Company, 1982), 2174.

[86] Sima Guang, *Zi Zhi Tong Jian* [*Comprehensive Mirror to Aid in Government*], vol. 248 (Kowloon: Chung Hwa Book Co., 1976), 8015–16; Benn, *Daily Life in Traditional China*, 16.

[87] Fang, *Wen Yuan Ying* Hua, vol. 429, p. 2174.

[88] Du Mu, "Hangzhou Xinzao Nantingzi Ji" ["Record of Newly Built South Pavilion in Hangzhou"], in *Quan Tang Wen Xin Bian* [*Complete Works of Tang New Edition*], vol. 753, ed. Zhou Shaoliang (Changchun: Jilin Science Literature Press, 2000), 8861–62.

[89] Gernet, *Buddhism in Chinese Society*, 139.

[90] Zhang, "Tangdai Siyuan Nubi Jieung Lueshuo," 182.

were reasonably successful, it would have been a generous liberation. However, it was very much an ambiguous act of emancipation.

When Wu Zong became emperor, he was confronted by a maze of environmental, socio-political, economic, and religious issues. It became clear from his proclivities that Wu Zong's immediate and primary concern was religion, and specifically, weakening the Buddhists and minimizing the significance they had in his empire. He was not gripped by compassion to free slaves. Their freedom was only one component of his strategy to crush his perceived enemies.

Wu Zong's treatment of the Buddhists was certainly aggressive. However, the outcome was short-lived. Xuan Zong ascended to the throne after Wu Zong's death. He killed monks who were particularly influential with Wu Zong and then issued an amnesty overturning Wu Zong's purge. Ennin states, "In mid-May of 846 there was an amnesty and an edict: 'Each state across the nation will build two monasteries, and each region is permitted to build three monasteries. Fifty monks will be placed in each monastery. The monks over fifty years old who were laicized last year will still be monks. Among them the ones older than eighty will be rewarded with 5 *guan* by the government.'"[91] In 847 an imperial edict referred to Wu Zong's work stating, "The reform was inappropriate."[92] The outcome of the abandoning of Wu Zong's policy was the rebuilding of monasteries and the return of the monks and nuns.

A web of irresolvable problems awaited Wu Zong as he ascended the throne. No doubt he had a range of options presented to him by his advisors designed to ameliorate one or more of the presenting issues. He chose to throttle the Buddhists. In the attempt, he liberated 150,000 people, but Buddhist influence continued unimpaired.

[91] Ennin, *Ru Tang Qiu Fa Xunli Xingji Jiaozhu*, 501.
[92] *Jiu Tang Shu*, [Old Book of Tang], vol. 18, p. 141.

CHAPTER 3

Environment and Enslavement in Highland Madagascar, 1500–1750: The Case for the Swahili Slave Export Trade Reassessed

Gwyn Campbell

Introduction

This chapter argues that recent claims of a major export in slaves to Muslim markets in the western Indian Ocean world (IOW) from 1500 to 1750 from the highlands of Madagascar are highly problematic given the environmental, economic, and demographic factors affecting the Malagasy highlands during the period concerned.

In 2003, Thomas Vernet published an article challenging the conventionally accepted view that the export trade in highland Malagasy slaves reached its height in the period 1770–1810. On the basis of new sources, Vernet claimed evidence for a large and hitherto unsuspected Swahili-run slave trade, based in Lamu, which shipped chiefly highland Malagasy from ports in north-west Madagascar to Muslim markets throughout the

I wish to acknowledge the support of the Social Science and Humanities Research Council of Canada, and Thomas Vernet for his comments on this chapter.

G. Campbell (✉)
McGill University, Montreal, QC, Canada

G. Campbell (ed.), *Bondage and the Environment in the Indian Ocean World*, Palgrave Series in Indian Ocean World Studies,
https://doi.org/10.1007/978-3-319-70028-1_3

western IOW.[1] This traffic was fuelled by population growth and internecine strife in the central plateau region of present-day Imerina as well as growing demand in Muslim countries for domestic workers, concubines, sailors, soldiers, guards, agricultural labourers, pearl divers, and palm plantation labour. Demand was accentuated by a high rate of emancipation, which was considered an act of piety for Muslims. The conventional view is that at the alleged height of the slave trade from the Malagasy highlands from 1770 to 1820, the number exported may have averaged between 700 and 1,336 a year—reaching an exceptional peak of 4,000 in 1820[2]—while in the seventeenth century, Muslim merchants shipped from Madagascar between 400 and 1,500 slaves annually.[3] However, on the basis of information from Turkish, Portuguese, and English sources, Vernet estimates that the Swahili exported an annual average of 2,000 Malagasy in the early 1500s, 3,000 by the mid-1600s, 4,000 by 1660, and over 4,000 between 1690 and the 1750s. This would give totals for slaves shipped directly from Malagasy shores to Muslim markets alone of around 200,000 for the sixteenth century, 180,000 from 1600 to 1660, and 360,000 from 1660 to 1750—or 740,000 for the entire period 1500–1750. The figure would be yet higher if exports of Malagasy slaves from the Comoro Islands were included. Only from the mid-1700s, when traders switched to East African sources, did the export of Malagasy slaves to Muslim markets decline.[4]

Vernet's claims greatly boost estimates for the Malagasy slave export trade in general, which, for this period, was hitherto believed to have been quantitatively limited and restricted largely to European slavers.[5] His

[1] Thomas Vernet, "Le commerce des esclaves sur la côte swahili, 1500–1750," *Azania* 38, no. 1 (2003): 69–97.

[2] Richard B. Allen, "Satisfying the 'Want for Labouring People': European Slave Trading in the Indian Ocean, 1500–1850," *Journal of World History* 21, no. 1 (2010): 68; Gwyn Campbell, "Madagascar and the Slave Trade, 1810–1895," *Journal of African History* 22, no. 2 (1981): 208; see also Alfred Grandidier and Guillaume Grandidier, *Histoire physique, naturelle et politique de Madagascar: ethnographie de Madagascar* (Paris: Imprimerie Nationale, 1908), 310, 325, footnote 2.

[3] James C. Armstrong, "Madagascar and the Slave Trade in the Seventeenth Century," *Omaly sy Anio* 17–20 (1983–1984), 216.

[4] Thomas Vernet, "Slave Trade and Slavery on the Swahili Coast (1500–1750)," in *Slavery, Islam and Diaspora*, ed. Paul Lovejoy, Behnaz A. Mirzai, and Ismael M. Montana (Trenton, NJ: Africa World Press 2009), 56–57.

[5] For example, see J. T. Hardyman, "The Madagascar Slave-Trade to the Americas (1632–1830)," in *Association Historique Internationale de l'Océan Indien Congrès, Océan*

article, first published in French versions,[6] was republished in English in 2009, and since then his thesis has gained wide acceptance.[7] This chapter reviews these factors before re-evaluating the nature of Vernet's sources and the conclusions to be drawn from them.

The Malagasy Context

The assertion of a "continuous and massive" trade in highland Malagasy slaves, facilitated by population growth and internecine warfare,[8] needs to be examined firstly from an environmental perspective. The high central plateau, 1,300 to 1,700 metres above sea level, possesses largely infertile lateritic soils except for volcanic pockets around Lake Itasy and in the Ankaratra mountains. The temperature and rainfall of the plateau are well below those of the east coast. It experiences two distinct seasons: hot and humid from late October to early April and dry for the remaining part of the year, with July and August being particularly cool and windy. The plateau is hit often by hail storms and occasionally, in the wet season, by

Indien et Méditerrannée: travaux du Sixième Colloque International d'Histoire Maritime et du Deuxième Congrès de l'Association Historique Internationale de l'Océan Indien, session de Lourenço Marques, 13–18 August 1962 (Lisbon: SEVPEN, 1964), 501–21; Virginia Bever Platt, "The East India Company and the Madagascar Slave Trade," *William and Mary Quarterly* 26, no. 4 (1969): 548–77; Armstrong, "Madagascar and the Slave Trade," 211–33; R. K. Kent, "Madagascar and the Islands of the Indian Ocean," in *Africa from the Sixteenth to the Eighteenth Century*, ed. Bethwell A. Ogot (Paris: UNESCO, 1992), 849–94.

6 Vernet, "Le commerce des esclaves"; Vernet, "Les réseaux de traite de l'Afrique orientale: côte swahili, Comores et nord-ouest de Madagascar (vers 1500–1750)," *Cahiers des anneaux de la mémoire* 9 (2006): 67–107.

7 Vernet, "Slave Trade and Slavery," 37–76; Rudolph T. Ware III, "Slavery in Islamic Africa," in *The Cambridge World History of Slavery: Volume 3, AD 1420–AD 1804*, ed. David Eltis and Stanley L. Engerman (Cambridge: Cambridge University Press, 2011), 76; Claude Allibert, "Le *Kitāb-i bahriyye* de Pīrī Re'īs et l'Océan Indien dans le contexte vohémarien: analyse des versions de 1521 et 1526," *Études Océan Indien* 46–47 (2011): 197–220; Henri Médard, "La plus ancienne et la plus récente des traites: panorama de la traite et de l'esclavage en Afrique orientale et dans l'Océan Indien," in *Traites et esclavage en Afrique orientale et dans l'Océan Indien*, eds. Henri Médard et al. (Paris: Karthala, 2013), 68; Richard B. Allen, *European Slave Trading in the Indian Ocean 1500–1850* (Athens, Ohio: Ohio University Press, 2014), 7.

8 Vernet, "Slave Trade and Slavery," 43; see also Vernet, "Le commerce des esclaves," 76; R. J. Barendse, *The Arabian Seas: The Indian Ocean World of the Seventeenth Century* (Armonk, N.Y.: M. E. Sharpe, 2002), 263–73.

cyclones from the east.[9] Low soil fertility, low temperatures, wind, hail, and locust attacks render cultivation more difficult and self-sufficiency more tenuous than in the lowlands.[10]

However, the period under review was the height of the Little Ice Age (LIA), when eastern Africa and Madagascar were affected by shifting regional and global environmental forces. Africa's basic climatic zones are to some degree decided by latitude. The northern and southern ends of the continent are characterized by Mediterranean-type dry summers and wetter winters. They border subtropical deserts: the Sahara in the north and the Namib coastal desert in the south-west. Between these deserts lies a wide tropical zone governed by the seasonal migration of the Intertropical Convergence Zone (ITCZ), which results in northern and southern belts of monsoonal climates with summer rains and winter drought. In between these belts is a humid equatorial zone characterized by a double rainfall maximum.[11] The ITCZ normally moves south to span latitudes 5°S–10°S during southern hemisphere summer, at which time atmospheric convection and thus precipitation in the south-west Indian Ocean increases.

During the LIA, the ITCZ shifted significantly more to the south in the austral summer, to below 12°S, to affect the northern reaches of Madagascar. This area thus experienced much the same LIA climate as coastal equatorial East Africa, where, from 1500 to 1800, ITCZ–ENSO interactions possibly triggered El Niño-like conditions and increased rainfall with postulated prosperity, agricultural expansion, and population

[9] Hugon, "Aperçu de mon dernier voyage à ancova de l'an 1808," British Library, London, Add. 18137, 11; Le Sage, "Mission to Madagascar," (1816), CO167/34, National Archives, Kew, UK, 102; Samuel Pasfield Oliver, *Madagascar: An Historical and Descriptive Account of the Island and its Former Dependencies*, vol. 1 (London: Macmillan, 1886), 450, and vol. 2, p. 3; Grandidier, *Histoire*, vol. 4 (Paris: Hachette, 1928), 6–8, 30–39.

[10] Hugon, "Aperçu," 11; Dumaine, "Voyage à la côte de l'ouest, autrement dite pays des Séclaves," January 1793, British Library, Add. 18128, 302; Le Sage, "Mission," 102; Grandidier, *Histoire* (1908), 363, and (1928), 6–7, 30–39, 51, 342; Oliver, *Madagascar*, vol. 2, pp. 3, 6; Maurice Bloch, *Placing the Dead: Tombs, Ancestral Villages and Kinship Organisation in Madagascar* (London: Seminar Press, 1971), 92–93; G. S. Chapus and G. Mondain, "Un chapitre inconnu: des rapports de Maurice et de Madagascar," *Bulletin de l'Académie Malgache* 30 (1951–52), 3–8; Raymond Decary, "Poids et mesures d'autrefois," *Bulletin de l'Académie Malgache* 35 (1957), 130; Alain Delivré, *L'histoire des rois d'Imerina: interprétation d'une tradition orale* (Paris: Klincksieck, 1974), 201; Daniel Couland, *Les Zafimaniry, un groupe ethnique de Madagascar à la poursuite de la forêt* (Tananarive: Imprimerie Fanontam-Boky-Malagasy, 1973), 164.

[11] Françoise Gasse, "Hydrological Changes in the African Tropics since the Last Glacial Maximum," *Quaternary Science Reviews* 19 (2000): 190.

growth. By contrast, regions further south, in common with subtropical southern Africa, experienced the reverse conditions.[12] The coral luminescence record confirms that from 1500 to 1800, north-east Madagascar experienced a significantly wetter climate than today,[13] while studies for eastern Madagascar generally indicate periods of relatively good rainfall from 1570 to 1594, 1617 to 1724, 1746 to 1762, and after 1771, broken by arid episodes from 1594 to 1617, 1724 to 1746, and 1785 to 1799.[14] For north-western Madagascar, the stalagmite record in Anjohibe Cave suggests that from 1700 to 1750, the region was affected by possibly the highest frequency of El Niño events experienced since 1550, with notable wet periods at 1620, 1675, 1725, and 1790 and dry intervals centred on around 1600, 1650, 1700, and 1770.[15] Historical writings confirm good rains and economic prosperity in 1613–1614[16] and in 1650,[17] while a universally cold period started around 1709.[18]

Southern Africa, and thus probably southern reaches of Madagascar, experienced a generally cold and dry climate, notably from 1690 to 1740 and between the 1790s and 1810s, with extreme droughts in the period around 1600 and in *ca.* 1700, when there was a depression of about 2°C

[12] J. M. Russell and T. C. Johnson, "Little Ice Age Drought in Equatorial Africa: Intertropical Convergence Zone Migrations and El Niño-Southern Oscillation Variability," *Geology* 35, no. 1 (2007): 23; see also Dirk Verschuren, "Decadal and Century-Scale Climate Variability in Tropical Africa during the Past 2000 Years," in *Past Climate Variability through Europe and Africa*, ed. R. W. Battarbee, Françoise Gasse, and Catherine E. Stickley (Dordrecht: Kluwer 2004), 149–53.

[13] Craig Alexander Grove, "Madagascar's Climate History Unlocked by Giant Corals" (PhD diss., Vrije Universiteit, Amsterdam, 2012), 12–13.

[14] Glen M. MacDonald and Roslyn A. Case, "Variations in the Pacific Decadal Oscillation over the Past Millennium," *Geophysical Research Letters* 32 (2005): L08703; Rosanne d'Arrigo and Rob Wilson, "On the Asian Expression of the PDO," *International Journal of Climatology* 26 (2006): 1607–17.

[15] George A. Brook et al., "A High-Resolution Proxy Record of Rainfall and ENSO since AD 1550 from Layering in Stalagmites from Anjohibe Cave, Madagascar," *The Holocene* 9, no. 6 (1999): 695–705.

[16] Luis Mariano, "Relation du voyage de découverte fait à l'île Saint-Laurent dans les années 1613–1614, par le capitaine Paulo Rodrigues da Costa et les pères jésuites Pedro Freire et Luis Mariano à bord de la caravelle Nossa Senhora da Esperança," in *Collections des ouvrages anciens concernant Madagascar*, vol. 2, eds. Alfred Grandidier and Guillaume Grandidier (Paris: Comité de Madagascar, 1904), 12.

[17] Robert Hunt, "Île d'Asada," in *Collections des ouvrages anciens concernant Madagascar*, vol. 3, ed. Alfred Grandidier and Guillaume Grandidier (Paris: Comité de Madagascar, 1905), 264–65.

[18] Jacques Galas et al., *Histoire de Vaucluse* (Avignon: A. Barthélemy, 1993), 10.

in mean annual temperature.[19] Taking Natal as a proxy, southern Madagascar in general probably experienced, for the period under review, good rainfall in 1496 and from 1588 to 1595, 1786 to 1796, and 1812 to 1819 but arid conditions from about 1320 to 1480, 1490 to 1570, 1675 to 1770, and after 1810, with droughts from around 1483 to 1487, 1527 to 1530, and 1543 to 1544 and in around 1580 (severe), 1688, and the 1790s (severe) to 1810s.[20] If conditions in Mozambique can be taken as a proxy for the central west coastal region of Madagascar, it experienced generally dry arid conditions from 1300 with severe droughts from *ca.* 1580 until the 1810s, notably in *ca.* 1700 and in the 1790s.[21] Possibly related is an epidemic that occurred in western Madagascar in 1701.[22]

Direct climatic indicators for the central plateau of Madagascar during the LIA are sparse, although it probably experienced a more arid dry period from 1300, notably from the 1790s to 1810s.[23] There are however references in oral traditions to three severe famines. The first, the *Tsimiofy* (lit. "do not peel"), which occurred during the reign of Andriamasinavalona (r. 1675–1710), was allegedly caused by violent winds that destroyed the rice crop. However, as the famine lasted "seven years,"[24] it is likely that other factors were also at work, notably high sulphur-rich volcanic activity from 1693 to 1696 that caused surface temperatures globally to fall by 0.2–0.3 °C, possibly from 1696 to 1699. The impact of volcanism was accentuated by the ENSO effect in 1692–1695 and 1702. Certainly the generalized extreme cold event centred around 1700 is confirmed for

[19] Verschuren, "Decadal and Century-Scale Climate Variability," 153; P. D. Tyson, W. Karien, K. Holmgren, and G. A. Heiss, "The Little Ice Age and Medieval Warming in South Africa," *South African Journal of Science* 96 (2000): 121.

[20] Clive Spinage, *African Ecology: Benchmarks and Historical Perspectives* (Berlin: Springer, 2012), 186–88; Verschuren, "Decadal and Century-Scale Climate Variability," 153.

[21] Anneli Ekblom, "Forest-Savanna Dynamics in the Coastal Lowland of Southern Mozambique since c. AD 1400," *The Holocene* 18, no. 8 (2008): 1247–57; Isla S. Castañeda, Josef P. Werne, and Thomas C. Johnson, "Wet and Arid Phases in the Southeast African Tropics since the Last Glacial Maximum," *Geology* 35, no. 9 (2007): 825; Verschuren, "Decadal and Century-Scale Climate Variability," 153.

[22] Grandidier, *Histoire* (1908), 643.

[23] F. Gasse and E. van Campo, "A 40,000-Yr Pollen and Diatom Record from Lake Tritrivakely, Madagascar, in the Southern Tropics," *Quaternary Research* 49, no. 3 (1998): 307; Verschuren, "Decadal and Century-Scale Climate Variability," 153.

[24] R. P. Callet, *Histoire des rois: Tantaran' Ny Andriana* (Tananarive: Éditions de la Librairie de Madagascar, 1974), 296–97; Gwyn Campbell, *An Economic History of Imperial Madagascar, 1750–1895: The Rise and Fall of an Island Empire* (Cambridge: Cambridge University Press, 2005), 24.

south-western Madagascar by coral records.[25] It is thus possible that the *Tsimiofy* occurred from *ca.* 1696 to *ca.* 1703. Another cold period started around 1709, and there was a marked ENSO event from 1715 to 1716.[26] Two other severe famines afflicted Imerina in the eighteenth century: the *Mavovava* (yellow mouth) in *ca.* 1747 and another, the "younger famine" (which caused less deaths than the *Mavovava*), which, if regional comparisons are taken into account, may have occurred in the years 1755–1756.[27] Of notable comparative significance are famine, which hit Mozambique from 1748 to 1751 and Cape Colony in 1747[28]; sulphuric volcanic activity from 1740 to 1744, in August 1739 in the wake of the massive explosion of Tarumai volcano in Japan, and from 1752 to 1756; and very strong ENSO effects in 1754 and 1759.[29]

It is against this background that the demographic history of the highlands needs to be discussed. Madagascar was first permanently settled in the eighth century on the north-west coast. Human activity becomes noticeable in the plateau interior only from the thirteenth and fourteenth centuries with cultivation of hillsides and swamplands. Around 1500, intensive wet rice cultivation was established in the area around Antananarivo, and significant population growth occurred. However, the first major drainage and irrigation projects to promote riziculture are traditionally considered to have commenced in the last third of the sixteenth century, with more sophisticated irrigated riziculture techniques being introduced during the eighteenth century.[30] Hierarchical highland polities

[25] Willie Soon et al., "Reconstructing Climatic and Environmental Changes of the Past 1000 Years: A Reappraisal," *Energy & Environment* 14, no. 2–3 (2003): 254; Verschuren, "Decadal and Century-Scale Climate Variability," 153; Tyson, Karien, Holmgren, and Heiss, "The Little Ice Age," 121.

[26] Stephen Codrington, *Planet Geography* (Sydney: Solid Star Press, 2011), 401.

[27] Callet, *Histoire des rois*, 404; see also H. C. V. Leibbrandt, *Precis of the Archives of the Cape of Good Hope. Journal, 1699–1732* (Cape Town: W. A. Richardson, 1896), 302; Spinage, *African Ecology*, table 4.1, p. 188; David Ernest Hutchins, *Cycles of Drought and Good Seasons in South Africa* (Wynberg: "Times" Office, 1889), 69.

[28] René J. Barendse, *Arabian Seas, 1700–1763*, vol. 1 (Leiden: Brill, 2009), 58; George McCall Theal, *History and Ethnography of Africa South of the Zambesi*, vol. 3 (New York: Cambridge University Press, 2010), 17.

[29] Candace Gudmundson, "El Niño and Climate Prediction," in *El Niño: Overview and Bibliography*, ed. A. M. Babkina (New York: Nova Science, 2003); Campbell, *Economic History*, 121.

[30] R. P. Malzac, *Histoire du Royaume Hova depuis ses origins jusqu'à sa fin* (Tananarive: Imprimerie Catholique, 1912), 18; David A. Burney, "Late Holocene Vegetational Change in Central Madagascar," *Quaternary Research* 28, no. 3 (1997): 141; Gasse and Campo, "A

emerged during the fourteenth century and became significant from about 1500; however, R.P. Malzac dates the first major defensive settlements to the mid-sixteenth century.[31] In the early 1600s, a cluster of eight villages in Ambohimanga would have held about 2,500 people, rising by 1700 to 3,000,[32] by which time a settlement pattern had emerged in Imerina comprising fortresses of 2–4 hectares (500–1,000 people) surrounded by subsidiary forts of 1–2 hectares (250–500 inhabitants) and small hamlets (with on average 7 people per house).[33]

The highland economy was dominated by riziculture in which yields were inferior to those of coastal wet rice and swidden because of low soil fertility, low temperatures, wind and hail, and locust attacks. It was essential for rice seeds and shoots to be nurtured in water, for not only did the top soil harden, crack, and crumble during the dry season, exposing the roots to a quick death in the sun, but plateau soil, devoid of phosphoric acid, chalk, potassium, and nitrogen, was low in fertility. Long-standing marshland contained a residue of fertility, but dry land needed to be assiduously broken, oxygenized, irrigated, and fertilized with humus or human or animal excrement before a reasonable yield could be expected. According to estimates, an average Merina family of four or five people, involving adults and children, working 330 days a year on one hectare of land, produced 2,000–2,500 kg of rice.[34]

40,000-Yr Pollen and Diatom Record," 307; Henry T. Wright, "Early State Dynamics as Political Experiment," *Journal of Anthropological Research* 62, no. 3 (2006): 313–14; M. Virah-Samwy, K. J. Willis, and L. Gillson, "Evidence for Drought and Forest Declines during the Recent Mega Faunal Extinctions in Madagascar," *Journal of Biogeography* 37, no. 3 (2010): 516; Robert E. Dewar and Henry T. Wright, "The Culture History of Madagascar," *Journal of World Prehistory* 7, no. 4 (1993): 426, 439, 447–48, 454; Raombana, "Histoires" (Archives de l'Académie Malgache, Antananarivo), 18–19; Gerald M. Berg, "Riziculture and the Founding of Monarchy in Imerina," *Journal of African History* 22, no. 3 (1981): 306; Anneli Ekblom, Paul Lane, Chantal Radimilahy, Jean-Aime Rakotoarisoa, Paul Sinclair, and Malika Virah-Sawmy, "Migration and Interaction between Madagascar and Eastern Africa, 500 BCE–1000 CE: The Archaeological Perspective," in *Early Exchange between Africa and the Wider Indian Ocean World*, ed. Gwyn Campbell (New York: Palgrave Macmillan, 2016), 195–230; Campbell, *Economic History*, 121.

[31] Malzac, *Histoire du Royaume Hova*, 31–32.

[32] Dewar and Wright, "The Culture History of Madagascar," 456.

[33] Wright, "Early State Dynamics," 313–14; Dewar and Wright, "The Culture History of Madagascar," 455–56.

[34] Mayeur, "Voyage au pays d'ancove, par le pays d'ancaye autrement dit des Baizangouzangoux," 1785, British Library, Add. 18128, 224; James Hastie, "Diary" (1817), 187, CO167-/34, National Archives, Kew; Hugon, "Aperçu," 11; Dumaine,

The manpower demands of both intensive riziculture and defensive settlements led to an enlargement of *fanompoana*, or unremunerated obligatory labour for the crown. *Fanompoana* first emerged under Andriandranolava, probably in the late fourteenth century, as a ritual honorific service restricted to a few privileged groups.[35] Succeeding sovereigns expanded the concept in order to gain control over scarce manpower resources, deemed essential in a sparsely populated and largely infertile plateau region.[36] Andrianjaka (r. *ca.* 1610–1630) regulated its structure through restricting clans to certain localities, and his example was followed by his successors, notably Andriamasinavalona, who, in the wake of the *Tsimiofy* famine, extended *fanompoana* to include all "first fruits" and the provision by Hova (the old term for the Merina) freemen of labour for state infrastructure. It was applied principally in a seven-year project to construct a dyke several metres deep and wide, flanking the river Ikopa for some 26 km of its course between Alasora and Andriantavy, and drain the surrounding Betsimitatatra marshland and convert it to rice fields. Subsequently, Andriamasinavalona divided the entire Merina male population into caste-based labour units and applied *fanompoana* to drain marshes over much of central Imerina and build defences around towns containing strategic stores of grain. By the late eighteenth century, the policy had succeeded: in 1777, Andriantsimarofy, King of Antananarivo, possessed a granary containing 4,000 to 5,000 kg of rice.[37] Such granary towns attracted ordinary cultivators seeking both protection and, in times of dearth, provisions.

"Voyage," 302; Le Sage, "Mission," 102; Grandidier, *Histoire* (1908), 30–39, and (1928), 3–7, 11, 51, 363, 342; Oliver, *Madagascar*, vol. 2, pp. 3, 6; Bloch, *Placing the Dead*, 92–93; Chapus and Mondain, "Un chapitre inconnu," 3–8; Decary, "Poids et mesures," 130; Delivré, *L'histoire des rois*, 201; Couland, *Les Zafimaniry*, 164.

[35] Jean-François Baré, "Remarques sur le vocabulaire monarchique sakalava du nord," in *Les souverains de Madagascar: l'histoire royale et ses résurgences contemporaines*, ed. Françoise Raison-Jourde (Paris: Karthala, 1983): 167; William Ellis, *History of Madagascar*, vol. 1 (London: Fisher, 1838), 175, 177–79.

[36] Campbell, *Economic History*, 120; J. P. Raison, "Perception et réalisation de l'espace dans la société merina," *Annales: économies, sociétés, civilisations* 32 (1977): 416–17, 423–25; Jacques Dez, "Développement économique et tradition à Madagascar," *Cahiers de l'institut de science économique appliquée* Suppl. 129 (1962): 79–108; Jacques Dez, "Le fokonolona malgache: institution désuète ou cellule de développement?" *Cahiers de l'ISEA: économie et sociologie rurales* (April 1965): 189–252; Jacques Dez, "Éléments pour une étude de l'économie agro-sylvo-pastorale de l'Imerina ancienne," *Terre Malgache* 8 (1970): 9–60.

[37] Mayeur, "Voyage au pays d'ancove, autrement dit des hovas ou Amboilamba dans l'intérieur des terres, Isle de Madagascar," 1777, British Library, Add. 18128, 156–57, 171.

The subsequent concentration of population facilitated the imposition by Andrianampoinimerina (r. *ca.* 1787–1810) of further limits on the geographical mobility of the population in order to control their labour for *fanompoana*.[38] He incorporated the territorially restricted population groups, or *foko*, into larger regional units called *toko*, six of which formed greater Imerina: Avaradrano, Marovatana, Ambodirano, Vakinisisaony, Vonizongo, and Vakinankaratra. The king claimed the labour of each male subject for six days each season (a total of 24 days a year) for "public works," notably in agriculture. By the end of the eighteenth century, *fanompoana* labour permitted the expansion of rice cultivation from one to two crops annually, the *vary vaky ambiaty* and *vary aloha*, thereby guaranteeing a regular agricultural surplus for the first time in Imerina's history.[39]

Highland Enslavement and Slave Exports

It is in the context of the fragility of the agricultural base of the highland economy, notably during the LIA, and intense all-year-round demand for labour that the issue of enslavement and the slave trade needs to be discussed.

Vernet is correct in assuming internecine conflict and slave raiding to have been major contributory factors to enslavement in the highlands.[40] The *Tantara*, or Merina royal traditions, state that slaves become numerous in Imerina chiefly as a result of war. However, they indicate that this first occurred in the reign of Andramasinavalona (r. *ca.* 1675–1710), well after 1500—the date when a massive and continuous efflux of highland

[38] Campbell, *Economic History*, 24; Grandidier, *Histoire* (1908), 846; Callet, *Histoire des rois*, 238–39, 275–76, 279–80, 282, 297–98, 304; Raombana, "Histoires," 247; Alfred Grandidier, "Souvenirs de voyages, 1865–1870," *Documents anciens sur Madagascar* (Tananarive: Association Malgache d'Archéologie, n.d.), 33; Conrad Keller, *Madagascar, Mauritius and the Other East African Islands* (London: S. Sonnenschein, 1901), 123; Dez, "Éléments," 39, 41.

[39] Raombana, "Histoires," 69; Callet, *Histoire des rois*, 276–77, 288, 823–27, 1111; Léon Ozoux, *Vieux principes d'économie rurale malgache* (Tananarive: Imprimerie Industrielle, 1926), 5; Raison, "Perception et réalisation," 417; Hubert Deschamps, "Tradition and Change in Madagascar, 1790–1870," in *Cambridge History of Africa*, vol. 5, ed. John E. Flint (Cambridge: Cambridge University Press, 1976), 400–02; Hubert Deschamps, *Histoire de Madagascar* (Paris: Berger-Levrault, 1972), 204, 209; Dez, "Le fokonolona malgache," 201.

[40] Vernet, "Slave Trade and Slavery," 42–43; see also Barendse, *The Arabian Seas*, 263–73.

slaves allegedly started.[41] The inflow of foreign coins and arms, indices of the volume of external exchange, also gained any magnitude only in the early and late seventeenth century, respectively.[42]

Moreover, any efflux of the Hova population of the order of 2,000 to 3,000 a year would have critically undermined the economy of the central plateau. Enslavement through warfare and raiding was common practice in the highlands, but the Portuguese observation cited by Vernet that victors in war were captured rather than killed cannot be interpreted as indicating a massive investment in the slave export trade.[43] Initially, non-combatants who submitted peacefully to conquering armies were spared, but captured male opponents were killed and their wives and children enslaved. "Blacks" from the south (possibly Betsileo) were also kidnapped and enslaved, while men found guilty of breaking royal ordinances were executed and their wives and children enslaved.[44] Forms of enslavement changed in the internecine wars that characterized Imerina from *ca.* 1720 to 1790, when fighting was effected with minimal casualties. Combatants, comprising spearmen and artillery, assisted by non-combatants, mostly relatives and slaves of the warriors, aimed to out-manoeuvre and "kidnap" enemy fighters. In comparison to European engagements involving similar forces and weapons, few combatants were killed or wounded. Indeed, fighting was suspended when crises arose—such as the arrival of a plague of locusts—that might threaten common agricultural interests. The aim was rather to ransom a captive back to his relatives, and only when the family was too poor to afford the ransom was the captive retained or sold as a slave.[45]

Warfare and slave raiding, like long-distance trade, were restricted to the dry season from mid-April to mid-November because overland travel was difficult, sometimes impossible, in the wet season—notably between the

[41] Callet, *Histoire des rois*, 321–22.

[42] Campbell, *Economic History*, 47; Callet, *Histoire des rois*, 360.

[43] Vernet, "Le commerce des esclaves," 76.

[44] Callet, *Histoire des rois*, 321–22.

[45] Mayeur, "Voyage au pays d'ancove," (1775), 62; Mayeur, "Voyage au pays d'ancove," (1785), 214–15, 229; Anon., "Mémoire historique et politique sur l'Isle de Madagascar," 1790, British Library, Add. 18126, 73; Grégoire Avine, Voyages aux isles de France d'Anjouan de Madagascar, de Mosambique, de Zanzibar et de la côte Coromandel (Paris: G. Durassié, 1961): 55; Raymond Decary, *Coutumes guerrières et organisation militaire chez les anciens Malgaches*, vol. 1 and 2 (Paris, Éditions Maritimes et d'Outremer, 1966).

interior and lowlands. In addition, highlanders travelling to the lowlands were highly vulnerable to malaria. A high incidence of the sickle cell trait, and resistance gained through exposure, offered some protection against malaria to coastal peoples, although it entailed high child mortality. By contrast, inhabitants of the traditionally malaria-free highlands possessed no acquired resistance or genetic defence against the disease, from which, during the summer months, they suffered as much as Europeans.[46] This became evident in the nineteenth century when, in attempting to conquer the lowlands, an estimated 25–50 per-cent of Merina soldiers died each year.[47] As great a percentage of Merina slave traders and slaves would have similarly perished had they attempted travel to the coast in the rainy season. Given these factors—the time-consuming and costly nature of enslavement and the 600 km overland trail to the north-west coast—it is inconceivable that slave caravans from the highlands, certainly those of the size alleged by the *Francis* (see below), reached the main north-western ports in March and April.

Claude Allibert suggests that highlanders could have been marched to the coast during the dry season, shipped to the Comoros, and there held in camps that, on the basis of the writings of the Ottoman admiral Pîrî Reis, might have formed slave-breeding centres. However, such camps, if established, would have experienced high rates of slave mortality due primarily to the exposure of highlanders to malaria once on the coast. Slaves were also highly vulnerable to other diseases. These included smallpox, which was present on the western Indian Ocean in the early seventeenth century.[48] Estimates for its impact on unvaccinated peoples in South and East Africa vary, but, for concentrated populations, mortality from smallpox varied from 40 to 70 per cent.[49] Of the 22 Malagasy slaves purchased

[46] J. J. Freeman, "Extracts from a Letter from the Rev. J. J. Freeman, Dated Tananarivou, Madagascar, October 23, 1827; Addressed to the Late Secretary [Burder?]," *Evangelical Magazine and Missionary Chronicle* (April 1828): 169; Campbell, *Economic History*, 146; see also Leonard Jan Bruce-Chwatt, *Essential Malariology* (London: Heinemann, 1980), 58–61.

[47] Gwyn Campbell, "Slavery and Fanompoana: The Structure of Forced Labour in Imerina (Madagascar), 1790–1861," *Journal of African History* 2, no. 3 (1988): 468–69; Campbell, *Economic History*, 148–49.

[48] William Foster, *The English Factories in India 1637–1641: A Calendar of Documents in the India Office, British Museum, and Public Record Office* (Oxford: Clarendon Press, 1912), xxiii–xxiv, 226, 294–95.

[49] Marc H. Dawson, "Disease and Population Decline of the Kikuyu of Kenya, 1890–1925," *African Historical Demography* vol. 2 (Edinburgh: University of Edinburgh, 1981): 127–28; Robert Ross, "Smallpox at the Cape of Good Hope in the Eighteenth Century," *African Historical Demography* vol. 1 (Edinburgh: University of Edinburgh, 1977): 416–28.

by the *Francis*, 13 (i.e. 59 per cent) developed smallpox on the voyage back to Surat, where they died shortly after arrival in November 1640.[50] Had there been a slave trade of the magnitude and regularity postulated, it might be expected that smallpox would have also appeared in the Malagasy highlands during the period under review. However, the evidence is that smallpox affected present-day Imerina only following increased exposure to foreign trade from the late eighteenth century and major epidemics occurring in 1817 and 1833–1835.[51]

A further reason for doubting the report of Hova (the old term for the Merina) slave caravans is the claim that their commanders also annually drove 10,000 cattle from the interior to the ports on the north-west coast. While animal husbandry was widespread in Madagascar, the main grazing lands were the southern and western plains, where cattle raising was the main occupation of the pastoral Bara, Mahafaly, Antandroy, and Sakalava, and the northern highlands occupied by the Tsimihety. Thus, in *ca*. 1711, Robert Drury commented of a prince in northern Menabe, in western Madagascar:

> The king's cattle are marked with a mark called Chemerango. Here are so many thousands of them that it is not known to two or three thousand how many he has. You may see oxen not able to walk for age and other for fatness. They never trouble themselves to milk the fourth part of the cows, though the place is populous and as large as some king's dominions ... He has eight thousand head of cattle of his own, and three hundred slaves.[52]

By contrast, the Hova were agriculturalists and possessed few cattle that they raised on uncultivated hilltops. Not until the imperial expansion of Imerina from the end of the eighteenth century did the Merina elite start to acquire large herds of cattle.[53]

[50] Foster, *English Factories*, xxiii–xxiv, 226, 294–95; see also Eric Axelson, *Portuguese in South-East Africa 1600–1700* (Johannesburg: Witwatersrand University Press, 1969), 121.

[51] Charles Theodore Hilsenberg and Wenceslaus Bojer, "A Sketch of the Province of Emerina, in the Island of Madagascar, and of the Huwa, its Inhabitants; Written during a Year's Residence," in *Botanical Miscellany*, vol. 3, ed. William Jackson Hooker (London: John Murray, 1833), 261; Freeman to Ellis, Tamatave, 3 October 1833, and Vohitsara, 11 October 1833, School of Oriental & African Studies, London, London Missionary Society Archives, Madagascar Incoming Letters, B4.F4.JC.

[52] Robert Drury, *Madagascar; or, Robert Drury's Journal, during Fifteen Years' Captivity on that Island*, ed. Pasfield Oliver (London: Fisher Unwin, 1890), 271; see also Mariano, "Relation," 10, 20.

[53] Campbell, *Economic History*, 29.

Thus, in traditional patterns of enslavement, most people who became slaves in the highlands were retained locally, forming the *Andevo*, the third of the three castes of Imerina (the Hova formed the second "free" caste below the noble *Andriana*). Those sold for export were chiefly condemned "unransomable" prisoners rejected by their communities.[54] Only from the mid-eighteenth century, after substantial population growth and the development of major internecine strife, did a steady and well-documented slave export trade develop from the highlands in the direction of the burgeoning plantation economy of the French-ruled Mascarene Islands of Bourbon (Réunion) and Mauritius.[55] The slave export trade to the Mascarenes peaked in the late eighteenth century, as is reflected in the formation of specific slave markets, such as the Friday market which emerged in Imamo by 1785, and a dramatic increase in the volume of imported coins that then became the predominant currency in the foreign trade of Madagascar.[56]

The Sources

The environmental and demographic constraints on slave exports indicated above prompt a re-examination of the seven principal sources used by Vernet.[57] The earliest is a passage from *Kitab-ı Bahriye* (*Book of Navigation*), written between 1511 and 1521 by Pîrî Reis (*ca.* 1470–1553), a celebrated Ottoman admiral and cartographer, in which he describes the island of Kaziçe (probably Ngazidja or Grande Comore), in the Comoro Archipelago:

> They treat their slaves like sheep and lambs.
> Some are old and some are young.
> There are men with a thousand slaves male and female. These are cared for like cattle.
> And their sons and daughters they constantly put up for sale.
> Sailors come hither and buy them. They fill their ships and take them away.
> They always sell them in Yemen though sometimes they take them as far as Cidde.[58]

[54] Callet, *Histoire des rois*, 360.
[55] Berg, "Riziculture," 308; Wright, "Early State Dynamics," 313–14.
[56] Grandidier, *Histoire* (1908), 261, 570; Campbell, *Economic History*, 47.
[57] Vernet, "Le commerce des esclaves," 76–78.
[58] Pîrî Reis, *The Book of Bahriye* (Istanbul: Boyut, 2013), 25.

Vernet interprets this passage as indicating that Comoro traders exported Malagasy slaves to the Comoro Islands, where they were held and even encouraged to breed before being sold to Swahili ships that called annually at the Comoros and Madagascar.[59] Allibert, who backs Vernet's thesis, comments that if slaves were bred for sale as suggested by Reis, it implies a far greater volume of trade—a thesis that appears confirmed by Portuguese accounts placing a higher value on a pregnant slave than on an older male.[60]

However, Reis's report is second-hand. His maritime voyages were limited to the Mediterranean until 1516, when he participated in the Ottoman seizure of Egypt, to where he returned in 1524 carrying the Ottoman Grand Vizier. Considerably later, in the 1640s, when he was admiral of the Ottoman fleet in Egypt that sailed in the Red and Arabian seas, he may have visited the Comoro Islands. However, as his description of the Comoros and Madagascar was written and revised by 1525, it must have been derived from second-hand sources. In that epoch slaves were trafficked on the Comoros. However, had these been highland Malagasy slaves, they would, as noted above, have suffered high rates of mortality due to malaria. In fact, in the same section of his book, Reis indicates that, rather than being Malagasy, the slaves were local, the subject of internecine strife between the different chiefs of the island who "hunt each other down and sell their prey."[61] Reis also indicates a vibrant slave trade in north-west Madagascar, where "The people of the cities [ports] go inland; and capturing them, they bring them down to the ships where they sell them. In this way they go on continuous slave hunts day and night."[62] However, this indicates that the slave raiders were Antalaotra (Islamized population of Madagascar) rather than Hova, while "inland" without any other qualification suggests peoples of the "hinterland" rather than those of a distant mountain interior.

Four of Vernet's sources, which he considers the most significant, are Portuguese. They all derive from Roman Catholic clerics or prominent lay Catholics who shared a common vision of Portuguese religious and political dominance, and a common loathing of Muslim influence, in the western Indian Ocean region. Indeed, as all these source documents were stored in Catholic archives in Europe, some were probably consulted by, and had a direct influence on, authors of later reports on the region. The

[59] Vernet, "Le commerce des esclaves," 77; Vernet, "Slave Trade and Slavery," 42.

[60] Allibert, "Le *Kitāb-i bahriyye.*"

[61] Reis, *Book of Bahriye*, 25.

[62] Reis, *Book of Bahriye*, 24.

earliest Portuguese source is Luis Mariano, a Jesuit priest based in Goa, who made four visits to Madagascar between 1613 and 1630 to proselytize the Malagasy. Vernet states that Mariano's 1616 and 1619 reports indicate that Swahili boats sailed on an almost annual basis from Pate to north-west Madagascar for slaves.[63] Mariano certainly found, to his dismay, that the major ports of the north-west coast were inhabited by Muslim converts under the influence of Swahili traders, chiefly from Malindi, Mombasa, Lamu, and Pate. He also discovered that hinterland Malagasy were impervious to Christianity.[64] Frustrated at his lack of success, Mariano lambasted the Muslims as the chief obstacle to his mission. He further accused them of conducting a large trade in captive Malagasy, notably children whom they shipped to Muslim markets for "infamous and repugnant purposes."[65]

However, Alfred and Guillaume Grandidier comment that Mariano is mistaken in attributing a uniquely Swahili origin to the Muslim traders of north-western Madagascar. The Antalaotra were, rather, a mixture of "Arabs from Yemen, Oman and the Persian Gulf, Persian, Indians from Kutch and Gujarat, and Swahili and Comorian metis."[66] Moreover, Mariano does not indicate that most slaves in the region were from the high plateau interior. In 1613, Mariano witnessed some Hova who had been brought for sale in Boina Bay "to Arabs from Malindi,"[67] but does not indicate that the slave traffic from the highlands was either large or regular. Indeed, Jean Gomez, a priest who accompanied Mariano to Madagascar, commented in 1620 that people from the north-western ports had stopped travelling into the highlands because one caravan had been attacked by Hova who had killed a large number of coastal traders and seized their goods and money.[68]

The second Portuguese account, dated 1630, states,

> To this island of Pate come ships, on their way from Mecca to the island of Saint Laurence [Madagascar] with sharifs, who are their cassizes [qasis], and who spread their sect there, and take back many Buques [Buki—Malagasy], pagan children, to bring them to Mecca.[69]

[63] Vernet, "Le commerce des esclaves," 78.

[64] Raymond K. Kent, *Early Kingdoms in Madagascar 1500–1700* (New York: Holt, Rinehart and Winston, 1970), 180.

[65] Mariano, quoted in Armstrong, "Madagascar and the Slave Trade," 213.

[66] Grandidier, *Histoire* (1908), 164.

[67] Grandidier and Grandidier, ed., *Collection*, vol. 2 (1904), 12–13.

[68] Jean Gomez, letter, "*Mazalagem [Boina], 1620*," in Grandidier and Grandidier, ed., *Collection*, vol. 2 (1904), 328.

[69] Quoted in Vernet, "Slave Trade and Slavery," 46.

This citation needs to be considered in context. It is an extract from one of a number of reports, dated 1625, 1629, 1630, and 1640, produced by the Portuguese Augustinian order as part of a propaganda campaign aiming to persuade the Catholic powers of Europe to act against Muslims in the western IOW. Their major premise for this was the alleged torture and execution of the "martyr" Christian queen Ketevan in 1624, one year after the Shah had permitted the Augustinians to establish a foundation in Shiraz, because she refused to convert to Islam. However, another central refrain of Augustinian anti-Muslim propaganda was the claim that Muslims were involved in a vast traffic in which they enslaved pagan children and converted them to Islam. It is in this light that the extract above should be interpreted.[70]

The third Portuguese account, dated 1663, is derived from Eric Axelson, who uses it to contend that:

> With the decline of Portuguese maritime power vessels from Pate were sailing almost unmolested to Madagascar, where they were exploiting a lucrative trade in slaves—it was estimated that they were transporting some 3,000 or 4,000 a year.[71]

Again, this citation needs to be placed in context. It forms part of a much larger report written by the Supervisor of the Misericórdia in Mozambique, a prominent Portuguese merchant, whose primary objective was to destroy Muslim commerce in the western IOW.[72] As local Portuguese forces proved incapable of stopping Muslim attacks on Portuguese posts and trade, the author appealed directly to authorities in Portugal to intervene, notably against Muscat and Pate, which were, at that time, allied in their fight to liberate the region from the Portuguese. For example, in 1661, Omani forces sacked Mombasa, where Augustinians were nurturing local

[70] Domingos do Espírito Santo, *Breve relaçam das christandades que os religiosos de N. Padre Sancto Agostinho tem a sua conta nas partes do Oriente, & do fruyto que nellas se faz, tiradas principalmente das cartas que nestes annos de lá se escrevem; em que se contem cousas muy notaveis* (Lisbon: por Antonio Alvarez, 1630); John Flannery, *The Mission of the Portuguese Augustinians to Persia and Beyond (1602–1747)* (Leiden: Brill, 2013), 197–222.

[71] Axelson, *Portuguese in South-East Africa*, 141; see also Vernet, "Le commerce des esclaves," 78.

[72] For a description of the workings of the Misericórde in colonial Portuguse settings, see e.g. Liam Matthew Brockey, *Portuguese Colonial Cities in the Early Modern World* (Farnham: Ashgate, 2008), 166.

converts to Catholicism, but left because they could not take Fort Jesus.[73] Again, the Portuguese at Mozambique resented the strong commercial and religious connections between Pate and north-west Madagascar. Mozambique Island frequently experienced major food shortages that its inhabitants sought to overcome through imports of provisions, chiefly from surrounding mainland African regions, India, the Comoros, and north-west Madagascar. In 1633, the Conde de Linhares even urged that two forts be established in Madagascar to provide provisions for Mozambique.[74] Three years later, when the Portuguese at Mozambique again sent to Madagascar for provisions, they found that local Muslims had granted refuge to Yusuf (Jeronimo Chingulia), the rebel King of Mombasa.[75] It was in this context that the Supervisor of the Misericórdia resorted to the familiar allegations of a vast slave trade in which "pagans" were being forcibly converted to Islam.[76] He thus urged Portuguese authorities to maintain a fleet of four ships:

> To cruise along the coast of the Island of São Lourenço [Madagascar] on the ships of the Muscat enemy and the Moors of Pate, who every year transport some three to four thousand slaves to make them Moors.[77]

The fourth Portuguese source used by Vernet is a letter written in 1667 by Father Manuel Barreto, Superior of the Jesuit College at Sena, to João Nunes da Cunha, the viceroy in Goa. Barreto desired to abolish corruption and fill the colonial administration with men who would create a Portuguese empire in Eastern Africa—from the Red Sea to the Cape—and its islands including Madagascar. Under Portuguese rule, commerce in the region would flourish, and the indigenous peoples would be converted to

[73] Teotonio R. De Souza, *Goa through the Ages: An Economic History*, vol. 2 (Goa: Goa University, 1999), 159; Augnet (website of the Order of St Augustine), http://www.augnet.org/?ipageid=685

[74] Axelson, *Portuguese in South-East Africa*, 101–02.

[75] C. H. Stigland, *The Land of Zinj: Being an Account of British East Africa, its Ancient History and Present Inhabitants* (London: Frank Cass, 1966), 20; Axelson, *Portuguese in South-East Africa*, 114; see also Grandidier, *Histoire* (1908), 519, footnote 1; Vernet, "Slave Trade and Slavery," 44–46.

[76] George McCall Theal, *Records of South-Eastern Africa*, vol. 5 (London: William Clowes, 1901), 144–45, 156–57.

[77] Quoted in Barendse, *Arabian Seas, 1700–1763*, vol. 3, 1242–43.

Christianity.[78] It is in this context that Barreto's plea concerning Madagascar should be viewed. Barreto called for the Portuguese crown to grant Bartolomeu Lopes a fleet of

> six well-armed royal ships manned by Portuguese troops, including his launch and several boats manned by Kaffirs from the other side [of Africa], he would be able to prevent all Muslim vessels from Mecca, Brava, and Mogadishu, from coming to this island [Madagascar] to buy Buques [Malagasy]. The Muslims purchase more than three thousand each year. To this end, they have in many ports, Muslim priests who in the course of the year go to buy and catechize them [into Islam] to the great opprobrium of the Christian name.[79]

Here Barreto is clearly restating the claim to Madagascar for Portugal and the Catholic faith. In 1654, colonists who had abandoned the French settlement in south-east Madagascar informed the Portuguese at Mozambique that with an army of 300 men they could conquer all of Madagascar.[80] Indeed, describing Mozambique Island as a graveyard for the Portuguese, Barreto recommended abandoning it for Madagascar, where provisions were abundant and from where ships could easily sail to Mozambique.[81] Moreover, to underscore the urgency of conquering Madagascar, Barreto restates the familiar Roman Catholic allegation that Muslim slavers were carrying away and converting thousands of heathen Malagasy every year. This was not a condemnation of the traffic in slaves. In fact, Portuguese missionaries encouraged the slave trade as a means of proselytizing captive pagans and preventing them from converting to Islam. In 1616, Mariano had purchased a young male slave in Madagascar,[82] and in 1657 in Ampaza (Faza on Pate Island), an Augustinian priest sent young converted slaves to Goa so that they could grow up without danger

[78] Manuel Barreto, "Moçambique e Madagascar: informação do estado e conquista dos rios de Cuama vulga e verdadeiramente chamados rios de Ouro," *Boletim da Sociedade de Geographia de Lisboa* 4, no. 1 (1883): 33–59; see also George McCall Theal, *History of Africa South of the Zambezi*, vol. 1 (London: George Allen & Unwin, 1916), 433–34.

[79] Manuel Barreto, "Moçambique e Madagascar," 55.

[80] Axelson, *Portuguese in South-East Africa*, 134.

[81] Axelson, *Portuguese in South-East Africa*, 132.

[82] Luis Mariano, letter, "Sadia, c. July 1616," in Grandidier and Grandidier, ed., *Collection*, vol. 2 (1904), 214.

to their faith.[83] In such a context, Barreto's comments about Madagascar should be seen in the same light as those of those of other Catholic clerics and the Supervisor of the Misericórdia—as religious propaganda, not as historically accurate observations.

It is in the light of the above discussion that the final two sources, both English, need to be considered. The first is a report from the ship *Francis* in 1640, which Verney cites as stating that "every year the Hova came down the river Mahavavy, south of the Bay of Boeny, forming 'caravans' of 10,000 heads of cattle and 2,000–3,000 captives to be sold in Boeny"[84] in north-west Madagascar.[85] In its original form, the report, which probably emanates from Henry Tyrrell, captain of the *Francis*, reads,

> There's a river ye So. Side of Matthewlodge [Mattelage] from where they trade wth. Ye Country people, called ye Hoves for cattle & slaves wth other provisions & sell em again to ye Portugalls at a set rate. Oxen a 2 R pr. Pe. Cowes at 1½ pr. Pe. 20 hens pr. R. Goats & Sheep at ½ a R Ye people kind but all comes from ye maine.
>
> Therefore if ye river from whence they have there trade were used wth like Truvck as they deal for as brass; beads & other comodityes, things wod. Be very cheap. To ys river comes ye Hoves wth there caffalo in March & Aprill with 10,000 head of Cattle & 2 or 3000 Slaves. Ye portuguize had agreed wth ye Matthewledgians not to go over at least they made us believe so.[86]

This extract raises a number of contextual issues. First, as already noted, it is unlikely that any Hova would have been found arriving in trade caravans from the highlands in March and April: they would never have traded cattle on such a scale and rarely, if ever, would have traded such quantities of slaves. Second, the report was obviously second-hand. The *Francis* was an English East India Company (EIC) ship of 100 tons and 30 crew, based at Surat and employed in trade chiefly off the coast of west India. In 1640 the Portuguese leased the *Francis* to sail with cloth and beads from Cambay to Mozambique. However, the EIC in Surat ordered it, while awaiting its return cargo from Mozambique, and possibly without the

[83] Augnet.
[84] Vernet, "Slave Trade and Slavery," 43.
[85] Armstrong, "Madagascar and the Slave Trade," 214.
[86] "A voyage in ye ship Frances from Mossambique for St. Lawrence," in Stephen Ellis, "Un texte du XVIIème siècle sur Madagascar," *Omaly sy Anio* 9 (1979), 157.

knowledge of the Portuguese, to sail to north-west Madagascar—its first and only visit to the island—which it reached on 30 May, at least a month after the slave caravan from the Malagasy highlands was supposed to have reached the north-west coast (Fig. 3.1).[87]

Third, the report derived from a Portuguese agent whom Tyrrell found, and questioned, at Boeny. The Portuguese were deeply resentful of their European rivals, notably the Protestant Dutch and English. Dutch ships had twice, in 1604 and 1608, attacked and attempted to seize Mozambique fort.[88] Since at least the 1610s, EIC ships had competed with the Portuguese for provisions in the region, notably in St Augustine Bay and Mohilla and Anjouan in the Comoro group, "where refreshing is plentiful and cheap."[89] They also competed for slaves. For example, a French ship that in mid-1708 visited Masselage stated that "the Dutch customarily send a ship there every year in the month of May to trade for slaves."[90] The English were possibly more aggressive. In January 1630, an EIC fleet in the Persian Gulf captured a Portuguese ship carrying 100 slaves, and in the following August another EIC fleet seized 126 Malagasy slaves and rice from a Portuguese vessel at Mohilla.[91] Unsurprisingly, when another EIC vessel, the *Blessing*, called at Mozambique in early 1637, Portuguese authorities there refused the captain permission to purchase slaves locally.[92] As Tyrrell's orders were to purchase as many Malagasy slaves as possible for the EIC posts at Surat and Bantam, it is highly improbable that Tyrrell's Portuguese informant would have offered his English rival accurate details about the structure of the Malagasy slave trade. Indeed, The *Francis* sailed from Madagascar with a cargo comprising chiefly rice and cattle, having purchased only 22 slaves. As the latter were very cheap, they were almost certainly from the immediate hinterland rather than from the

[87] Foster, *English Factories*, xxiii–xxiv, 226, 294–95; see also Axelson, *Portuguese in South-East Africa*, 121.

[88] Malyn Newitt, *A History of Portuguese Overseas Expansion 1400–1668* (London: Routledge, 2005), 191.

[89] Fursland and Brockedon to Fitzherbert, Batavia, 6 Oct. 1621, quoted in Foster, *English Factories*, 290; Anon., "Early Notices of Madagascar from the Old Voyagers, Part III: Hamilton's description of Madagascar, 1744," *Antananarivo Annual and Madagascar Magazine* 16 (1892), 497; see also Newitt, *History of Portuguese Overseas Expansion*, 207.

[90] "Le mémoire de Parat sur Madagascar (1714)," ed. J. C. Hebert, *Bulletin de l'Académie Malgache* 56, no. 1–2 (1978): 50.

[91] W. Noel Sainsbury, ed., "East Indies and Persia, 1630–34," *Calendar of State Papers, Colonial Series* (London: Eyre and Spottiswoode, 1892), 4, 6; Foster, *English Factories*, 310.

[92] Foster, *English Factories*, 39.

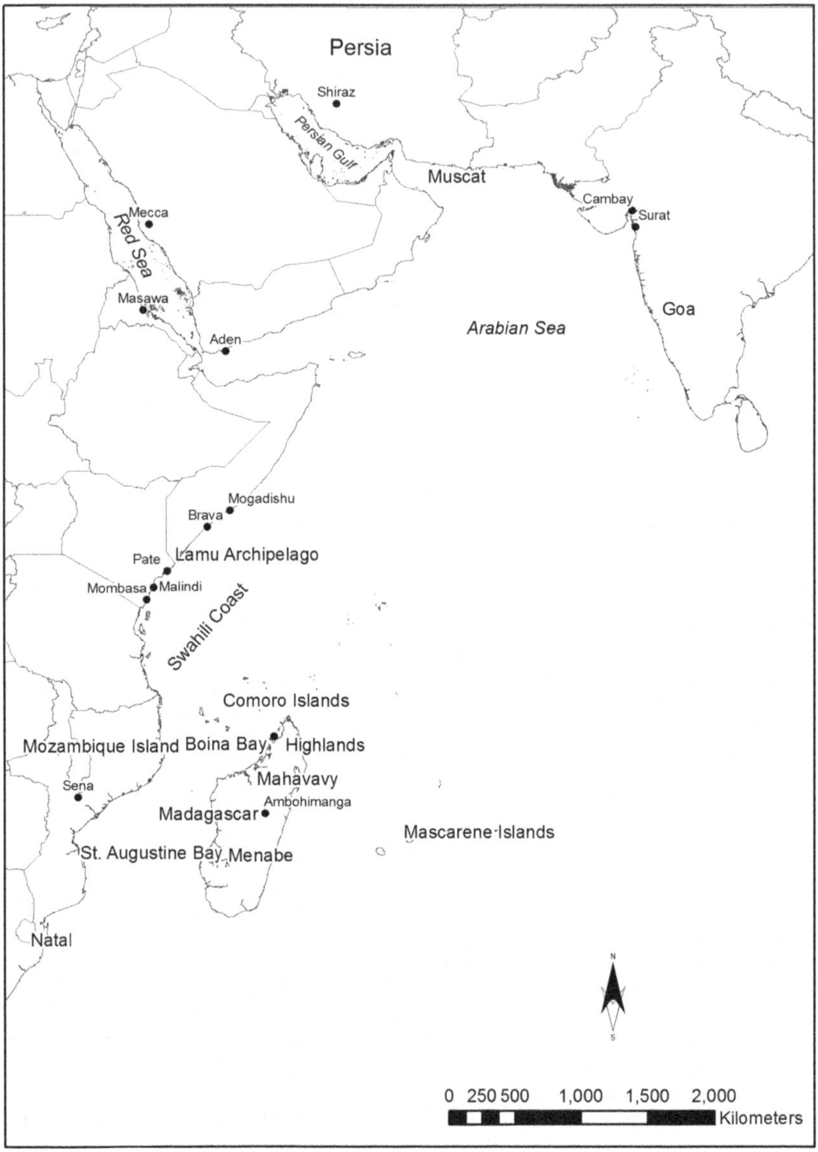

Fig. 3.1 The Western Indian Ocean World

distant highland interior.[93] Tyrrell nevertheless commended the commercial potential of Madagascar to his superiors in Surat and later sent a report of his Madagascar visit to the EIC headquarters in London.

The second English source is Nicholas Buckeridge (*ca.* 1625–*ca.* 1690), an "Anglo-Oriental merchant and adventurer,"[94] who sailed extensively in East African waters, visiting both Madagascar and the Swahili coats. In a 1663 report commissioned by the EIC, Buckeridge reported of Pate:

> Pate is an Island nere ye equinoctial Line which hath Safe harbours but dangerous to bee com to: inhabited by Arrabbans & other Mahometans. But ye Portugalls have ye cheife trade there & did take customs of all others that did trade thither. & ffrom Mombasse & this plave they did usualie carie to Indea A greater quantitie of Eliphants Teeth then they doe from Mosambiq, besides A good quantitie of Bees Wax & Tortice Shells, some Civett & ambergreece also procurable here. Ye goods that vend here are Cheiflie callicoes, but of A Different & finer sort than at either mombas or Mosambiq. Alsoe Iron, clth, Lead &c. will vend here more than at any other place on these Coasts. These people have A trade to Maselage & Assada on ye Island of St. Lawrence cheiflie for slaves but some sugar & provisons they have from assada, but noe great quantitie. But of Slaves they yearly bring from Masselage about 2000 head & Slaves are Cheap & plentifullie to be had at all places on those Coats but where ye Portugalls came.[95]

Vernet considers Buckeridge's account of Pate to be decisive evidence that a minimum of 2,000 highland Merina slaves were shipped through Pate annually, although he sides with the more inflated Portuguese estimate of 3,000–4,000.[96] Again, however, this account cannot be taken at face value. Buckeridge, who hailed from an elite High Anglican and pro-royalist background in England, joined the EIC in the early 1640s and was based in Surat, from where he supervised a voyage to Basra in 1644. He subsequently became the EIC agent in Persia, where he compiled a number of

[93] Foster, *English Factories*, xxiii–xxiv, 226, 294–95.

[94] Historical Manuscripts Commission, *Manuscripts of J. Eliot Hodgkin, Esq., F.S.A., of Richmond, Surrey* (London: printed by Eyre and Spottiswoode, 1897), 8.

[95] Report, 15 July 1663, in Nicholas Buckeridge, *Journal and Letter Book of Nicholas Buckeridge 1651–1654*, ed. John R. Jenson (Minneapolis: University of Minnesota Press, 1973), 45–46.

[96] Vernet, "Slave Trade and Slavery," 44.

commercial reports, and spent a number of years at Aden.[97] Buckeridge would have heard of Tyrrell's Madagascar visit first-hand when at Surat in 1840 and would also have learned of attempts, led by John Smart in St Augustine Bay from March 1645 to May 1646 and five years later by Robert Hunt on Nosy Be ("Assada"), to found English settlements in Madagascar.[98] In 1651 and 1653, the EIC Directors, who had also read Tyrrell's report and accounts of the Madagascar expeditions of Smart and Hunt, dispatched Buckeridge to explore the commercial potential of the Swahili coast and Madagascar.[99] In Mozambique in 1651, Buckeridge met the governor and leading traders. He also twice encountered and debated with Jesuits priests, who claimed the right to a percentage of the trade of the town.[100] On his second voyage from mid-December 1652 to 30 November 1653, he visited Anjouan, St Augustine's Bay, Sena, Mozambique—where he held a conference with Mozambique traders with strong commercial relations with north-west Madagascar[101]—Zanzibar, and Pate before returning to Bassein.[102]

[97] Buckeridge, *Journal and Letter Book*, 3–4; Historical Manuscripts Commission, *Manuscripts of J. Eliot Hodgkin*, 3, 7–8; Foster, *English Factories*, 21, footnote 1; see also Philip Anderson, *The English in Western India: Being the Early History of the Factory at Surat, of Bombay, and the Subordinate Factories on the Western Coast, from the Earliest Period until the Commencement of the Eighteenth Century. Drawn from Authentic Works and Original Documents* (London, Smith, Elder and Co., 1856), 34; Frederick Charles Danvers, *List of Marine Records of the Late East India Company, and of Subsequent Date, Presented in the Record Department of the India Office, London* (London, printed by Eyre and Spottiswoode: 1896), 4; Steven C. A. Pincus, *Protestantism and Patriotism: Ideologies and the Making of English Foreign Policy, 1650–1668* (Cambridge: Cambridge University Press, 1996), 349, footnote 41; Richard Grassby, *Kinship and Capitalism: Marriage, Family, and Business in the English Speaking World, 1580–1740* (New York: Cambridge University Press, 2001), 213; P. E. McCullough, "Buckeridge, John (d. 1631), Bishop of Ely," *Oxford Dictionary of National Biography*, http://www.oxforddnb.com/view/printable/3854

[98] Foster, *English Factories*, 272–73; Foster, "An English Settlement in Madagascar in 1645–6," *English Historical Review* 27 (1912): 239–50; Robert Brenner, *Merchants and Revolution: Commercial Change, Political Conflict, and London's Overseas Traders, 1550–1653* (Princeton, N.J.: Princeton University Press, 1993), 171–81; Alison Games, "Oceans, Migrants, and the Character of Empires: English Colonial Schemes in the Seventeenth Century *History Cooperative* (2003), http://webdoc.sub.gwdg.de/ebook/p/2005/history_cooperative/www.historycooperative.org/proceedings/seascapes/games.html

[99] Anderson, *English in Western India*, 34; Danvers, *List of Marine Records*, 4.

[100] Buckeridge, *Journal and Letter Book*, 25.

[101] Buckeridge, *Journal and Letter Book*, 62.

[102] Buckeridge, *Journal and Letter Book*, 11, 49–84.

However, in the reports of his 1651 and 1653 missions, which were specifically on the region's commercial potential, Buckeridge makes no mention of slaves. His first allusion to slaves occurs in a report compiled for the EIC in July 1663, following his retirement to England and ten years after his second voyage to East African waters. It would appear that before composing this, he consulted Tyrrell's report—filed at the EIC headquarters in London—of his 1640 voyage to Madagascar, and possibly interviewed Tyrrell himself. He may also have remembered allegations against Muslim traders in the region made by the Jesuits he had met on his own earlier visits to Mozambique. It is most likely that these are the reports that influenced him to allude to a slave trade from Madagascar to Pate and to estimate the number of slaves involved—the lower end of the number given by the Portuguese to Tyrrell.

Further, Buckeridge's account confirms the significant and consistent presence of Portuguese traders from Mozambique in the Comoros and ports of north-west Madagascar.[103] Portuguese commercial activity there was such that local Comoro and Malagasy commercial agents learned to speak Portuguese.[104] Portugal's European rivals were also actively trading for slaves and provisions in the Comoros and north-west Madagascar, as were European pirates from about 1680 to 1720.[105] However, the indications from eyewitness sources and visiting vessels are that most slaves exported from western Madagascar were a result of raids launched by

[103] Report, 15 July 1663, in Buckeridge, *Journal and Letter Book*, 44.

[104] Walter Payton, "A Journall of All Principall Matters Passed in the Twelfth Voyage to the East India, Observed by mee Walter Payton, in the Good Ship the Expedition: The Captaine whereof was M. Christopher Newport, Being Set Out, Anno 1612," in Samuel Purchas, *Hakluytus Posthumus or Purchas his Pilgrimes Containing a History of the World in Sea Voyages and Lande Travells by Englishmen and Others*, vol. 4 (Glasgow: James MacLehose and Sons, 1905), 183; Grandidier, *Histoire* (1908), 310–20; William Foster, ed., *The Embassy of Sir Thomas Roe to the Court of the Great Mogul, 1615–1619*, vol. 1 (London: Hakluyt Society, 1899), 19; Malyn Newitt, *A History of Mozambique* (Bloomington: Indiana University Press, 1995), 173.

[105] See e.g. Grandidier, *Histoire*, vol. 1 (1908), 310–20; "Richard Boothby, A Brief Discovery or Description of the Most Famous Island of Madagascar, or St. Laurence, in Asia, Near unto the East-Indies," in *A Collection of Voyages and Travels*, vol. 2, ed. Thomas Osborne (London: Thomas Osborne, 1745), 629; "Early Notices of Madagascar," 497; J. M. Filliot, *La traite des esclaves vers les Mascareignes* (Paris: ORSTOM, 1974), 24, 27–28, 31; R. E. P. Wastell, "British Imperial Policy in Relation to Madagascar, 1810–96" (PhD diss., London University, 1944), 34; Malyn Newitt, *The Comoro Islands: Struggle against Dependency in the Indian Ocean* (Boulder, Colorado: Westview, 1984), 18–20; Samuel Copland, *A History of the Island of Madagascar* (London: Burton & Smith, 1822), 152–54.

Sakalava in the western lowlands rather than warfare in the central highlands. As the population of the western plains was predominantly pastoral, and population density was low, the opportunity to obtain large and regular numbers of slaves for export was limited.[106]

Buckeridge also underscores the heavy disruption to trade caused by Portuguese–Muslim conflict along the Swahili coast:

> all these places were nere ruined by wars with ye Portugalls which the doe cheiflie manage by devieing these Petty Princes & Governors & imploying their forces against one an other. Nor doe the Portugalls suffer any they can hinder to trade to any of these places espetialie ye Arabians & Indeans whom they doe punish with Confiscation of vesaile & goods (if not captiivitie of their persons) if they are taken tradeing there.[107]

This conflict centred on the traffic in slaves: "Commodities they [the Portuguese] doe not suffer Mohametans to trade espetalie in slaves".[108]

Pate, the focus of considerable European commercial interest, was particularly embroiled in the Portuguese–Muslim conflict:

> Ye Portugalls had a ffactorie fformerly at Ampasa A Toune on ye Island of Pate where they had A Church & Preists & severall families resided there which I saw anno 1653 but in anno 1660 ye Arabbs of Muscatt were there with A fleete & destroyed or captivated all they found there & tooke 6 Saile of Shipps from them that usualie traded betene Indea & those parts. & what altercation hath been thereby occasioned since I know not.[109]

This strife, which severely disrupted trade, would appear to undermine the case for Pate being a major recipient and distribution centre for highland

[106] Mariano, "Relation," 9–10, 14, 22, 68; Drury, *Madagascar*, 230–31, 236, 258–68; R. K. Kent, "The Sakalava, Origins of the First Malagasy Empire," *Revue française d'histoire d'outre-mer* 55, no. 199 (1968): 167, 170–73, 178, 181–82; Grandidier, *Histoire* (1908), 306, 359, 503–04, 517; H. C. V. Leibbrandt, *Precis of the Archives of the Cape of Good Hope, January 1659–May 1662: Riebeeck's Journal, &c.*, part 3 (Cape Town: W. A. Richards, 1897), 81–82; James C. Armstrong, "Madagascar and the Slave Trade in the Seventeenth Century" *Omaly sy Anio* 17–20 (1983–84): 219–20; Entry, 8 December 1663, *Cape Monthly Magazine* 3, no. 13 (1880): 136.

[107] Report, 15 July 1663, in Buckeridge, *Journal and Letter Book*, 45.

[108] Report, 15 July 1663, in Buckeridge, *Journal and Letter Book*, 45–46.

[109] Report, 15 July 1663, in Buckeridge, *Journal and Letter Book*, 46; see also Alexander Hamilton, *A New Account of the East Indies*, vol. 1 (London: printed for C. Hitch and A. Millar, 1744), 12.

Malagasy slaves. Indeed, Abdul el Zein, recording traditions of slave acquisition in Lamu, states that its inhabitants began to acquire slaves in any number only in the late eighteenth century, when they established agricultural plantations on the East African mainland. These slaves were non-Muslims chiefly from present-day Tanzania and were acquired predominantly in Bagamoyo and Zanzibar.[110] Comorian slaves, generally skilled sailors employed on boats, were already Muslim, so they were considered superior to African slaves, were permitted to build their own mosque, and lived as an endogamous group, often fetching marriage partners from the Comoros and maintaining contact with them through the Swahili maritime trade network.[111]

Madagascar was thus clearly not the sole or principal source of slaves. Buckeridge notes that, provided they could avoid the Portuguese, Muslim traders found plentiful supplies of cheap slaves throughout East African waters. Indeed, Richard Allen estimates that the number of slaves shipped by non-Europeans from East Africa (excluding Madagascar) to regions outside the immediate Swahili coastal region amounted to 80,000–87,500 (800–875 p.a.) in the 1500s, 76,900–84,400 (769–844 p.a.) in the 1600s, and 144,000–180,000 (1440–1800 p.a.) in the 1700s.[112] Even if inflated, such numbers, and the fact that "non-Europeans" here means Muslim traders, would indicate that the supply of East African slaves easily satisfied demand from the Swahili coastal states, with a considerable surplus being shipped to Muslim markets in the Red Sea and Persian Gulf regions. In addition, there existed from at least the early sixteenth century a significant trade, via Massawa, Zeila, and Mocha, in Christian Ethiopian slaves, who were preferred to neighbouring "pagans," to Muslim markets in the Red Sea, Arabian Sea, and Persian Gulf regions.[113] The trade continued to be vibrant into the nineteenth century. Abdul Sheriff has estimated the

[110] Abdul Hamid M. el Zein, *The Sacred Meadows: A Structural Analysis of Religious Symbolism in an East African Town* (Evanston, Ill.: Northwestern University Press, 1974), 27.

[111] el Zein, *Sacred Meadows*, 28–30.

[112] Allen, "Satisfying the 'Want for Labouring People'," 56–57, 68.

[113] "The Voyage of Sir Francis Alvarez, a Portugall Priest, Made unto the Court of Prete Janni, the Great Christian Emperor of Ethiopia," in Purchas, *Hakluytus Posthumus*, 101, 205; Francisco Alvarez, *Narrative of the Portuguese Embassy to Abyssinia during the Years 1520–1527 by Father Francisco Alvrez*, trans. and ed. Lord Stanley of Alderley (London: Hakluyt Society, 1881), 302, 347; Hamilton, *New Account of the East Indies*, vol. 1, pp. 15–16, 24–25.

annual import of African slaves into Muscat in the period 1827–1831 to have been about 1,400 to 1,700, of which 75 per cent came from East Africa and the remainder from Ethiopia.[114] All in all, this calls into question the assertion of a large demand for Malagasy slaves, notably highland Malagasy, the transport costs and mortality rates for whom would necessarily have been significantly higher than for slaves from East Africa, which, because of its closer proximity to Persian Gulf markets, had lower slave mortality rates of between 10 and 20 per cent.[115]

Indeed, other sources appear to confirm this. In July 1614, William Keeling, an EIC captain who spoke good Arabic, visited the Comoros and upon enquiry reported that the north-west coast between 13°S and 16°S provided 400 slaves annually to traders who arrived each year from "Magadoxa [Mogadishu], Pata [Pate], Brava, and that coast [i.e. the Swahili coast]" and to the Portuguese.[116] Eighty years later, in 1694 a VOC captain stated that by August that year, Muslim traders had already shipped from north-west Madagascar to the Red Sea (Aden?) the 400 slaves held in readiness for export.[117] This would suggest that exports of Malagasy slaves into the Muslim trade network of the western Indian Ocean amounted to between 5 and 10 per cent of the figures postulated by Vernet. Moreover, there were certain years, such as 1710 at the start of a universally cold period, when it appears that few if any slaves were available in north-west Madagascar.[118]

[114] Abdul Sheriff, "The Slave Trade and its Fallout in the Persian Gulf," in *Abolition and its Aftermath in Indian Ocean Africa and Asia*, ed. Gwyn Campbell (London: Routledge, 2005), 104–06.

[115] Guillaume Grandidier and Raymond Decary, *Histoire physique, naturelle et politique de Madagascar*, vol. 5, no. 3, fasc. 1 (Tananarive: Imprimerie Officielle, 1958), 206; Foster, *English Factories*, xxiii–xxiv, 226, 294–95; Thomas M. Ricks, "Slaves and Slave Traders in the Persian Gulf, 18th and 19th Centuries: An Assessment," *Slavery & Abolition* 9, no. 3 (1988): 67; Allen, "Satisfying the 'Want for Labouring People'," 59.

[116] Quote from Keeling's journal, entry for 23 July 1614, in Michael Strachan and Boies Penrose, eds., *The East India Company Journals of Captain William Keeling and Master Thomas Bonner, 1615–1617* (Minneapolis: University of Minnesota Press, 1971), 90; see also Strachan and Penrose, *The East India Company*, 91–92; J. K. Laughton, "Keeling, William (1577/8–1620)," rev. G. G. Harris, *Oxford Dictionary of National Biography* (Oxford: Oxford University Press, 2004), http://www.oxforddnb.com/view/article/15243

[117] Armstrong, "Madagascar and the Slave Trade," 215.

[118] Grandidier, *Histoire* (1908), 498.

Summary

The thesis that from 1500 to 1750 there was a large, consistent Swahili-run trade in Malagasy highland slaves via ports in north-west Madagascar to Muslim markets throughout the western IOW cannot be sustained. It incorrectly assumes that during that period, highland Madagascar possessed a population surplus from which large numbers could be annually taken for export as slaves. The high central plateau is an infertile region that was afflicted with locust plagues and a series of adverse LIA climatic events that induced three major famines. Cultivation, focused on riziculture, demanded enormous labour inputs, while self-sufficiency was achieved only in the late eighteenth century following state-imposed *fanompoana*, or forced labour, in order to create and maintain an expanding network of dykes to drain and convert marshland into rice fields. War and armed raids produced slaves. However, most of the enslaved were ransomed back to their home communities or retained as slave labour in the highlands. Those exported as slaves from the highlands were chiefly criminals. Their numbers were insufficient to sustain the proposition that there existed a large and consistent slave export trade, and they could not have travelled to the north-west coast to be shipped out in March and April as suggested without suffering high mortality due to malaria, to which highlanders were as vulnerable as Europeans and which prevailed in the lowlands from late October to early April.

Moreover, a closer examination of the sources used reveals that they cannot be taken at face value. The main Portuguese sources emanate from Roman Catholic missionaries and a top Catholic layperson aggrieved at their inability to successfully propagate Catholicism in a region dominated by their religious archenemies—the Muslims. Their greatest propaganda tool, which aimed to persuade authorities in Europe to send sufficient military force to create a Portuguese empire throughout the region and then financially back Catholic proselytization, was to emphasize the Muslim threat, and specifically to exaggerate and decry a Muslim-run slave trade in which captive pagans were forced to convert to Islam and subjected to sexual exploitation by their Muslim masters. The two English sources used are based largely on these and similar Portuguese reports.

The historical evidence rather points to considerable commercial rivalry and hostility in East African waters, including the Comoros and north-west Madagascar, both inter-European and between Europeans and

Muslims. In the period under review, this rivalry, which for commercial and religious reasons focused particularly on the slave trade, heavily disrupted commercial traffic in the region. This, and the questionable nature of the sources used, renders highly improbable the claim for an immense and regular trade in highland Malagasy slaves to Muslim markets throughout the western Indian Ocean. Indeed, the evidence indicates that most slave exports from north-west Madagascar were from the hinterland, rather than highland interior, and a more realistic indication of the average number annually exported into the Muslim commercial network would be around 400, the lower end of James Armstrong's estimate for the seventeenth century.

CHAPTER 4

Volcanoes, Refugees, and Raiders: The 1765 Macaturin Eruption and the Rise of the Iranun

James Francis Warren

Introduction

This chapter is an investigation of the ways in which the unpredictable power of nature and the rise of the Iranun, as long-distance, saltwater slavers, changed the course of South-East Asian history. I discuss the effects of the eruption of Mount Macaturin on a people and a region. Too little is known about the extraordinary history of the eruption of Macaturin Volcano 1765—the most devastating explosion in centuries—and the remarkable events that occurred over the following several decades in Mindanao, the Sulu Archipelago, and north Borneo, including harvest failure, famine, social dislocations, and displacement. As far as is possible, I reconstruct what actually happened to the Iranun/Maranao people of the Rio Grande Valley in the period 1765–1790, and I try to make a balanced assessment concerning the probable links between the cataclysmic eruption of the Macaturin Volcano and its social and material impacts on the Iranun/Maranao, who lived in its shadow.

J.F. Warren (✉)
Murdoch University, Murdoch, WA, Australia

G. Campbell (ed.), *Bondage and the Environment in the Indian Ocean World*, Palgrave Series in Indian Ocean World Studies,
https://doi.org/10.1007/978-3-319-70028-1_4

Social scientists have recently turned to the "natural hazard" or the "post-event" forms of "disaster" as significant subjects of interest, and these topics have been approached from a variety of standpoints and emphases.[1] Geological hazards occur irregularly in time and space and cause negative impacts on humans and the environment. Periodically, South-East Asian peoples have been decimated by geologic and climate-related natural hazards, including volcanic eruptions, earthquakes, and tsunamis, typhoons, floods, droughts, and associated famine and disease. Although climate scientists are clarifying the natural and human-induced contributions to climate change and the variation in weather systems operating in the region, little is known about the historical basis of the social, cultural, and economic impacts of natural hazards beyond the linking of major Asian political events to natural disasters.[2] The social and economic impacts of stochastic events such as volcanic eruptions, earthquakes, and tsunamis are yet to be explored.

Terry Cannon argues that "*hazards* are natural but ... disasters are not."[3] The stress here is on the vulnerability of a people that makes it possible for a hazard to become a disaster. Given that "the repercussions of similar hazard events tend to have very different impacts on one community or society compared to another," Cannon argues that the factors that turn a geologic hazard, like a volcanic eruption, into a natural disaster require comparative analyses that focus on various economic and political systems and on the way these factors structure societies.[4] Geologic hazards,

[1] Guillermo R. Balce and Emmanuel G. Ramos, "Geologic Hazards in the Philippines: A Definition and Overview," in *Geologic Hazards and Disaster Preparedness Systems*, ed. Ann Varley (Manila: Philippines Institute of Volcanology and Seismology, 1987), 1–21; Greg Clancey, "Symposium on Natural Disaster in Asian History, Culture and Memory," unpublished draft (National University of Singapore, December 2004), 1.

[2] Greg Bankoff, *Cultures of Disaster* (London: Routledge Curzon, 2002); Alfred Crosby, *Ecological Imperialism: The Biological Expansion of Europe* (New York: Cambridge University Press, 1986); Jarrod Diamond, *Collapse* (New York: Viking Books, 2005); Brian Fagan, *Floods, Famines and Emperors: El Niño and the Fate of Civilizations* (New York: Basic Books, 1999).

[3] Terry Cannon, "Vulnerability Analysis and the Explanation of 'Natural Disasters'," in *Disasters, Development and Environment*, ed. Ann Varley (Chichester: John Wiley and Sons, 1994), 13–30 (author's emphasis); de Boer and Sanders, *Volcanoes in Human History*, 138, 146; Clive Oppenheimer, *Eruptions that Shook the World* (Cambridge: Cambridge University Press, 2011), 310–11.

[4] Piers Blaikie et al., *At Risk: Natural Hazards, People's Vulnerability and Disasters* (London: Routledge, 1994).

such as volcanic eruptions, are considered disastrous precisely because of the consequences they have for groups with particular cultural, ecological, and institutional arrangements at specific moments in time and space. Consequently, these cultural-ecological arrangements play an active role in determining whether, and to what extent, a natural event like a volcanic eruption is considered a catastrophe.[5]

The temporal dimension, or long-view history, is crucially important in aiding the understanding of the character and development of particular natural disasters. Such events unfold historically—even seemingly sudden, non-rhythmic occurrences like cataclysmic volcanic eruptions. The pre-conditions for such disasters ("root causes" and "dynamic pressures") would have been forming over a long period. Indeed, Anthony Oliver-Smith treats the Peruvian earthquake of 1970 as having "root causes" that reach back 500 years to the trauma of the Spanish conquest of the Inca Empire and the ensuing biological and social collapse that led to a loss of traditional methods of coping with environmental risk.[6]

At the southernmost end of the Philippines' volcanic belt, the Macturin cone provided one of the mid-eighteenth century's least-known geophysical disasters. The eruption produced a lava flow that stretched for more than 20 km, destroying scores of Iranun/Maranao settlements; the volcanic gases and poisonous smoke caused the death of thousands of Iranun people. Large stretches of the Rio Grande Valley were inundated by lava flows that contained huge boulders and abundant lava tubes and caused numerous dry river and stream beds. The event coincided with the arrival of the first English East India Company traders, who penetrated the unchartered coastal waters of south-western Mindanao and the Sulu Archipelago in search of trade goods to take back to China.[7]

Nearly three decades passed before the story of the eruption and the effects on Iranun/Maranao society and the environment were generally known. At this time, data on piracy and slave raiding began to be assembled, shared, and studied by rival colonial powers who were concerned

[5] William Torry, "Anthropological Studies in Hazardous Environments: Past Trends and New Horizons," *Current Anthropology* 20, no. 30 (1979): 517–40, 532.

[6] Anthony Oliver-Smith, "Peru's Five-Hundred Year Earthquake," in *Disasters, Development and Environment*, ed. Ann Varley (Chichester: John Wiley and Sons, 1994) 3–48; Blaikie et al., *At Risk*, 55.

[7] James Francis Warren, *Iranun and Balangingi: Globalisation, Maritime Raiding and the Birth of Ethnicity* (Quezon City: New Day Publishers, 2001), 22–23.

about the sudden emergence and devastating impact of the "Lanun."[8] Government officials, naval officers, and newspaper columnists speculated on possible reasons for their rise. However, it was an astute East India Company country trader, Thomas Forrest, who informed his superiors and readers about the significance of the year 1765, when south-western Mindanao suffered a massive volcanic eruption, when crops failed, when hunger threatened the Rio Grande Valley, and when famine and disease stalked the southern regions of the great island.[9]

The question is whether the eruption of Mount Macaturin eventually caused the Iranun diaspora of the 1770s and 1780s and the social dislocation and long-distance maritime raiding of the following years. Did its influence extend as late as the 1790s—the last years of the great Iranun diaspora? We need to understand the way in which this specific Iranun society worked and changed over time in the face of the sudden destruction of its homeland.

Philippine Volcanoes and Mount Macaturin

The Philippines is located on the borders of a volcanic and earthquake belt known as the "circum-Pacific belt" or "ring of fire," which comprises a near-continuous series of oceanic trenches, volcanic arcs, volcanic belts, and/or tectonic plates.[10] Across the centuries, the Philippine Archipelago has experienced numerous volcanic eruptions and earthquakes. Indeed, the islands owe their fertility to monsoon rains and the chemically rich grey ash of active, dormant, and extinct volcanoes. There are 220 quaternary volcanoes in the country, with 21 classified as having been active during historic times. Five of these 21 active volcanoes (Mayon, Taal, Bulusan, Canlaon, and Hibok-Hibok) have relatively short repose periods (8–50 years between eruptions), and they have been the most active in the geological record.[11] Philippine eruptive phenomena range from lava flows

[8] Warren, *Iranun and Balangingi*, 1, 53.

[9] Thomas Forrest, *A Voyage to New Guinea and the Moluccas from Balambangan: Including an Account of Maguindanao, Sooloo and Other Islands* (London: C. Scott, 1779), 192–93; Melvin Mednick, "Some Problems of Moro History and Political Organisation," *Philippine Sociological Review* 5 (1957): 39–52, 43.

[10] "Ring of Fire," *U.S. Geological Survey Earthquakes Visual Glossary*, 2009, http://earthquake.usgs.gov/learn/glossary/?term=Ring of Fire

[11] Raymundo S. Punongbayan, "Disaster Preparedness Systems for Natural Hazards in the Philippines: An Assessment," in *Geologic Hazards and Disaster Preparedness Systems* (Quezon

to pyroclastic ejections and gas emissions.[12] The fall and flow of hot and/or cold materials during volcanic eruptions have caused major changes to the Philippine environment and ground disruption due to intrusion or explosive actions. These volcanic phenomena have also caused environmental changes, with the size of the area affected being dependent on the magnitude of the event.[13]

The Jesuits' dedication to volcanology in the Philippines forms one of their most important scientific contributions. From the sixteenth century to the eighteenth, they studied single volcanoes and eruptions and wrote about the probable causes of these events. Towards the end of the nineteenth century, the Spanish Jesuits established a seismological section within the newly established weather bureau to continually study and investigate geophysical data relating to volcanological phenomena.[14] The Spanish had documented volcanic events in the Philippines since their occupation of the country. These records note that there had been a large eruption on Mindanao on 4 January 1641 that almost overwhelmed the ships of a Spanish squadron off the south coast of the island because of the amount of the ash ejected.[15]

Indeed, Mount Macaturin has a grim and terrible history that includes the 1641 and 1765 eruptions. The late nineteenth-century Spanish seismologists considered Macaturin a mysterious volcano because of its inaccessible location in Muslim-controlled territory, but they reported three eruptions in the nineteenth century: the first on 20 January 1840, the second on 1 November 1856, and the third just before the earthquake of 8 December 1871. At the beginning of the twentieth century, the Jesuit Miguel Saderra Maso noted that the actual location and life of the volcano were still not known, although there was evidence of Macaturin's past work. He noted:

City: Philippine Institute of Volcanology and Seismology, 1987), 77–101, 78.

[12] Jerome B. Cruz, "Volcanism and Associated Hazards in the Philippine Setting," in *Geologic Hazards and Disaster Preparedness Systems* (Quezon City: Philippine Institute of Volcanology and Seismology, 1987), 22–37, 22.

[13] Balce and Ramos, "Geologic Hazards in the Philippines," 12.

[14] See Miguel Saderra Maso, *Historia del observatorio de Manila fundado y dirigio por los padres de la mision de la Compañia de Jesus de Filipinas 1865–1915* (Manila: E. C. McCullough Co., 1915); W. C. Repetti, SJ, *The Manila Observatory* (Washington: Georgetown University, n.d.); Benjamin Defensor, "What Ails the Weather Bureau?," *Manila Chronicle*, 18 July 1960, pp. 1, 11.

[15] *Description of the Philippines Compiled by the Bureau of Insular Affairs, War Department* (Manila: Government Printing Office, 1903), 37.

> Various authors say that volcanic bombs thrown up by this volcano may be seen near the port of Polloc, which is at a distance of 7 leagues [about 21 miles] from the volcano. Many volcanic rocks may be seen lying on the coral banks which extend along the coast from Polloc to Malabang. In this military post, which is famous for the springs which issue from cellular basalt mixed with clay beds, there may be seen a kind of porous or scoriaceous detritus, probably due to some relatively recent eruption.[16]

Iranun: Early History

Ruurdje Laarhaven notes that the Iranun were linked through kinship and alliances to the Maguindanao.[17] They were Maranao-speaking and culturally and linguistically related to the Maguindanao.[18] However, the pejorative Spanish label "moro" tended to obscure both the nature and extent of internal differentiation between Iranun, especially those based in other parts of South-East Asia and other ethnic groups of Mindanao and Sulu. Early Spanish and Dutch references to the Iranun attest to their encampments in the densely populated Lake Lanao basin, their military prowess and warlike nature, and their willingness to engage in slave raiding for the Sultanate of Cotabato.[19]

The Iranun originally inhabited coastal stretches around the mouth of the Pulangi, Polok (Polluc) harbour, and the eastern shore of Illana Bay. By the start of the seventeenth century, thousands of Iranun had also

[16] Miguel Saderra Maso, SJ, *Volcanos and Seismic Centers of the Philippine Archipelago* (Washington: Government Printing Office, 1904), 30–31.

[17] Ruurdje Laarhoven, "Lords of the Great River: The Maguindanao Port and Polity during the Seventeenth Century," in *The Southeast Asian Port and Polity: Rise and Demise*, ed. J. Kathirithamby-Wells and John Villiers (Singapore: Singapore University Press, 1990), 161–86, 166. The word *Maguindanao* is derived from the root *danao*, which means inundation by sea, river, or lake. *Maguindanao* means "that which has been inundated." It is the most appropriate term that could have been designated to describe the broad lowland of the Rio Grande de Cotabato because of the often flooded condition in which this intermontane basin and neighbouring river valleys are found. It was the Maguindanao, people of the flood plain, who lent their name to the vast island, which the Spaniards shortened and corrupted to Mindanao. See Najeeb M. Saleeby, *Studies in Moro History, Law and Religion* (Manila: Bureau of Public Printing, 1905), 13.

[18] Melvin Mednick, "Encampment of the Lake: The Social Organisation of a Moslem Philippine (Moro) People" (PhD diss., University of Chicago, 1965), 18; personal communication, David Barradas, 7 February 1974.

[19] Laarhoven, "Lords of the Great River," 166; Reynaldo Clemena Ileto, *Maguindanao 1860–1888: The Career of Dato Uto of Buayan* (Ithaca: Cornell University, 1971), 4–6.

migrated inland to the lake and plateau region at the south-western corner of the Tiruray Highlands. This coast and bay, whose shoreline constituted a continuous line of impenetrable mangroves and swamps, was readily linked to the great lake behind it, which the Iranun considered their stronghold and home.

This "homeland"—a vast area of lakes, dense forests, rivers, and mangrove swamps—was a significant obstacle to travel and, until the second half of the nineteenth century, a natural barrier between the Iranun and the Spanish. Laarhoven notes that the geography of Mindanao facilitated the maintenance of the Cotabato Sultanate.[20] Its prime down-river coastal position, with access to harbours and major sea routes, coupled with its central position surrounded by large mountains in the Pulangi river valley, provided a favourable environment in terms of the economy, production, and natural security. Laarhoven estimates that at the beginning of the eighteenth century, there would have been 90,000–100,000 Iranun and Samal in the inland and coastal areas.[21]

The valley of the Rio Grande river, above which towered several imposing volcanic peaks, possessed some of the most spectacularly fertile soils in the Philippines thanks to regular monsoon rains and volcanic eruptions. While their earlier eruptions may have caused disaster on a large scale, they also had a beneficial effect: the chemically rich volcanic ash made the soil extremely fertile, so the inhabitants were able to grow rice crops intensively. The result was steady population growth in the Rio Grande Valley. In the mid-eighteenth century, Mount Macaturin's untamed forces had far greater potential for destruction than ever before. The burgeoning human population meant larger settlements and more densely packed riverbeds and shorelines around the lake, which increased the risk of a major catastrophe due to a volcanic eruption.

Several English witnesses provide accounts that invite us into the lives of the Iranun when Macaturin threatened annihilation. One is William Dampier, who, for seven months during 1686 and 1687, lived among the Maguindanao and Iranun. Dampier, who crossed three oceans and four continents aboard a British privateering vessel, the *Cygnet*, was a gifted navigator and an exceptional observer and commentator about the places

[20] Ruurdje Laarhoven, *Triumph of Moro Diplomacy: The Maguindanao Sultanate in the 17th Century* (Quezon City: New Day Publishers, 1989), 182.

[21] Laarhoven, *Triumph of Moro Diplomacy*, 109–10.

and people he visited.[22] He wrote about the Maranao society he encountered in meticulous detail and with a certain amount of detachment. Slave raiding in this earlier period often extended overland into the interior of the large island of Mindanao. For many Maranao, Iranun, and Maguindanao, slaving was a temporary occupation to which they resorted only to acquire captives to work their lands and help them settle down. These raids were conducted by small bands of men rather than the large numbers of raiders and long ships usually deployed in the Iranun maritime raiding of the late eighteenth century.

In his account, Dampier claimed that the Iranun/Maranao society was prosperous and stable with a developed commercial life. He noted that they built "good and serviceable ships and barks for the sea." The Iranun traded with the Maguindanao in beeswax and gold from the interior of Mindanao and received in return "their calicos, muslins and China silk."[23] While Dampier wrote that the Iranun were a "martial race" that had built some "ships of war," he made no mention of any marauding propensities or of the terrifying attacks on the coasts of South-East Asia that became characteristic a century later. Why, then, after a century of relative obscurity, did Iranun warrior-aristocrats take to the seas of South-East Asia to make such a dramatic impact on the population and the region? Before the end of the eighteenth century, scant documentation is available regarding the exodus of the Iranun/Maranao from the lake region to the southern coast of Mindanao. Until that time they had remained little known to Spanish invaders, European travellers, and traders because of their up-country agricultural orientation. When Dampier visited Cotabato in 1686, he described the insularity of the Iranun/Maranao thus: "The Hilanoones live in the heart of the country, they have little or no commerce by sea, yet they have proe's [*sic*] that row with 12 or 14 oars apiece. They enjoy the benefit of the gold mines; and with their gold buy foreign commodities off the Mindanao people."[24]

It is difficult to get a clear and accurate picture of what transpired at the time of the 1765 disaster. Nevertheless, one of the richest sources on the Iranun was written by the English East India Company captain and explorer Thomas Forrest, who lived on the island of Balambangan and

[22] William Dampier, *A New Voyage Round the World* (London, 1697; repr. with an introduction by Sir Albert Fray, London: A & C Black, 1937).

[23] Dampier, *A New Voyage Round the World*, 227–28.

[24] Dampier, *A New Voyage Round the World*, 333.

who, for several years, sailed to the east with the Maguindanao and Bugis as an explorer-cum-political emissary. What is fascinating about his published account of the voyage of the *Tartar Galley*, which occurred over 20 months from 1774 to 1776, is that it is more than tales of plundering expeditions and the sea battles fought by the Iranun and Bugis against the Dutch. He also includes meticulous descriptions of the new lands and new people they encountered en route. Forrest vividly describes the coasts and islands of Mindanao, Sulu, and Sulawesi and parts of the Moluccas and New Guinea, and catalogues their rivers, harbours, and exotic fauna and flora, as well as the culture and society of the inhabitants he met along the way—especially those of Maguindanao and Iranun.[25]

In the 1760s, Forrest had taken part in the initial English East India Company voyages of discovery and trade to the Mindanao-Sulu region, and he recorded in his diary a description of the Maguindanao and Iranun and their material culture as well as the fauna and flora of the island.[26] He actually learned the languages of the Maguindanao and Iranun and studied their history, customs, nautical skills, and boat building. He subsequently wrote a remarkable account of their way of life in the mid-1770s—a decade after the Macaturin eruption and precisely when everything was in flux socially, politically, and economically because of the advent of the China trade.[27]

There is a paradox between the rise of the Sulu Sultanate, increased Iranun maritime raiding, and the opening and imminent decline of China at the hands of Europe. These events all took place at much the same time (the second half of the eighteenth century) as the Macaturin eruption and the introduction of tea from China to Europe.[28] By the end of the eighteenth century, Britain's insatiable desire for tea and the impact of the 1765 eruption on the Iranun were to change the face of Asian history and the shape and future destinies of both Sulu and China.[29]

It was the scale and impact of the largely unchronicled volcanic disaster in this isolated part of Mindanao that probably accounted for its legendary

[25] Forrest, *A Voyage to New Guinea and the Moluccas*, 174–337.

[26] Forrest, *A Voyage to New Guinea and the Moluccas*, 174–337.

[27] "The Memorial of Thomas Forrest to the Court of Directors of the East India Company", 29 June 1779, 21 July 1779, East India Company Court Minutes, vol. 88, F. 188. East India Company Office Records, British Library.

[28] James Francis Warren, *The Sulu Zone, the World Capitalist Economy and the Historical Imagination* (Amsterdam: VU University Press/CASA, 1998), 15.

[29] Warren, *The Sulu Zone, the World Capitalist Economy and the Historical Imagination*, 25–26; Peter Burke, *History and Social Theory* (Oxford: Polity Press, 1992), 152.

role in some Iranun folklore. The 1765 incident drove thousands of families out of the Rio Grande Valley to take up residence in the Sulu Islands and beyond. Ninety years after Dampier's visit to Lake Lanao, it was Thomas Forrest who put his finger on the link between the disastrous eruption of Mount Macaturin and the key events that followed: namely, the migration of the Iranun and their sudden rise to prominence as saltwater slavers.

Forrest wrote of a Plinian eruption and pyroclastic flows that released large volumes of superheated gas and debris, which rolled towards the sea with percussive force. Lahar from the Macaturin Volcano destroyed towns on its flanks, and hail storms of ash and stone battered the countryside. Forrest hiked across stretches of the coastal landscape of south-west Mindanao that had been dramatically altered by this recent event. In May 1775, he noted in a journal entry the impact the dreadful eruption had made on the now "dry harbour" of Tubug, which had become a principal place for assembling Iranun raiding vessels:

> Large stones loaded many places, even at the sea side, and at Tubug near Pulo Ebus, I have seen fresh springs burnt out, (at low water) from amongst the black stones, of many tons weight, in various parts of the dry harbour. I was told that a river was formerly there, where is not the least appearance of one now. At present there seems to be a good deal of mold intermixt with the black sand, which is favourable for vegetation and the country hereabouts is now covered with long grass called lalang. In some places are reeds eighteen foot [5.5 m] high.[30]

The scene of utter devastation and the way in which Mindanao's surface had been transformed by the age-old geophysical process of volcanism could still not account for the scenes of mass death and the catalogue of wretchedness and misery that emerged after the eruption struck: famine, epidemic, broken families, violent death, and displacement. Obviously, there are gaps in our knowledge of landscape history and environmental change as well as the emergence of refugees and raiders during a turbulent period in the latter half of the eighteenth century that Forrest and others cannot fill in their journal notes. Given the fragmentary nature of the evidence, a great deal must be inferred.

[30] Forrest, *A Voyage to New Guinea and the Moluccas*, 192–93.

The Eruption and Flight

The 1765 geophysical event that caused violent mass deaths and resulted in acute hunger, disease, and widespread property loss to Iranun communities across south-west Mindanao was, by any definition, a disaster. It was not hard for Forrest to think how appalling things must have been during the days of the eruption. Far from being ancient history in his time, it was an event so extreme, so destructive, and so enduring that its effects are still being felt today. Indeed, it is almost inevitable that such a prodigious eruption would have found its way into Iranun oral tradition precisely because it was so vast and destructive.

Prior to the eruption, condensed clouds of steam must have hung over the summit of Mount Macaturin—the volcano that had erupted several times in the previous centuries. But, as with any eruption, the greatest dangers were the pyroclastic flows that could roar down the slopes at up to 100 km per hour, enveloping villages at will, followed by huge explosions that brought ash and debris down upon the surrounding areas. Blistering gas and huge ballistic fragments incinerated houses, triggered chaotic evacuations, and, within hours, pushed the death toll into the hundreds or thousands.

The mountain explosion created panic among the Iranun inhabitants of the valley. Men, women, and children ran pell-mell seeking safety, driving before them their frightened domestic animals. Some women and the elderly were trapped in their smouldering homes, blanketed by ash several feet deep. Elsewhere there was a veritable pandemonium, with scenes of chaos and panic as residents scrambled to escape the heat, clouds of poisonous gas, and pyroclastic flows in what Forrest considered the mountain's most violent eruption in a century.

As the eighteenth-century Iranun population of south-west Mindanao had expanded, the farmers had pushed higher up the volcanic slopes, employing dry rice cultivation methods. But, in 1765, some of these pioneering settlers found themselves within the radius of a permanent danger area around Mount Macaturin—a zone of at least 11–34 km, depending on the direction and development of volcanic activity. The volcanic eruption of 1765 destroyed vast tracts of agricultural land, which were buried by volcanic ejecta, and agricultural productivity was further reduced by volcanic ash that smothered and singed vegetation.[31]

[31] Punongbayan, "Disaster Preparedness Systems FOR Natural Hazards in the Philippines," 81.

Thousands of hectares of arable land were rendered useless, and reclamation was particularly difficult because there were so many large boulders in the deposits. Peter Boomgaard has suggested that volcanic eruptions of the magnitude of the 1765 disaster would have contributed to mortality indirectly by destroying crops, livestock, and boats, which directly affected subsistence and nutrition.[32]

Rampaging nature and the growth of the population were responsible for the onset of famine and starvation. Here the cause of the hunger must not centre on ownership and exchange but rather on people pushed beyond the brink who suddenly lacked sustenance. Starvation had not been a total stranger to the mid-eighteenth century Iranun. Throughout their earlier history, famines, usually triggered by drought, had exacted a terrible toll. But no recent famine had caused more suffering and death than the one that ravaged south-west Mindanao for several long years beginning shortly after the eruption of 1765. From this point of view, Forrest's brief, albeit fascinating, account of the eruption points to unrelieved deprivation and a decade of extraordinary suffering among the Iranun.

The Iranun region devastated by the Macaturin volcanic eruption became a wasteland. Obviously, to judge from Forrest's account, based on the experiences of eyewitnesses and his own observations, there was no way to rejuvenate the area and assist victims of the disaster to forge new local livelihoods. Suffice it to say that women, children, and the elderly have historically been the least able to anticipate, cope with, resist, and recover from the impact of such an event.[33]

When the fundamental equation that balances population density with the carrying capacity of land shifts as suddenly as it did in 1765, people have no choice but to disperse and settle elsewhere.[34] With their homes gone, famine victims in some devastated Iranun villages sought shelter with more distant relatives, friends, or even distant neighbours like the down-river Maguindanao and the Taosug of the Sulu Archipelago. Those refugees who were able fled to the Sulu Islands in the south or to the

[32] Peter Boomgaard, "Crisis Mortality in Seventeenth Century Indonesia," in *Asian Population History*, ed. Ts'ui-jung Liu et al. (Oxford: Oxford University Press, 2001), 193–97.

[33] Blaikie et al., *At Risk*, 218.

[34] Brian Fagan, *Floods, Famines and Emperors: El Niño and the Fate of Civilizations* (New York: Basic Books, 1999), xviii.

north coast of Borneo in the west, where food and good fortune were still readily available. The women and the elderly, who lacked the means or strength to leave in order to rebuild their lives in the aftermath of the devastating eruption, faced a grim future.

Resettlement

As Melvin Mednick notes, the Iranun/Maranao became segmented. In 1776, Forrest listed no fewer than 33 sultanates that divided up a population of some 61,000 persons. Forrest in his account noted that the Iranun or "Illanon" resided near the "great Lanao" lake and that, "on the banks of the Lano, are no fewer than 17 titled Rajahs, 16 who take the title of Sultan, besides those on the coast."[35] It was clear that the region's fragmented political and social institutions were ill equipped to deal with the displaced population or to rebuild shattered economies.

The displaced Iranun established settlements and mooring sites along the river mouths of the south-east coast of Illana Bay from Punta Flechas to Polloc and along the west coast as far as Sindangan Bay.[36] Their cruisers were fitted out in the vicinity of the Bay of Illigan at Larapan in the north and at Tubug and Tukurun in the south. Forrest described the harbour of Tubug as the chief point for the assemblage of marauding flotillas.[37] Behind the mangrove-screened wall of Illana Bay and in its innumerable inlets, the Iranun lived, built, and repaired their *prahus* (sailing craft) with impunity. Unable to return to the devastated plateau basin, they evaded capture by the Spanish in these new, out-of-reach settlements protected by mangrove swamps and a narrow river. The ingenious construction of bamboo scaffolds and track systems enabled them to rapidly remove their raiding craft from the water. In trees along the coast of the bay, the Iranun built a series of camouflaged watchtowers as

[35] Mednick, "Some Problems of Moro History and Political Organisation," 42; Forrest, *A Voyage to New Guinea and the Moluccas*, 276.

[36] Tomas de Comyn, *State of the Philippines in 1810, Being a Historical Statistical and Descriptive Account of the Interesting Portion on the Indian Archipelago* (Manila: Filipiniana Book Guild, 1969), 132; Saleeby, *Studies in Moro History, Law and Religion*, 15.

[37] "No. 7 Governor General a Señor Secretario de Estado, 4 June 1896," Archivo General de Indias, Filipinas 510, Seville, Spain; Forrest, *A Voyage to New Guinea and the Moluccas*, 193.

another line of defence. From the lake, in their *prahus*, they also pushed a passage northwards via a small river that ran into the sea.[38]

Less than two decades after the volcanic eruption in 1765 and the subsequent advent and rapid growth of the China trade, whole groups of displaced Iranun, dissatisfied with their fate and life under the Maguindanao Sultanate, moved to the Sulu Archipelago, where they became subject to the sultan at Jolo, the seat of the sultanate.[39] They had primarily left Mindanao to escape from the heavy burden of ecological disaster under the expectation that they were going to something better. These Iranun settled on the north and east coast of Jolo and on Basilan and established clientage relationships with Tausog *datus* (chiefs or aristocrats). In particular, Basilan, which is an island of considerable size located directly between southern Mindanao and the Balangingi cluster of islets and Jolo, became a place of frequent resort for the Iranun. Some of its Samal Yakan inhabitants likewise became slave raiders from their birth, and it was not unusual for them to identify themselves with the Iranun as maritime marauders. James Rennell noted in one of his journal footnote entries that "some of these Illanians [from Mindanao] are now settled at Basseelan of Sooloo from whence they fit out their privateers."[40] Several years later, Alexander Dalrymple, annotating and amending his colleague's journal, added at the foot of the page, "Many Illanos are now settled in Sooloo Dominions."[41]

In 1774, Forrest described many of Jolo's inhabitants as *orang Illano*: those "who live in a quarter by themselves."[42] The Iranun lived in distinct colonies or wards with their own headman, *nakodahs* (captain sailing craft), and blacksmiths, along with Visayan *renegados* (renegades), who had acquired the necessary expertise for successful slave raiding in South-East Asia. In this regard, Iranun blacksmiths developed their industrial art to a degree unparalleled by their Tausog or Maguindanao neighbours, and

[38] "Commander Blake to Rear Admiral F. Maitland, 13 August 1838," Admiralty 125/133-Sulu Piracy, National Archives, Kew, London, England; "The Illanoons," *Singapore Free Press and Mercantile Advertiser*, 13 December 1838.

[39] Warren, *Iranun and Balangingi*, 33.

[40] James Rennell, *Journal of a Voyage to the Sooloo Islands and the Northwest Coast of Borneo, from and to Madras with Descriptions of the Islands, 1762–1763* (British Museum, Add. 19299, April 1762–February 1763), 38.

[41] James Rennell, *Journal of a Voyage to the Sooloo Islands*, 38.

[42] Forrest, *A Voyage to New Guinea and the Moluccas*, 193.

at the beginning of the nineteenth century, they introduced important changes in the smithing industry and brass manufacture at Jolo.[43]

At the end of the eighteenth century, the Iranun population of the Sulu Sultanate experienced rapid economic growth, as Tausog *datus* heavily involved in the China trade became "silent partners" in maritime raiding expeditions, now more interested in the redistributive and exchange aspects of slave raiding. *Datus* advanced the outfits and, equally importantly, provided a market at Jolo for captives and all spoils taken. Iranun expansion was in character with the profound social and economic changes then occurring in the global economy of the Sulu Sultanate. Spanish reports covering the period 1774–1795 show the ever-increasing opportunities for the Iranun to settle, trade, and equip vessels at Jolo,[44] and it became Tausog practice to invite groups of Iranun to settle in the Sulu Archipelago for the purpose of slave raiding.

In 1814, J. Hunt listed the Iranun communities in the Sulu Archipelago from which slaving attacks were launched against the Visayas and Sulawesi. By then several thousand resettled Iranun were exclusively engaged in maritime raiding and slaving. The other "piratical establishments" mentioned by Hunt were predominantly Samal-speaking communities, notably on the small islands of Balangingi and Tunkil, which clearly also had many Iranun residents (Table 4.1).[45]

After the 1830s, while the entire Sulu group was still occasionally subject to visits from the Iranun during their extended voyages, they were no longer in the habit of stopping over at any permanent settlements except those at Basilan, Jolo, and Balangingi. The Iranun generally obtained their supplies of ammunition either at Jolo or by trafficking with other Iranun communities and bases, which were in communication with various small Dutch settlements on the coasts of Borneo and the adjacent islands.[46]

[43] See Arturo Garin y Sociats, "Memoria sobre el Archipielago de Jolo," *Boletin de la Sociedad Geografica de Madrid* 10 (1881): 110–33, 161–97, 193.

[44] "Juan Cencilli a Señor conde de Aranda," 16 April 1774, Archivo Historico Nacional, Estado 2845, caja 2; Jose Montero y Vidal, *Historia General de Filipinas desde el descubrimiento de diches Islas hasta nuestras dias*, vol. 2 (Madrid: Imprenta de M. Tello, 1888), 311.

[45] J. Hunt, "Some Particulars Relating to Sulo in the Archipelago of Felicia," in *Notices of the Indian Archipelago and Adjacent Countries*, ed. J. H. Moor (London: Cass, 1967), 50–51, 57–60.

[46] Warren, *Iranun and Balangingi*, 52.

Table 4.1 The Iranun communities in the Sulu Archipelago (1814)

Location	*Kampong/ ward*	*Jurisdiction/ leadership*	*Size of male population*	*Number of raiding prahus*
Jolo Town	Subyon	Datu Molok Datu Bukon	600	20
Jolo Island	Bual	Raja Muda Buling Panglima Daud	1500	30
Tulayan Island			300	
Pilas Island			300	10
Pangutaran Island	Bait-Bait	Orang Kaya Malik	Many Iranun	
Basilan			300	
Balangingi Island			Many Iranun	
Tunkil Island		Datu Timbing	500	10
Tawi-Tawi		Maguindanao Datu	300	10
Palawan	Babuyan		600	20

Source: The table is based on information compiled by J. Hunt for Stamford Raffles. I have listed only those communities designated as Iranun. J.Hunt, 'Some Particulars Relating to Sulu', pp. 50–51, 57–60

By the 1790s, the Iranun were strong enough to drive the Tausog out of Tempasuk in north Borneo and establish their own sphere of local influence.[47] Thereafter, the commanders and crew of the Tempasuk raiding vessels were no longer authorized by a commission of the Sultan of Sulu to capture the trading vessels of hostile states. After 1790, The Iranun of Tempasuk were independent mercenaries or privateers because their *joangas* (raiding vessels) were private ships in private ownership as opposed to raiding ships belonging to wealthy Tausog sponsors or to the Sultan of Sulu. No sovereign or state owned the *joangas* of Tempasuk.[48]

The Iranun Impact

The 1765 Macaturin disaster had a long-term effect on the Iranun and, as revealed through this reconstruction of the natural disaster, on the population and history of South-East Asia. The terrors of the sudden harsh

[47] Vicente Barrantes, *Guerras piraticas de Filipinas contra Mindanaos y Joloanos* (Madrid: Imprenta de Manuel H. Herrandez, 1878), 160.

[48] Warren, *Iranun and Balangingi*, 60–64.

presence of these well-armed alien Iranun raiders live on to this day in the oral recollections, reminiscences, popular folk epics, and drama of the victims' descendants in the Philippines, Indonesia, and Malaysia.[49]

Until recently, villages in many parts of eastern Indonesia, particularly Buton, were either situated well inland or, if on the coast, on steep cliffs with extremely difficult access—this being the historical legacy of defence against the threat of Iranun and Balangingi marauders and slave raiders.[50] Robert Barnes, in his classic study of Lamalera, notes that the village is really a "twin settlement," with a lower one (Lamalera Bawah) on the beach and an upper one (Lamalera Atas) on a nearby cliff top for protection from earlier maritime slave raids. Such villages in eyrie-like settings were usually palisaded, but in this case (as at Tira, the site of Michael Southon's fieldwork in Buton), the main defence was inaccessibility.[51] Christian Heersink also notes that on Salayer, most of the nineteenth-century settlements were situated in the interior. Here the northern and southern extremities of the island were the least safe and suffered most from Iranun piracy, while the alluvial west coast became the prominent zone of security and trade.[52] Evidence has also emerged supporting the fear and dread of the Iranun in the Java Sea. Kurt Stenross, researching the traditional sailing boats of Madura, accidentally stumbled upon people on the north coast in a small isolated village who still had terrifying memories of the Iranun.[53]

[49] See Charles Frake, "Abu Sayyaf Displays of Violence and the Proliferation of Contested Identities among Philippine Muslims," *American Anthropologist* 100, no. 1: (1998): 41–54; Benedict Sandin, *The Sea Dayaks of Borneo before White Rajah Rule* (London: Macmillan, 1967), 63–65, 127; Esther Velthoen, "Wanderers, Robbers and Bad Folk: The Politics of Violence, Protection and Trade in Eastern Sulawesi 1750–1850," in *The Last Stand of Autonomous States, 1750–1870: Responses to Modernity in the Diverse Worlds of Southeast Asia and Korea*, ed. Anthony Reid (London: Macmillan, 1997), 367–88; Warren, *The Sulu Zone, the World Capitalist Economy and the Historical Imagination* (Amsterdam: VU University Press/CASA, 1998).

[50] Michael Southon, *The Navel of the Perahu: Meaning and Values in the Maritime Trading Economy of a Butonese Village* (Canberra: Department of Anthropology, Australian National University, 1995), 22.

[51] Robert Barnes, *Sea Hunters of Indonesia: Fishers and Weavers of Lamalera* (New York: Oxford University Press, 1996), 44; Southon, *The Navel of the Perahu.*

[52] Christian Heersink, "Environmental Adaptations in Southern Sulawesi," in *Environmental Challenges in Southeast Asia*, ed. Victor T. King (London: Gunon, 1998), 95–120, 103–04.

[53] Report of interview with Daruji conducted on 4 January 2000, in Stenross to Warren, personal correspondence, 8 March 2000.

Scattered along the coastline of the Philippine Archipelago, it is still possible to find, albeit with difficulty, the remnants of the century-long presence of these raiders: an old stone watchtower, a crumbling church-cum fortress, or the ruins of a Spanish fort and cemetery. The remains of such neglected sites and fortifications, primarily concentrated along the coasts of llocos, Catanduane, Albay, Cebu, Leyte, and Samar, bear silent witness to the advent of growing affluence in the Sulu Sultanate and deep despair, displacement, and dispersion of people captured and sold by the well-armed Iranun.[54]

Whether as warriors, traders, or pioneering colonizers, the Iranun reached almost every part of South-East Asia. In the course of their sudden expansion, the Iranun raiders and settlers had a dramatic impact on the environment by mining tin, cutting down forests, harvesting *tripang* (sea cucumber) on a hitherto unknown scale, and kidnapping and enslaving humans. This process of Iranun expansion was known in the Malayo-Muslim world by the term *merantau*—also the name of a leading Iranun leader in the 1840s on the north-west coast of Borneo—meaning "to go on a quest or journey": a journey to seek good fortune or, to put it another way in Malay, to *makan angin*, literally to eat the wind.

Conclusion

The dramatic Iranun expansion westwards across South-East Asia and southwards to the fabled Spice Islands in the 1780s and 1790s is still not fully understood. What made many Iranun abandon their heartland in south-western Mindanao? When and why were the Iranun settlements in the present-day Philippines, Indonesia, and Malaysia established? The refugees from the 1765 Macaturin volcanic eruption became saltwater raiders who regularly travelled further than most South-East Asians had ever gone before, establishing a vast network of bases and communications. Such events marked the beginning of the period known to late eighteenth-century Europeans as the Iranun Age.

[54] James F. Warren, *The Sulu Zone 1768–1898: The Dynamics of External Trade, Slavery and Ethnicity in the Transformation of a Southeast Asian Maritime State* (Singapore: Singapore University Press, 1981); Rene B. Javellana, SJ, *Wood and Stone for God's Greater Glory: Jesuit Art and Architecture in the Philippines* (Manila: Ateneo de Manila Press, 1991), 61–110; and Rene B. Javellana, *Fortress of Empire: Spanish Colonial Fortifications of the Philippines 1565–1898* (Manila: Bookmark, 1997).

The initial fate of the Iranun was decided in large measure by the uncontrollable forces of nature. The displacement, relocation, and expansion surrounding the 1765 Macaturin eruption subsequently had far-reaching and often terrifying impacts upon the coastal populations of South-East Asia and the course of regional history. Greg Clancey notes that natural catastrophes are sudden breaches not only in the "normality of nature but also in the normality of social organisation and potential control."[55] In terms of large-scale interruptions to social reproduction and disaster causation, my approach here attests to an abiding historical interest in how individuals and social systems both respond to and perform under acute and chronic stress.[56] In this chapter, I have examined the enduring region-changing effects of the catastrophic eruption of Mount Macaturin and the immense events that followed. These events were generally so destructive and enduring that they are still being felt by institutions today, and they continue to exert an effect on the nature and growth of populations in parts of South-East Asia.

The initial migration of thousands of distraught people from the lake region of south-western Mindanao across a strait to the neighbouring island of Borneo, landing at Tempasuk, was first mentioned in a short paragraph on the pages of Thomas Forrest's history of reconnaissance and travel.[57] Many were farmers whose homeland had become a devastated, boulder-strewn volcanic landscape with no empty arable land. Fishers were better off, but they needed boats and nets. At first the newcomers had a negligible impact on the traditional societies who had lived on the north-west coast of the large island for centuries, but within decades a major transformation occurred. More Macaturin migrants arrived, and they took control of much of north-west Borneo, utilizing European arms and the seafaring technology of large-scale maritime raiding developed in southern Mindanao and the Sulu Archipelago as a direct consequence of the impact of the China trade on Tausog economy and society.

Fortified Iranun/Maranao communities and bases with an aristocracy linked by kinships and long-distance maritime raiding soon emerged and

[55] Clancey, "Symposium on Natural Disaster in Asian History, Culture and Memory," 5.

[56] See Pitirim A. Sorokin, *Manual Society and Calamity the Effects of War, Revolution, Famine, Pestilence, upon Human Mind, Behavior, Social Organisation and Cultural Life* (New York: Simon and Schuster, 1976); and Greg Bankoff, George Frerks, and Dorothea Hilhorst, eds., *Mapping Vulnerability Disasters, Development and People* (London: Earthscan, 2004).

[57] Forrest, *A Voyage to New. Guinea and the Moluccas*, 192–93.

spread to the straits of Malacca. By the early nineteenth century, the once large number of Iranun refugees and adventurers that Forrest had chronicled in passing had developed into a loosely unified confederation with a new culture: one centred on maritime raiding and slaving triggered by the rise of the Sulu Sultanate and its ferocious desire for captives and slaves. Forrest noted that some of the Iranun/Maranao migrants and warriors who had formerly held a lowly rank within the traditional hierarchy of the Maguindanao and Tausog involved in the China trade became men of power and prowess—both master and lord—when they became "Iranun."

In their remorseless search for captives and slaves, the Iranun brought the "border arcs," or moving frontiers of the margins of the Malayo-Muslim world and of the various colonial peripheries, home to the centres, striking back at the various empires' hearts around Batavia and Singapore, in the straits of Malacca and Manila Bay, and beyond, reaching right across the top of northern Australia. Their mobility, kinship, and diplomatic connections as well as their capacity to either protect or disrupt trade enabled the Iranun to forge region-wide links through the establishment of distant settlements and forward bases of operation. Their powerful fluid political confederation could make or break local states and destroy colonial trade networks and population centres.[58] In addition to being fierce warriors and slavers of popular stereotype, the Iranun were master artisans (e.g. smiths, shipwrights, weavers, and carvers), shrewd traders, and fearless, intrepid explorers. Their raiding and trading activities stimulated political, demographic, environmental, and social changes right across South-East Asia and in Europe and China.

Volcanic eruptions were capable of shaking the very foundations of some regions to their core on an intermittent basis, revealing in the process the complex inner workings and institutional moorings that underpin particular societies. Obviously, volcanic eruptions have been, like wars, droughts, and epidemics, potent events in the history of Mindanao and the Sulu Archipelago. Forrest understood the unfolding significance of the catastrophic eruption of Mount Macaturin. He readily recognized how the fate of a people has often been decided by the uncontrollable and unpredictable power of a geophysical hazard like a volcanic eruption. Forrest succinctly described the devastation and terrible effects caused by the 1765 Macaturin eruption upon the course of Iranun history in this manner:

[58] Warren, *The Sulu Zone 1768–1898*, 150–51; Mednick, "Encampment of the Lake," 47.

> About ten years ago, one of the mountains, six or seven miles inland from their [Iranun] part of the coast, broke out into fire and smoke … It ejected such a quantity of stones and black sand, as covered a great part of the circumjacent country, for several feet perpendicular. During the eruption of the volcano the black sand was driven to Mindanao, the ashes as far as Sooloo which is about forty leagues distant; and the Illanon districts suffered so much that many colonies went to Sooloo, even to Tampasook and Tawarran, on the west coast of Borneo in search of a better country, where many of them live at [*sic*] this day.[59]

Volcanic eruptions of the magnitude described here also led to an increase in slavery and raiding in other places and eras. The biggest and most influential volcanic eruption in recorded human history occurred on 10 April 1815, when the Indonesian island of Sumbawa became ground zero for a worldwide climatic catastrophe.[60] The cataclysmic eruption of Mount Tambora affected agricultural and trade-based societies on opposite sides of the world, from Native American tribes like the warlike Comanche to starving Chinese farmers, who were all confronted with the outbreak of unprecedented climatic volatility in a short period of time. Past volcanic eruptions had also caused widespread harvest failure, mass death, and migration.[61] But people the world over affected by this extraordinary 1815 eruption faced a subsistence crisis of unparalleled proportions. The Indonesian islands of Sumbawa, Lombok, Bali, Sulawesi, and eastern Java

[59] Forrest, *A Voyage to New Guinea and the Moluccas*, 192–93; Mednick, "Some Problems of Moro History and political Organisation," 43.

[60] Henry Stommel and Elizabeth Stommel, *Volcano Weather: The Story of 1816, the Year without a Summer* (Newport: Seven Seas Press, 1983); Jelle Zeilinga de Boer and Donald Theodore Sanders, *Volcanoes in Human History: The Far-Reaching Effects of Major Eruptions* (Princeton: Princeton University Press, 2002), 138–56; Clive Oppenheimer, *Eruptions that Shook the World* (Cambridge: Cambridge University Press, 2011), 295–319; Richard B. Stothers, "The Great Tambora Eruption in 1815 and its Aftermath," *Science* 224, no. 4654 (15 June 1984): 1191–98; Gillen D'Arcy Wood, "The Volcano Lover: Climate, Colonialism, and the Slave Trade in Raffle's 'History of Java' (1817)," *Journal for Early Modern Cultural Studies* 8, no. 2 (Fall–Winter 2008): 33–55; Michael Greshko, "201 Years Ago, this Volcano Caused a Climate Catastrophe," *National Geographic*, 8 April 2016, http://news.nationalgeographic.com/2016/04/160408-tambora-eruption-volcano-anniversary-indonesia-science/; "Red Skies over Tambora," *Tempo* (12 April 2015): 48–58.

[61] See Peter Boomgaard, "Human Capital, Slavery and Low Rates of Economic and Population Growth in Indonesia, 1600–1910," in *The Structure of Slavery in Indian Ocean Africa and Asia*, ed. Gwyn Campbell (London: Cass, 2004), 83–95, 89.

were the worst affected, as possibly more than 70,000 people were killed outright or died from disease and hunger in the months that followed. Among the islands the impacts were prolonged and complex, as multitudes of starving people sold their children into slavery.[62] The eruption also caused major weather changes that summer in China, where huge floods in the valleys of the Yangtze and Yellow Rivers destroyed Chinese farmers' rice paddies, forcing thousands of children into child slavery.[63]

The Tambora eruption, like the Macaturin eruption, enabled the consolidation of indigenous systems of slavery and raiding. In North America, 1816 became known as "the year without a summer." The traumatic effects of the Tambora eruption on Euro-American agricultural communities, when their crops failed and they were forced to migrate, is well documented. What is less well known is that this enormous eruption caused a major shift in the economies of Native American peoples in the trans-Mississippi west.[64] The Comanche abandoned agriculture and took to the horse and slave raiding. Their traditional way of life changed because of climatic variability and the disruption of agriculture. Just as for the Iranun, the question of adaptability was crucial for the Comanche, and they gave up the hoe for the horse and gun.[65] They attacked Spanish and French colonies on horseback from Louisiana to Texas and New Mexico and took captives on an unprecedented scale. The captives were adopted into the tribe, ransomed back to their own tribe, or sold to other tribes or foreigners. One of the far-reaching effects of the Mount Tambora eruption was that these emergent mounted warriors and hunters quickly became known and feared as "The Lords of the South Plains."[66]

[62] De Boer and Sanders, *Volcanoes in Human History*, 38, 146; Oppenheimer, *Eruptions that Shook the World*, 310–11.

[63] De Boer and Sanders, *Volcanoes in Human History*, 148; Shuji Cao, Yushang Li, and Bin Yang, "Mt. Tambora, Climatic Changes, and China's Decline in the Nineteenth Century," *Journal of World History* 23, no. 3 (September 2012): 590, 597–99, 607. Greshko, "201 Years ago, this Volcano Caused a Climate Catastrophe," *National Geographic* (8 April 2016).

[64] Gary Dixon, *Slavery in the West: The Untold Story of the Slavery of Native Americans in the West* (n.p.: Xlibris Corporation, 2011), 88; Juliana Barr, "Captivity, Gender and Social Control in the Texas-Louisiana Borderlands," in *Major Problems in the History of North American Borderlands*, ed. Pekka Hamalainen and Benjamin H. Johnson (Boston: Wadsworth, 2012), 68–82; Ernest Wallace and E. Adamson Hoebel, *The Comanches: Lords of the South Plains* (Norman: University of Oklahoma Press, 1952), 261–66.

[65] Preston Holder, *The Hoe and the Horse and the Plains: A Study of Cultural Development among North American Indians* (Lincoln: University of Nebraska Press, 1974).

[66] Wallace and Hoebel, *The Comanches*, 261–66.

CHAPTER 5

The Environment and Slave Resistance in the Cape Colony

Nigel Worden

Introduction

It has long been recognized that small-scale desertion and individualized forms of slave resistance were more prevalent in the Cape Colony of southern Africa than large-scale revolts.[1] The reasons usually given for this draw from comparative studies of transatlantic slavery: slaves did not form an overwhelming majority of the colony's population and were approximately equal in number to settlers; they were scattered across small-scale farmsteads rather than concentrated on large plantations; and they lacked a strongly unifying cultural background since there was only limited creolization and they originated from diverse regions across the Indian Ocean world.

[1] Most notably in R. Ross, *Cape of Torments: Slavery and Resistance in South Africa* (London: Routledge, 1983) and N. Worden, *Slavery in Dutch South Africa* (Cambridge: Cambridge University Press, 1985), Chap. 9. For a more recent summary, see N. Worden, "Revolt in Cape Colony Slave Society," in *Resisting Bondage in Indian Ocean Africa and Asia*, ed. E. Alpers, G. Campbell, and M. Salman (London: Routledge, 2007), 10–23.

The translations in the chapter are the ones we used in Worden and Groenewald, *Trials of Slavery*

N. Worden (✉)
University of Cape Town, Cape Town, South Africa

G. Campbell (ed.), *Bondage and the Environment in the Indian Ocean World*, Palgrave Series in Indian Ocean World Studies,
https://doi.org/10.1007/978-3-319-70028-1_5

The environment is another factor behind this phenomenon in the Cape Colony. This chapter explores the ways in which the Cape's physical environment provided opportunities for individual or small-scale group desertion during the heyday of Dutch colonial rule. In so doing, it identifies a slave geography—that is, a landscape perceived and used by slaves in ways different from those of their owners and of the indigenous Khoi and San herders and hunter-gatherers of the Cape. This landscape included the areas around and between settler farms, mountains in the heart of the colony, and coastal and maritime environments as well as a vast and unknown terrestrial and maritime hinterland. Over the course of the eighteenth century, however, the ways in which slaves used this landscape changed. The chapter ends by proposing that the colony's first (and only) collective slave uprising in 1808 and the forms of slave resistance in the subsequent final three decades of slavery marked a new development: the attempted conquest of those parts of the Cape's environment hitherto dominated by the slave owners rather than an escape from it. The environment itself did not change, but in 1808, for a brief period, the ways in which slaves used and perceived it did.

A Slave Geography

Studies of Cape slavery have usually focused on the lived experience of bondage in settler households and farmsteads. But around and between these locales were spaces that slaves associated with different kinds of experiences. The settlement at Table Bay established by the Dutch East India Company (VOC) in 1652 was followed by the rapid occupation of the south-western Cape and the destruction of Khoisan hunting and herding economies. However, intensive arable agriculture of the kind used in Europe and in parts of the colonial Americas was swiftly found to be inappropriate. Extensive agriculture was used for the production of grain along the western slopes of the Table Mountain range and on the farms of Stellenbosch and Paarl, established on the other side of the sandy Cape flats in the 1670s.[2] The introduction of viticulture in the late seventeenth century and the extension of settler farming to the Drakenstein, Franschhoek, and Wagenmaker valleys did little to change the pattern of large farms scattered across the landscape of the arable regions with sandy wastelands, hills, *kloofs* (valleys), and mountains between them (Fig. 5.1).

[2] L. Guelke, "Freehold Farmers and Frontier Settlers, 1657–1780," in *The Shaping of South African Society, 1652–1840*, 2nd ed., ed. R. Elphick and H. Giliomee (Cape Town: Maskew Miller Longman, 1989), 69–73.

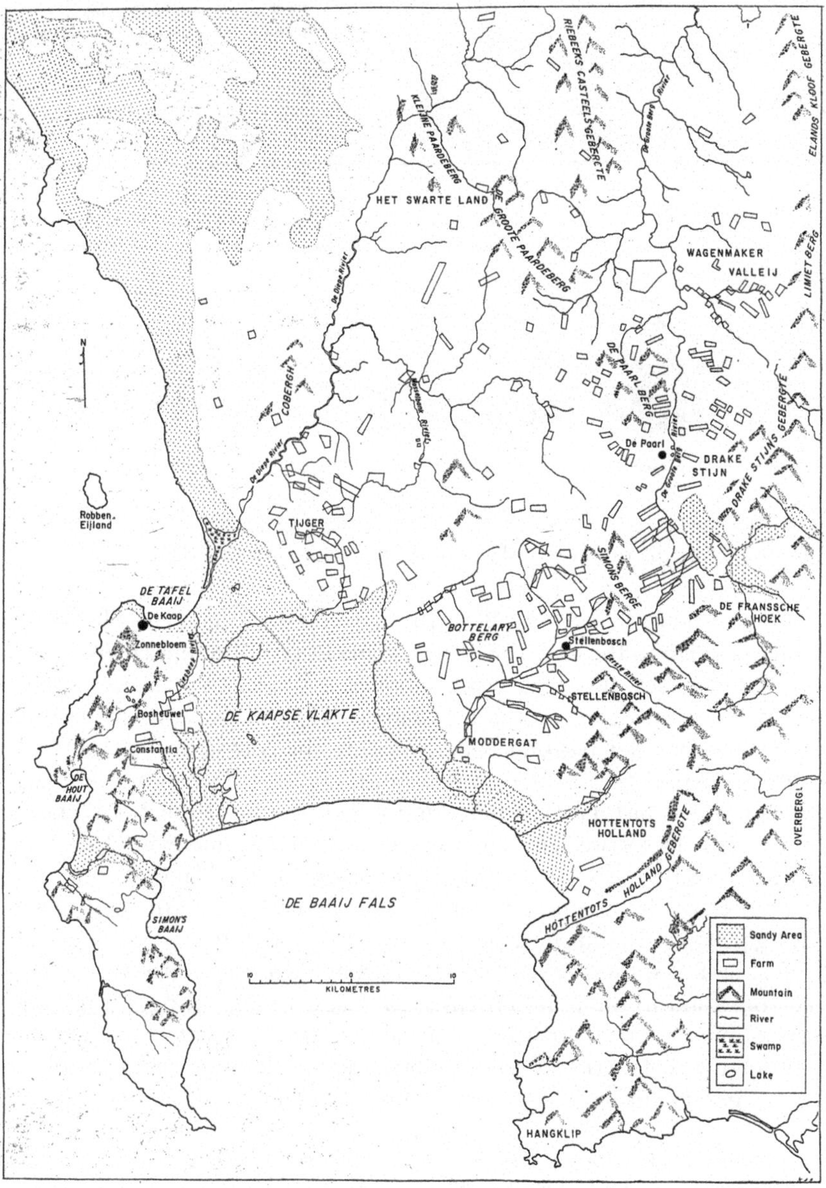

Fig. 5.1 The environment of the south-western Cape, *ca.* 1750

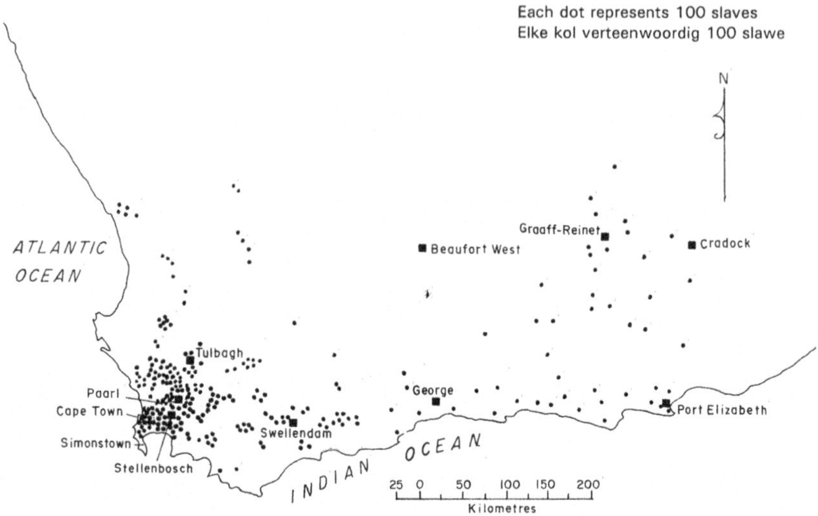

Fig. 5.2 The distribution of slaves in the Cape Colony, *ca.* 1820

Once settlers crossed the Berg River to the north and the Hottentots Holland range to the east, they entered an entirely different environment. Low rainfall, lack of regular water supplies, and infertile soils made pastoralism the only viable economic strategy. The "Cape frontier" is often associated with this pastoral zone, where in the course of the eighteenth century, competition over land, livestock, and water supplies dominated the continued history of conflict and dispossession of the Khoisan, and where some historians have located the roots of a distinctive trekboer settler society.[3] By the 1780s, settler pastoralists occupied lands hundreds of kilometres to the north and east of the original Cape settlement, where they grazed cattle and sheep across extensive pasturage.

As Fig. 5.2 shows, slaves were used primarily in the south-western arable zones and in the port settlement of Cape Town, although they were also found scattered across the pastoral regions of the colony, where they were used as field workers and herdsmen alongside dispossessed Khoi servants and indentured labourers. Thus, slaves lived in an environment

[3] S. Newton-King, *Masters and Servants on the Cape Eastern Frontier, 1760–1803* (Cambridge: Cambridge University Press, 1999); N. Penn, *The Forgotten Frontier: Colonist and Khoisan on the Cape's Northern Frontier in the Eighteenth Century* (Athens, OH, and Cape Town: Ohio University Press and Double Storey Books, 2005).

where settler control was widespread and intensifying but by no means absolute. Between the nodes of colonial concentration—the port and garrison, the farmsteads, and Company outposts—were regions that were much less firmly controlled by the VOC and its settlers. The boundary lines drawn on the Company's maps concealed the fact that its control over land well within these boundaries, indeed in the heart of the colony, was often tenuous. It was these liminal regions that provided the basis for an alternative slave use of the landscape.

The records of the Company are littered with complaints of slaves missing either temporarily or permanently. Slave desertion was a problem for the Company from the start: the first recorded escape was in March 1655, when Anthony of Madagascar ran away and was never seen again. In fact, the Company journal is filled with accounts of the escapes of the first group of imported slaves from Guinea and Angola.[4]

There were two environmental factors favouring slave desertion. One was the limited range of colonial settlement on the tip of a vast continent, which provided deserters with a ready hinterland into which they could disappear. Although many did just that, survival in such an alien environment was not easy. In comparison to military deserters, who could often get work on more remote farms, slave absconders were more visible and less likely to be sheltered from the authorities by colonists.[5] Slave deserters were dependent on the indigenous Khoi and San for directions, food, and water supplies, and without this assistance in the harsh and unfamiliar environment of the Cape hinterland, they were as helpless as their settler owners. And the Company knew this full well. From early on it offered Khoi rewards for the return of slaves, and as a result, as Robert Ross has observed, "even the sight of someone they suspected of being a Khoi, in the distance, was enough to send a party of [slave] fugitives into a panic."[6] But it was not only Company inducements that alienated runaways from the local inhabitants. As Nigel Penn has stressed in his study of the Bokkeveld and Roggeveld regions to the north of the colony, desperate and starving runaways were likely to be just as disruptive to the hunter and herding economies of the pastoral zones as were settler raiders.[7] Indeed,

[4] J. Armstrong and N. Worden, "The Slaves, 1652–1834," in *The Shaping of South African Society, 1652–1840*, 157.

[5] N. Penn, "Fugitives on the Cape Frontier, c.1680–1770," in *Rogues, Rebels and Runaways: Eighteenth-Century Cape Characters* (Cape Town: David Philip, 1999), 79–81.

[6] Ross, *Cape of Torments*, 44.

[7] Penn, "Fugitives," 75–77.

there was no reason why the Khoi should recognize any essential difference between them. Feelings of solidarity as victims of settler colonialism were likely to emerge only after the Khoi had been fully proletarianized and worked alongside slaves as labourers on settler farms; before then, slaves were yet another destructive element from an alien and intruding world. The slave landscape was not that of the Khoi; nonetheless, slave desertions were constant. As the despairing Cape Council of Policy observed to the VOC's directorate in 1680, "neither hunger nor thirst, the fury of ravenous animals, nor the murderous hordes of the more distant Hottentots [Khoi], nor finally the certainty of death, will suffice to shake their wicked resolution."[8]

However, there was another option available to slaves seeking desertion. Instead of escaping the colony and venturing into unknown regions, runaway slaves could attempt to live independently within its geographical boundaries. They were fugitives and outlaws and always subject to recapture, but they could attempt to evade this by surviving in a world with which they were familiar and where they could maintain links with fellow slaves and sympathizers. Many made such a bid, aided by the environmental features of the settler colony. There are regular reports in the judicial records of the first half of the eighteenth century of runaway slaves who survived in and around the settler farms of the south-western core of the colony, sometimes for lengthy periods of time. Karel Schoeman records how a group of slaves remained in the Stellenbosch area in 1704, eating slaughtered livestock and food raided from vineyards, orchards, and vegetable plots.[9] A few years later, in 1726, a group of at least five runaways hid out in the Blauuwberg and Tijgerberg hills and dunes within sight of Cape Town, and from there they raided farms over a period of six months.[10]

Perhaps the most striking testimony of the possibilities of survival for runaways within the boundaries of the colony comes from a case that appeared before the Cape courts in 1749. Reijnier of Madagascar was charged with having attacked his owner with a knife, after which he

[8] Cited in Worden, *Slavery in Dutch South Africa*, 125.

[9] K. Schoeman, *Early Slavery at the Cape of Good Hope, 1652–1717* (Pretoria: Protea, 2007), 277.

[10] Western Cape Regional Archives, Cape Town (hereafter CA), CJ 785, pp. 12–55. A selection of these records is transcribed and translated in N. Worden and G. Groenewald, *Trials of Slavery: Selected Documents Concerning Slaves from the Criminal Records of the Council of Justice at the Cape of Good Hope, 1705–1794* (Cape Town: Van Riebeeck Society, 2005), 101–06.

immediately took flight to above the ploughed fields of his owner, where he remained for about one week, and thereafter went from there to the mountains at Franschhoek and remained at that place, as best he can remember, for a period of sixteen years, after which he proceeded from there to the mountain situated above the farm of Jan du Busson, where he likewise remained some years, and where he was also, a few days ago, captured by some Europeans, taken to Stellenbosch and then delivered into the hands of justice. The confessant declares finally that during all of this time he had not spoken to a single person, and had maintained himself with dassies and fish which he caught with implements he manufactured to this end.[11]

It may well be that Reijnier's claim to have not spoken to a single person was designed to protect anyone who had assisted him from prosecution by the authorities. Nonetheless, his partner on the farm, the slave Manika of Bengal, now 60 years old, declared that "in all these years she has never heard from him, nor did she know if he was alive or dead."[12] It seems that Reijnier was able to survive for some 20 years in the Franschhoek Valley and the mountains that surrounded it.

The mountainous landscape of the south-western Cape provided the means by which Reijnier avoided detection for so long. The Franschhoek range remains still today a secluded region surrounding fertile vineyards and farms. Other mountains were even closer to the heart of the colony. Paarl, for example, a major arable and vineyard region in sight of Cape Town, was dominated by hills with caves and *kloofs* (secluded valleys) in which slave deserters could hide for lengthy periods. In 1706, for example, Ari, a runaway slave from near the Berg River in the north of the colony, headed back to the heart of the settler colony, where he met up with other absconders "near the Paarl Diamont" and raided vineyards and farms in the area. In 1713, Jacob of Madagascar fled from his owner with a gun, and he survived for five years in the Paarl Mountains near his owner's farm.[13] In the 1740s, runaways continued to group in the area, and in 1748 the Company organized a special commando to deal with the "vagabonds and runaways on the Paarl mountain."[14]

But the mountain that both symbolically and literally formed the core of the runaway slave's world was Table Mountain. Towering over the

[11] CA, CJ 357, pp. 7–9.
[12] Testimony of Manika van Bengal, 9 January 1749, CA, 1/STB 3/8, unpaginated.
[13] CA, CJ 784, pp. 102–03.
[14] Penn, "Fugitives," 90; Stellenbosch Dagregister, 2 July 1748, CA, C 653, pp. 508–09.

small colonial port and dominating the landscape of settler farms along its eastern slopes, in the eighteenth century the mountain represented not so much the picturesque as the untamed, unknown, and fearsome.[15] Abraham Bogaert, a visitor who climbed to its summit in 1702, found it "much toil and no little danger" and was repelled by its "unpleasant precipices" and "valleys beset with thick jungles and horrible caves," while others described it as "an unpleasant landscape," and the view from its top as one of "horror."[16] Visitors recorded fantasies of "fearsome lights" and remarkably shaped stones.[17] The mountain was a region in the heart of the colony that remained outside the control of the colonists. This was most evident in its occupation by runaway slaves. Their fires could be seen from the streets of the town, a constant reminder to the settlers of their vulnerability.

Yet for runaway slaves, the mountain represented not vulnerability but refuge and freedom. Slaves of the town and the farms nearby were familiar with its topography. The collection of timber from its slopes was a regular chore given to them, while slave women from Cape Town used Platteklip Gorge, where the mountain streams cascaded down towards Table Bay, for the equally relentless task of washing clothes.[18] There, as O. F. Mentzel reported, they could "avail themselves of the services of fugitive slaves. In return for some bread, meat and fish these outlaws would cut the trees down in the valley, hew them to shape and drag them up to the crest of the mountain."[19] In 1736, the authorities in Cape Town lamented the dangers of links between slaves working on the market gardens and homesteads above the town and the "runaways who live in groups on the

[15] N. Vergunst, *Hoerikwaggo: Images of Table Mountain* (Cape Town: South African National Gallery, 2000).

[16] R. Raven-Hart, *Cape of Good Hope 1652–1702: The First Fifty Years of Dutch Colonization as Seen by Callers*, vol. 2 (Cape Town: Balkema, 1971), 477; F. Valentyn, *Description of the Cape of Good Hope*, vol. 1, ed. P. Serton et al. (Cape Town: Van Riebeeck Society, 1971), 59.

[17] Raven-Hart, *Cape of Good Hope*, vol. 2, pp. 351, 372; P. Kolb, *The Present State of the Cape of Good Hope*, vol. 2 (New York: Johnson Reprint Corporation, 1968), 13; Valentyn, *Description*, vol. 1, p. 61; O. F. Mentzel, *Description of the Cape of Good Hope*, vol. 1, ed. H. Mandelbrote (Cape Town: Van Riebeeck Society, 1921), 93.

[18] E. G. Jordan, "'Unrelenting Toil': Expanding Archaeological Interpretations of the Female Slave Experience," *Slavery & Abolition* 26, no. 2 (2005), 217–32; E. G. Jordan, "It all Comes out in the Wash: Engendering Archaeological Interpretations of Slavery," in *Women and Slavery*, vol. 1, *Africa, the Indian Ocean World, and the Medieval North Atlantic*, ed. G. Campbell, S. Miers, and J. C. Miller (Athens: Ohio University Press, 2007), 335–57.

[19] Mentzel, *Description*, vol. 1, p. 90.

mountain."[20] In 1760, after these connections were revealed following the murder of a settler family by mountain runaways, town slaves were forbidden to go into the hills in search of firewood and a strict curfew was imposed.[21]

The mountain was a favoured destination for runaways for good reason. It was familiar and close to the core of the colony but also relatively secure. As Mentzel commented, "the presence of escaped slaves was well known, but pursuit was dangerous, since the hunted men could hide themselves among the rocks and stone their pursuers to death."[22] The runaways were more familiar with the mountain's terrain, its numerous caves and shelters, than were the soldiers sent out to capture them. Moreover, the steep slopes and crevices made the area unsuitable for mounted patrols. The trial of the 1760 murderers gives an indication of the difficulties encountered by these soldiers:

> [S]ince some time ago certain of the slaves of both the inhabitants of this town and those of the country areas ran away from their owners and, some earlier, some later, got together with one another at Table mountain until they reached the number of fifteen men. That this company was then, as soon as their hiding place became known, attacked in a cavern of the Windberg by some burghers who used every means possible to overpower them and to take them prisoner, without, however, it having been possible at this time on account of their vehement resistance, both through shooting with a pistol and by throwing down heavy stones on the aforesaid burghers; and because this company of runaway slaves escaped by fleeing after one of them, by the name of Cardoes, was shot dead by the said burghers, and then retreated to the back of Table mountain to a different cavern.[23]

Nonetheless, the mountain was neither sufficiently secure nor rich enough in resources to provide a permanent place of safety for runaways. It seems rather to have been a place of initial refuge before they moved to a more permanent hideaway. The mountain chain stretched down from Table Bay to the coast of False Bay, behind the farms of Constantia, and runaways could pass along it without crossing the boundaries of farms and settlements

[20] CA, CJ 340, pp. 97–103, 223–25.

[21] Resolutions of the Council of Policy, 15 July–19 August 1760, CA, C 138, pp. 348–74. These restrictions were withdrawn after the murderers were caught.

[22] Mentzel, *Description*, vol. 1, p. 90.

[23] CA, CJ 789, pp. 269–70. Documentation for this case is partially transcribed in Worden and Groenewald, *Trials of Slavery*, 355–84.

until they reached the wide shoreline of the latter. This gave them access to the sandy dunes of the Cape Flats, a region unoccupied by colonists, and thence along the coastline to more remote places of refuge (see Fig. 5.1).

There are numerous cases of such journeys recorded in the judicial records, revealing the existence of an underground runaway network that linked the urban and peri-urban core of the colony via Table Mountain to the wider Cape Peninsula and beyond. For example, a concerted commando operation following an attempt by slaves to set fire to the thatched roofs of Cape Town in 1736 caught a number of runaways on Table Mountain, near the Company outpost at Simon's Town on False Bay and further around the False Bay coastline.[24] In 1755, two Cape Town slave runaways were caught by soldiers at Simon's Town and confessed that they had survived for several months in the dune areas of the coast by relying on marine resources and food exchanged with slave shepherds in the area.[25]

The ultimate destination for many of these slaves was another notable place in slave topography: the rocky outcrops of Hangklip, which lay at the easternmost tip of False Bay. Hangklip was the closest the Cape came to possessing a maroon community, although, as Robert Ross has argued, there is no evidence that a self-reproducing population of runaway slaves ever emerged there.[26] Nonetheless, we know that a community of runaways existed at Hangklip from the 1720s until the 1750s, in the the late decades of the eighteenth century, and right up to slave emancipation in the 1830s. The geographical position of the region made it a perfect place of refuge. It jutted out into False Bay and so provided a lookout point from which commandos coming across the Cape Flats or ships in the bay could be readily spotted. It also contained a warren of underground tunnels and caves into which the runaways could escape and hide. Rich fishing and marine resources along the coast provided ample subsistence. Furthermore, the main route from Cape Town crossing the Hottentots Holland Mountains towards Swellendam and the eastern regions of the colony passed close by, and wagon convoys including food, ammunition supplies, and droves of cattle and sheep made an ideal target for highway

[24] Worden and Groenewald, *Trials of Slavery*, 139–43.

[25] Worden and Groenewald, *Trials of Slavery*, 330–36.

[26] Ross, *Cape of Torments*, 54–72, 122–24. Ross gives the fullest account of Hangklip (pp. 54–72) and an appendix listing 50 maroons known to have lived there (pp. 122–24).

holdups. Hangklip was also linked to the runaway network of the wider colony. In 1736, Cape Town was set on fire by runaways led by Leander van Bugis, who was based at Hangklip. In 1785, Sapdilje, a slave from Thailand, deserted from his owner near Simon's Town and fled to Hangklip, from where he led raids on the Company meat stores and warehouses at Simon's Town that he knew well.[27]

The reason why all of these cases—and there are many more like them—come to our notice is that the slaves involved were recaptured and put on trial. We of course will never know how many slaves successfully got away. However, as Nigel Penn has pointed out, the problem with desertion into these "peri-urban retreats" was that slaves were drawn back to settler areas for food and supplies and thus risked recapture.[28] Penn argues that the growth of the colony in the later eighteenth century made it more difficult for runaway groups to survive in its heartlands—although, as we have seen, Hangklip may have been the exception to this. Instead, he shows how areas beyond the Cedarberg Mountains, such as the Bokkeveld and Roggeveld, which were within striking range of the livestock herds of both settler and Khoi pastoralists, became favoured places to hide out.[29] Penn even suggests that runaway slaves, together with deserted soldiers and other refugees from the core of the colony, were "the true pioneers of the colonial frontier."[30]

The Global Hinterland

Not all runaway slaves sought refuge within the colony. Although, as I argued earlier, escape into an unknown landscape beyond its borders was filled with peril, slaves continued to abscond with the goal of reaching a specific destination elsewhere. Slaves, particularly the many who had been forcibly brought to the colony from elsewhere, saw the Cape as part of a wide geographical network. Some attempted to return to their homelands, or at least to parts of the network with which they were more familiar. In 1714, Tromp van Madagascar led a group of 11 slaves seeking to escape to "the Portuguese territory," a reference to Delagoa Bay in

[27] CA, CJ 795, pp. 239–49.

[28] Penn, "Fugitives," 76.

[29] N. Penn, "Droster Gangs of the Bokkeveld and Roggeveld, 1770–1800," in *Rogues, Rebels and Runaways: Eighteenth-Century Cape Characters*, ed. N. Penn (Cape Town: David Philip, 1999), 147–66.

[30] Penn, "Fugitives," 98.

south-east Africa.[31] Slaves from the south-west Indian Ocean region were particularly likely to try to get back home. In 1709, one Malagasy slave not only gathered fellow Malagasies to do so but also sought to persuade an unwilling South-East Asian slave to join them "to go to their country."[32]

More realistically, others sought to get not to their birthplace but to societies in the African hinterland where they believed they would be welcomed. The frequency with which runaways aimed to reach "Kaffirland" (the Xhosa communities of the eastern parts of southern Africa) suggests that knowledge of the possibilities of incorporation into such societies had penetrated back to the colony.[33] The lure of these welcoming communities was considerable. In 1738, a slave working on an outlying cattle post was approached by a runaway who asked him where the path lay that would lead him and his fellow fugitives "to the Caffers."[34] In 1751, a group of 12 Cape Town slaves and a convict ran away, inspired by the news "that there had recently been a group of slaves who had also taken flight and who had arrived safely at a free village of blacks and even on Madagascar."[35] In 1786, another group of Muslim slaves from Cape Town, believing that they were protected by talismans provided by a spiritual leader, "sought to get, in accordance with their planned scheme, far into the interior to the Caffers."[36] An alternative route was to go north into Namaqualand, although the danger of conflicts with local San and Khoi was more prevalent in these regions.[37]

However, the Cape's hinterland was not solely terrestrial; it was also maritime, and this presented other opportunities. Some of the first slaves brought to the colony from Guinea, reported the commander Jan van Riebeeck, "have even planned to seize a boat which belonged to one of the ships, and which was being used for fishing in the river or near the beach ... they intended to make off with it, searching for their country by sailing along the coast."[38] Throughout the eighteenth century, slaves attempted to get away in boats they purloined, without (as far as is known)

[31] Nationaal Archief, The Hague (hereafter NA), VOC 4073, p. 584, Q. 5.

[32] 3 October 1709, CA, 1/STB 18/155, unpaginated.

[33] For examples of such incorporations, see Ross, *Cape of Torments*, 81–95.

[34] CA, CJ 786, p. 194.

[35] CA, CJ 788, p. 59.

[36] CA, CJ 795, p. 377.

[37] For an example, see Worden and Groenewald, *Trials of Slavery*, 21–42.

[38] H. Thom, ed., *Journal of Jan van Riebeeck*, vol. 2 (Cape Town: Van Riebeeck Society, 1954), 355, cited in Schoeman, *Early Slavery*, 65.

much success. Moses van Balij tried in 1738 to make for the sea on a stolen boat he found at Vishoek (near Hangklip), but he could not control the vessel in rough waters and ended up stranded on a rock, from which he was rescued and captured.[39] In 1746, a group of slave charcoal burners from Hout Bay raided a farm on the shore:

> from where they took out of the workhouse a sail and a mast, as well as a rope which belonged to the boat that stood at that place on the beach … they also stole some mealies, water melons, pumpkins and apples, which fruit, after all of them put it in the aforementioned sail, they carried to the boat on the beach, in which was also to be found four ropes, a small anchor with its accessories and two small cod nets. After this, all nine of them brought the aforementioned boat to water, climbed into it and took off with it by day from the shore and, after being on the sea with good weather for four days and four nights, came on shore at Hangklip because of a storm, which caused the boat to leak, for which reason they remained there for two days.[40]

It is not clear where they were heading, although from Hangklip they tried to walk overland to "Kaffirland." Nearly 40 years later, the slave Cesar van Bengal also set out from Hout Bay in a fishing boat but was driven ashore in False Bay by strong winds.[41] A more ambitious plan was hatched by South-East Asian slave convicts on Robben Island in 1751 to overpower the crew of the island's supply ship and sail it to the East Indies.[42]

An alternative to sailing themselves was for runaways to stow away on a passing ship. In 1735, Januriij van Nagapatnam told his colleagues that rather than trying to escape overland, he would prefer to "go to the Cape to see if he could escape on an English ship."[43] His specific identification of an English vessel suggests that he was aware of the English East India Company's trading activities in his homeland on the Coromandel Coast and hoped to get back there. Stowing away was risky, however, since it entailed lurking unobserved in Cape Town and sneaking on board a ship in Table Bay harbour under the nose of the authorities. Certain "safe houses"

[39] CA, CJ 786, pp. 220–22.
[40] CA, CJ 354, part 2, pp. 464–67.
[41] CA, CJ 2488, pp. 56–57.
[42] P. Truter, "The Robben Island Rebellion of 1751: A Study of Convict Experience at the Cape of Good Hope," *Kronos: Journal of Cape History* 31 (2005): 42.
[43] CA, CJ 786, pp. 216–17.

were known in the town, such as the tavern and brothel in the 1730s called (with delicious irony) the *Schotse Tempel* and run by Clara Tant.[44] We will never know how many succeeded in getting away, but records of unsuccessful ventures suggest that it was certainly possible.

An intriguing example that took place in 1750 is that of Jan, a Cape-born slave. He later told the authorities that he "had been to the Cape with the wagon of his master, that he remained behind here, and that moreover he had got aboard ship with an English boat, and could not return again to shore, and thus got to the fatherland and to Zeeland, where he had married a woman with children." His mistake was to then enlist as a sailor with the VOC, claiming to be from St Helena (which would explain his darker skin). On arrival back in Cape Town, he was recognized and arrested.[45] A remarkably similar case took place two years later, when Jacob van de Caab escaped to Holland aboard a VOC vessel but then enlisted in the Company as a ship's boy and was found out when his ship arrived in Table Bay.[46]

Neither Jan nor Jacob was seeking a return to a world he had previously known, since both were born in the Cape. But their escape bids reveal that the maritime frontiers of the Cape provided a runaway slave with as good an opportunity for a new life as did the African hinterland.

Capturing the Settler Environment

The examples of slave resistance hitherto described have all shown ways in which slaves tried to escape from an environment dominated by their owners and to survive in a different landscape, either within the formal boundaries of the colony or outside it. These methods predominated throughout the seventeenth and eighteenth centuries. During this time, there were no attempts by slaves to take control over the settler environment created by their owners. There were attacks by runaways on farmsteads and burglaries in Cape Town houses, but these were temporary invasions, usually in search of food and sometimes also guns and ammunition. It was rare for the main house to be broken into; usually, slaves stole from the fields, storehouses, or barns.[47]

[44] NA, VOC 4135, p. 153. This emerged when a runaway was captured in a raid after neighbours complained of riotous noise.

[45] CA, CJ 33, pp. 31–33.

[46] CA, CJ 34, pp. 28–30.

[47] Worden, *Slavery in Dutch South Africa*, pp. 124–26.

Considerable alarm was caused in 1760, however, when a group of runaways from Table Mountain broke into a settler house on the outskirts of Cape Town and murdered the family. One of them, Achilles van de West Cust, had lived there and wanted revenge on his owner, but the prime goal was to obtain supplies in order to flee from the colony.[48] Their actions were characteristic of the topography we have been discussing: they assembled on the mountain and then crept into the garden of the house at night. Achilles then entered the house and warned the other slaves of what was about to happen. The lights were doused and the gang broke in, grabbed what they needed, killed the owners, and fled. They escaped to the dunes near the coast, a typical example of the interstitial spaces in the Cape's slave geography, where they survived for several weeks on food and tobacco provided by fellow slaves from a nearby farm and made preparations to escape beyond the colony's borders "to the Caffers' land."[49] But escape was impossible, and they were hunted down and killed. They remained part of the shadowy and dangerous underworld that fringed the settlement, but it was becoming increasingly difficult to survive inside the space of the colony, certainly in its heartlands.[50]

However, in 1808, a challenge of a different sort arose. In that year, slaves, led by a slave from Cape Town called Louis of Mauritius, attacked over 30 of the prosperous grain farms of the Koeberg and Swartland region immediately to the north of the town, took the farmers prisoner, and persuaded the slave labourers to join them.[51] Louis declared that

> it was his intention to make all the slaves rebel, to make himself master of all the wagons, horses, powder, arms and ammunition of the farmers, afterwards to join all together at the place where General Janssens encamped with the Dutch troops, to keep all the wagons collected at the Salt River, there to arm the slaves and on Friday in the night about 12 o'clock to go to Cape Town, there to surprise the lines and the Castle, to take possession of the magazines, to break open the prisons, to let loose all the prisoners and

[48] Worden and Groenewald, *Trials of Slavery*, pp. 355–84; M. Cairns, "The Smuts Family Murders: 14.7.1760," *Cabo* 2, no. 3 (1980): 13–16.

[49] CA, CJ 373, testimony of Isaac van Boegies, 4 August 1760.

[50] For further analysis of this case, see N. Worden, "Forgotten Revolutionaries: Slave Resistance at the Cape, 1760–1808," in *L'Atlantique révolutionnaire: une perspective ibéro-américaine*, ed. C. Thibaud et al. (Les Perséides: Bécherel, 2013), pp. 421–46.

[51] Their route and the names of farms attacked are mapped out in CJ 515, C, Marsch route der rebellen onder commando van Louis, pp. 40–41.

> to set at liberty all slaves, who then undoubtedly would assist, after which Lewis would make himself the chief or king.[52]

The Swartland uprising was swiftly defeated by government troops before it got to the town, and the whole event was over in 48 hours. But it marked a significant change of slave tactics, as this was a deliberate bid to capture and overturn core colonial space rather than to escape from it.

This change in tactics was clear in Louis's goal of conquering Cape Town, releasing all slaves and prisoners, and taking over power. He was driven to this after being told by Irish soldiers that "there were no slaves in our country, neither in England, or in Scotland or in America," to which Louis responded, "it was a bad thing that there should be slaves here."[53] The Cape rebels were clearly influenced by the revolutionary ideas sweeping the Atlantic world at this time, marked by the slave revolts of Haiti and elsewhere in the Caribbean and Latin America, as well as the class solidarity of Atlantic soldiers, sailors, and slaves. It was notable that they were supported by Khoi workers who laboured alongside the slaves on the grain farms of the region. By the early nineteenth century, independent Khoi pastoralists had been eradicated from the settler colony, and the remaining Khoi were forced to survive by working as farm labourers alongside slaves. Class solidarity was replacing the earlier conflicts between slave runaways and Khoi herders.[54]

This revolutionary ideology was evident in the ways in which the rebels behaved during their 48 hours of activity. Contrary to some historians who describe the revolt as disorganized anarchy, I have argued elsewhere that a close examination of features such as language, gesture, dress, and the use of space reveals ritualized cultural reversals that gave the revolt particular meaning.[55] Here I will focus on the spatial implications of the uprising.

[52] CA, CJ 516, pp. 76–77.

[53] CA, CJ 516, pp. 201–02.

[54] For a development of this argument, see N. Ulrich, "Abolition from Below: The 1808 Revolt in the Cape Colony," in *"Humanitarian Intervention" and Changing Labour Relations: Long-Term Consequences of the British Act on the Abolition of the Slave Trade, 1807*, ed. M. van der Linden (Leiden and Boston: Brill, 2010), 193–222.

[55] N. Worden, "'Armed with Swords and Ostrich Feathers': Militarism and Cultural Revolution in the Cape Slave Uprising of 1808," in *War, Empire and Slavery, 1770–1830*, ed. R. Bessel, N. Guyatt, and J. Rendall (London: Palgrave, 2010), pp.121–38.

Instead of running away from farms or raiding them for food, the 1808 rebels deliberately invaded and occupied them in order to overturn the social order. Some farmers fled or took refuge nearby but were captured when they were betrayed by their slaves or by the barking of dogs, after which they were humiliatingly returned to their own farms as captives on foot and driven by slaves on horseback.[56] Thus slaves occupied the farm while settlers sought refuge in the surrounding terrain, where they were hunted down by slaves in ways that exactly replicated the tactics used against slave runaways in the past; indeed, hiding in the bushes had in the past been an action of the runaway slave, not the settler farmer. When Matt van der Spuij was driven back to his farmhouse together with his wife and children, he was accosted by Goliath, one of the rebel slaves. Goliath told him, in language reversing the stereotypes of settler discourse about slaves, that skulking in the dark was "the work of a scoundrel, and honest people stay in their homes."[57] The farmer had become the runaway scoundrel, hiding like a slave deserter in the bushes of the marginal environment.

The rebel slaves gave commands while holding a *sjambok*, the quintessential symbol of the slave owner or overseer. Hendrick Grijling, an overseer on the farm of Jan Dreijer, was ordered by "a slave unknown to him, holding a large sjambok in his hand," to give him wine, and when Grijling provided the wine in a small glass, the slave said that "if he did not give him a large glass he would take the sjambok to him."[58] The slave's demand for a large glass suggests more than a desire for extra drink; it was an assertion of his authority and a reversal of the norms whereby slaves would be given tots of wine in amounts and at times chosen by the overseer. It was also a transgression of space. Grijling had been standing on the *stoep* (veranda) at the front of the house while the rebels had assembled the farm slaves in the farmyard. August's demand for a wine glass, and a large one at that, required Grijling to break the boundary between the "slave" werf and the "farmer" house and wine cellar.

Such spatial invasions were a source of particular concern to the authorities at the subsequent trial of the rebels. A frequent pattern was that the

[56] CA, CJ 515, p. 327. On the humiliation (and pain, if the horses kicked their heels) of being driven in front of a horse, see G. Sekoto's depiction of *Prisoner Being Led in Front of a Mounted Horse* (ball point on paper, *ca.* 1960), reproduced in G. Sekoto, *My Life and Work* (Johannesburg: Viva Books, 1995), p. 6.

[57] CA, CJ 515, p. 327.

[58] CA, CJ 515, pp. 286–87; CA, CJ 516, p. 662.

rebels assembled all the slaves and their owners in the farmyard, while horses and wagons were commandeered. Orders were given to hand over guns and ammunition and sometimes cash and food. Some slaves then entered the homestead itself, often (as above) with a standoff on the *stoep* at its entrance. At one level this was a plundering of the houses for food, ammunition, cash, and clothing, all of which were of practical use to the rebels.[59] They also rifled through the contents of desks and chests to take or destroy papers and documents that represented the legal powers of property that landowners and slaveholders held over them.[60] But the very passing over the threshold of the settler homestead without permission of the owner was a violation of the spatial order of the farmstead.[61] Particularly outrageous was the invasion of bedrooms, the most private of spaces. Adriaan Louw was lying sick in his room when the slave Geduld burst in, walked twice around his bed, and then stood at its foot pointing a gun at him and ordered him to be tied up.[62] Conversely, accusations that Pieter Theron, a Cape Town free burgher, was associated with the rebels were strengthened by the evidence of three farm slaves whom he had asked for a glass of water in the back kitchen and who had been seen standing on the *stoep* rather than entering the main part of the house.[63] In this case, a freeman was suspect because of his physical location in "slave space."

The most evident episode of the slaves' spatial invasion of settler farms in the uprising was an intriguing episode of deliberate deception. The rebellion began at *Vogelgezang*, the farm of Petrus Louw, in the Swartland. Louis presented himself there in disguise with "long white trousers, a blue jacket with red collar and gold epaulettes on the shoulders," dress that bore a striking resemblance to widely circulated images of Toussaint l'Ouverture, leader of the successful Haitian slave revolt. He was introduced by the Irishmen who accompanied him as a visiting Spanish naval captain who was travelling in the area "for pleasure" but who spoke no Dutch and therefore remained silent.[64] Petrus Louw was away from the farm, but his wife offered them accommoda-

[59] Ross, *Cape of Torments*, 102. Ross argues that "property rather than persons suffered most."

[60] CA, CJ 802, p. 749; CA, CJ 515, p. 265; CA, CJ 515, p. 330.

[61] P. Scully, *The Bouquet of Freedom: Social and Economic Relations in the Stellenbosch District, c.1870–1900* (Cape Town: Centre for African Studies, University of Cape Town, 1990), pp. 80–82.

[62] CA, CJ 515, p. 248.

[63] CA, CJ 516, pp. 737–38, 745.

[64] CA, CJ 516, pp. 25, 133.

tion and supper. As she later recalled, "this pretended black Captain first went and sat down in my hall at a table, and afterwards the others also—they then asked for a bottle of wine, and I ordered it to be given them ... the pretended Spanish Captain sat opposite me and my children and they were served by one of the slaves they had brought with them."[65] The slave thus became the guest of honour at the settler table and was subsequently offered overnight hospitality in the heart of the settler colony.

The events of 1808, in contrast to those of the eighteenth century, thus represented an attempt by slaves to occupy and control the physical world of their owners rather than escape from it. It was an appropriate time for this new tactic. A new and revolutionary ideology was sweeping through the Atlantic world, including the port of Cape Town. Political upheavals in the colony included the transfer of power from the VOC to the British in 1795, its return to the Dutch, and its second conquest by the British in 1806. It is perhaps no coincidence that the 1808 rebels came from the same region in which the Battle of Blouberg took place, where they would have observed the Dutch forces defeated. Significantly, Louis ordered the rebels to assemble "at the place where General Janssens encamped with the Dutch troops" two years earlier, a symbolic place in colonial topography.

The rebels' bid was unsuccessful. The 1808 rebellion was crushed, and colonial order restored. The leaders were executed, and their followers sent back to their farms for "domestic correction." But memory of the events of that momentous day had an important impact on slave perceptions and responses. Another uprising took place in 1825 on a farm in the Bokkeveld led by a slave named Galant and supported by both slave and Khoi labourers. They killed the farmer and his family and threatened to take over other farms in the area, but they were ultimately captured before they could carry this threat into effect. However, at their subsequent trial, the fiscal claimed that "the pernicious poison of strife and discontent infused into the minds of the slaves" in 1808 had not been quelled: "... still, however, the fire of discontent at the general hope of a general freedom appears to have been smouldering under the ashes, so that the smallest blast of wind is but necessary to make the flame burst out again more violently than ever."[66]

[65] CA, CJ 515, pp. 60–61, 64.

[66] G. Theal, *Records of the Cape Colony*, vol. 20 (London: Clowes Printers for the Government of the Cape Colony, 1904), pp. 314–15.

Galant and his accomplices, like the rebels of 1808, were convinced that their owners were acting against the wishes of the new British government. In 1807, the slave trade was abolished, leading to rising expectations among the 1808 rebels of total freedom. In 1825, the government was implementing "reforms" in its slave colonies. In 1823, regulations were passed which included restrictions on the severity of punishment allowed, minimum requirements for food and clothing, limitations on hours of work, and a ban on the sale of young slave children without their mothers.[67] Slaves were now entitled to report to the authorities if their owners failed to comply with such measures, and Galant had indeed complained to the Worcester magistrate that his master had unfairly and excessively punished him.

Moreover, Galant was not alone in this. Although the might of the colonial authorities and militia ensured that slave ambitions of total freedom were crushed, many slaves were now taking advantage of ameliorative legislation and complained to the authorities to demand better treatment and sometimes manumission.[68] Slaves continued to run away and seek refuge at Hangklip or beyond the colony, but it was increasingly difficult to survive in such circumstances. The limited rights that amelioration offered provided the possibility of redress within the heart of the settler colony rather than a precarious existence on its margins or outside it. The offices of the local slave protectors now became the main sites of slave assertiveness, sometimes successfully and to the humiliation of their owners.[69] Although their complaints were often dismissed as "frivolous" and unwarranted, the fact that they continued to come to the offices and courts of the authorities indicated the changing slave perception of the colonial environment as one in which their claims could and should be heard.[70]

[67] P. G. Warnich, "Die Toepassing en invloed van slawewetgewing in die landdrosdistrik Tulbagh-Worcester, 1816–1830" (PhD diss., University of Stellenbosch, 1988), Chaps. 2–3.

[68] J. Mason, "The Slaves and their Protectors: Reforming Resistance in a Slave Society: The Cape Colony, 1826–1834," *Journal of Southern African Studies* 17, no. 1 (1991): 103–28.

[69] For example, J. Mason, "Hendrik Albertus and his Ex-Slave Mey: A Drama in Three Acts," *Journal of African History* 31 (1990): 423–45.

[70] For the argument that Cape slaves by the nineteenth century developed a network of communication which took advantage of the structures of the ameliorative system, see F. Vernal, "Discourse Networks in South African Slave Society," *African Historical Review* 43, no. 2 (2011): 1–36.

Incorporation into this spatial environment was not, of course, tantamount to equality. After general slave emancipation in the 1830s, most slaves remained locked into a settler economy as impoverished farm labourers. Some therefore continued to practise the ways of their predecessors, moving to the remoter mountains and coastlines and eking out a marginal existence through fishing, hunting, or subsistence farming.[71] Thus even then-freed slaves continued to carve out and occupy a distinct environment for themselves, removed from that of the settlers.

In contrast to some of the other landscapes studied in this book, the physical landscape of the Cape Colony in the late eighteenth and early nineteenth centuries was not marked by significant environmental or climate changes. Rather, the ways in which slaves used and perceived it were transformed. Instead of fleeing from the farms and homesteads of their owners to seek refuge elsewhere, in 1808 and subsequently, slaves staked their claims within the physical environment of the settler colony.

[71] N. Worden, "Adjusting to Emancipation: Freed Slaves and Farmers in the Mid-Nineteenth Century Western Cape," in *The Angry Divide: Social and Economic History of the Western Cape*, ed. W. James and M. Simons (Cape Town: David Philip, 1989), pp. 31–39; P. Scully, *Liberating the Family? Gender and British Slave Emancipation in the Rural Western Cape, South Africa, 1823–53* (Portsmouth, NH: Heinemann, 1997).

CHAPTER 6

A Local View on Global Climate and Migration Patterns: The Impact of Cyclones and Drought on the Routier Family and Their Slaves in Île Bourbon (Réunion), 1770–1820

Sue Peabody

Family history is generally the purview of genealogists, amateur and professional, who scour censuses and church and civil records to trace a lineage back through the centuries to identify forbears and collateral relatives. Historians typically eschew such research as too limited or personal to shed light on wider questions of global historical significance. Yet a marriage between the two approaches—the dense history of a carefully selected family, situated within a social, political, and even environmental context—can produce a fascinating, rich narrative that sheds light on both the particular and the general.

S. Peabody (✉)
Washington State University, Vancouver, Washington, USA

G. Campbell (ed.), *Bondage and the Environment in the Indian Ocean World*, Palgrave Series in Indian Ocean World Studies,
https://doi.org/10.1007/978-3-319-70028-1_6

In this chapter, as part of a larger project,[1] I trace the history of a family held in slavery in France's Indian Ocean colony of Île Bourbon (now the department of Réunion), with particular attention to what is known about the island's environmental history, to see how global forces may have impacted the particular experiences of slaves, their masters, and related free people of colour. Climate, natural disasters (e.g. droughts and cyclones), and disease impacted the island in a series of identifiable crises. To investigate how these crises affected the lives of particular colonists and enslaved people, I have adopted a method of tracing the presence and absence of individuals listed in the census records over many decades. By paying careful attention to the names, ages, and origins of individual slaves, it is possible to discern apparent fluctuations in fertility and mortality as well as slaveholders' strategies in purchasing and selling their slaves. While my project focuses on a single family—that of Furcy, a man whose celebrated lawsuit for freedom drew widespread attention in France and England in the first half of the nineteenth century[2]—this methodology of tracing particular slaves on colonial census records over decades could be employed on a much larger scale to deduce patterns of bonded migration out of Africa, Madagascar, and South Asia in the eighteenth and nineteenth centuries.

Geography and Climate of Réunion

Réunion, located about 700 km to the east of Madagascar at the latitude of 21°S, is 2512 square km in area and possesses the only active volcano and the highest mountain along the Indian Ocean coastline.[3] Prevailing winds flow from the east, full of moisture, dumping 6000–7000 mm of rain annually along the wettest part of the coast, while the mountains cast a rain shadow to the west, especially at the lower altitudes of the coastline (see Fig. 6.1). The coast drops sharply into the sea, with only a few coral reefs on parts of the sheltered west side. The island possesses only one small

[1] Sue Peabody, *Madeleine's Children: Family, Freedom, Secrets, and Lies in France's Indian Ocean Colonies* (New York: Oxford University Press, 2017).

[2] Sue Peabody, "Furcy, la question raciale et le 'sol libre de France': une micro-histoire," *Annales: histoire, sciences sociales* 64, no. 6 (2009): 1305–34.

[3] "All round the island the land slopes steeply upwards towards the centre, 61 per cent of the land surface being above 1000 m." Anthony S. Cheke and Julian P. Hume, *Lost Land of the Dodo: The Ecological History of Mauritius, Réunion and Rodrigues* (London: A & C Black, 2010), 13–14.

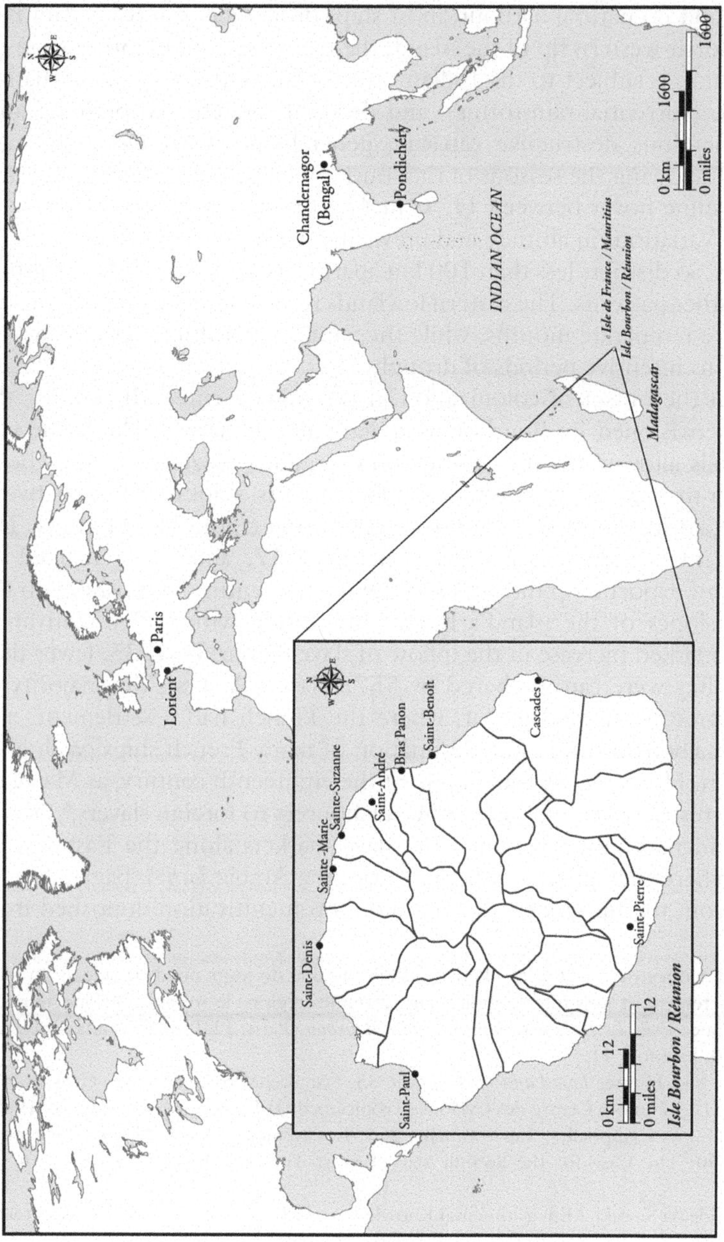

Fig. 6.1 Map of Isle Bourbon/Réunion in relation to the Indian Ocean world

lagoon and no natural harbour; most ships drop anchor in Saint Paul Bay, at the north-western tip of the island, the best point of disembarkation.

Réunion is subject to the Indian Ocean climate patterns of alternating *avalasses* (torrential rainstorms)[4] and drought, punctuated by hurricanes of tremendous destructive capacity, generally between December and April.[5] Following the seasons of the southern hemisphere, temperatures at the coastline hover between 17 °C in August and 30 °C from January to March. Variations in altitude and prevailing wind patterns produce micro-climates, so districts less than 100 km apart can experience radically different weather patterns. The eastern lowlands receive frequent rainfall during the more temperate months, while the western coastline is especially vulnerable to intensive periods of drought.

From the outset of colonization in the mid-seventeenth century, the French envisioned Île Bourbon as an agricultural factory. The fertile volcanic soils allowed for the production of grains (maize, rice, and wheat) and later manioc and potatoes to supply residents, ships transiting between Europe, Africa, and Asia, and the neighbouring colony of Mauritius. The French introduced coffee production in 1717, and this remained an important export crop through the eighteenth century, flourishing on the coastal slopes of the island's mountains. Along with coffee cultivation came a marked increase in the inflow of slave labour; by 1735, fewer than 600 whites were outnumbered by 7573 slaves. At first, the majority of slaves came from Madagascar, where the French had a settlement, and occasionally from India, the destination of many French ships passing Île Bourbon. However, by the middle of the eighteenth century, as Malagasy masters restricted the sale of enslaved labourers to foreign slavers,[6] French traders increasingly frequented the slave markets along the East African coast, where they purchased *cafres* (from the Arabic *kafir*): pagan African slaves from the hinterland.[7] Clove and cotton cultivation flourished from

[4] Ève describes the avalasse as "Période d'une dizaine de jours pendant laquelle la pluie tombe fortement et sans interruption. Dans certaines régions le sol est alors lavé jusqu'au tuf." Prosper Ève, *Le corps des esclaves de l'île Bourbon* (Paris: PUPS—PU Paris Sorbonne, 2012), 70, footnote 41.

[5] Cheke and Hume, *Lost Land of the Dodo*, 35. See also Mireille Mayoka, *Cyclones à La Réunion* (La Réunion: Centre des Cyclones Tropicaux de la Réunion, Météo France, 1998).

[6] See Gwyn Campbell, "Environment and Enslavement in Highland Madagascar, 1500–1750: The Case for the Swahili Slave Export Trade Reassessed," Chap. 3 of this volume.

[7] See Hideaki Suzuki, "Environmental Knowledge and Resistance by Slave Transporters in the Nineteenth-Century Western Indian Ocean," Chap. 9 of this volume.

the 1790s until the 1850s, when local production was eclipsed by other producers closer to the African mainland. Sugar cultivation started relatively late in Île Bourbon, accelerating during and after the British occupation of 1810–1815.[8]

Maize—grown to feed the slaves in Île Bourbon and Île de France—and wheat and rice—primarily destined for their masters' tables—were planted at the outset of the rainy season in December. The first coffee harvest began in March, with three subsequent crops collected from March to July. From the nineteenth century, the sugar harvest took place during the dry season, from July to November.[9] Yet cyclones, a persistent feature of Île Bourbon history, periodically threatened agricultural production. Hot, moist air brings torrential rains for one to three days, inundating the land and causing flash floods, while powerful winds of speeds of up to 250 km per hour wreak destruction upon structures and exfoliate trees.[10] As the French replaced indigenous forests, naturally selected over aeons to survive these storms, with agriculture, the destructive force of cyclones eroded the fertile topsoil and damaged crops.

Emmanuel Garnier and Jeremy Desarthe, tracing the impact of 89 cyclones in Réunion and Mauritius between 1656 and 2007, indicate that the period from 1750 to 1849 saw an unusually severe series of storms.[11] Nonetheless, they conclude that, "[c]ontrary to what one might think ... hurricanes did not cause exceptional loss of life [in the Mascarenes] ... Generally human losses from cyclones did not exceed fifty dead."[12] Immediate deaths in the Mascarene Islands generally resulted from the

[8] Cheke and Hume, *Lost Land of the Dodo*, 104–09. Alessandro Stanziani ("The Cyclone, the Meteorologist, the Planter, and the Indentured Immigrant: The Strange Story of Selective Cyclone Damage in Réunion Island, 1840s–1870s," Chap. 7 of this volume) shows the resilience of sugarcane to cyclones.

[9] Ève, *Le corps des esclaves de l'île Bourbon*, 246–53.

[10] Cheke and Hume, *Lost Land of the Dodo*, 35; S. Kamala Devi, "History of Cyclones on the Coromandel Coast with Special Reference to Tamil Nadu, 1800–1900" (PhD diss., Bharathidasan University, 2011), 12–18, 34–48, 76–77.

[11] Emmanuel Garnier and Jérôme Desarthe, "Cyclones and Societies in the Mascarene Islands 17th–20th Centuries," *American Journal of Climate Change* 2 (2013), no. 1: 3–4. http://www.scirp.org/journal/ajcc. By their own admission, these records are incomplete and tend to be more systematic and reliable after 1789. The archives that Garnier and Desarthe consulted consist primarily of correspondence of the Directors of the English East India Company, 1700–1789, and then the Archives of the Colonial Office and Secretary of State for the Colonies, including Series C, 2C, 11C, 20C, 22C, 2C, and M of the Archives Départementales de La Réunion (hereafter ADR), in detail before about 1820.

[12] Garnier and Desarthe, "Cyclones and Societies," 8.

collapse of houses or drowning in flooded rivers or ravines. This picture contrasts with Devi's findings for the Coromandel Coast of India, where denser settlement, and especially a different geography, made the population more vulnerable to the immediate devastation of storm surges and flooding.[13] However, Prosper Ève's study of cyclones and weather patterns between 1770 and 1818 makes it clear that the devastating effects of severe weather could be felt not in the hours, days, or weeks following the inundations, but in their impact on future harvests, mainly in the destruction of seed intended for sowing (Table 6.1).[14]

The Routier Family and Their Slaves

The effects of periodic storms and droughts can be examined through the lens of one plantation family. Charles Gabriel Routier was born to a French East India Company officer and his wife in Île de France (today, Mauritius) in 1731 and would be the only one of their six children to survive infancy. In 1752, Charles married Françoise Panon, who was from a prominent early settler family of Île Bourbon, and he subsequently fathered two children.[15] Françoise died in 1764, and Charles wasted no time in marrying Marie Anne Ursule Desblottières, his first wife's cousin, who gave birth to three sons in three years. By 1768, Charles had moved his growing family to Saint Denis, Île Bourbon.[16]

In 1770–1772, a series of severe cyclones and drought afflicted Île Bourbon and Mauritius.[17] These weather disturbances and the economic upheaval they caused may well have prompted Charles and his family to

[13] Devi, "History of Cyclones on the Coromandel Coast," 78–79.

[14] Ève, *Le corps des esclaves de l'île Bourbon*, 65–75.

[15] Françoise Marguerite (born 1753 in Port Bourbon, Île de France) and Marie Elie (born 1761 in Saint Denis). Lucien-Jacques-Camille Ricquebourg, *Dictionnaire généalogique des familles de l'île Bourbon (La Réunion) 1665–1810*, vol. 3 (Mayenne: Imprimérie de la Manutention, 1983), 2589.

[16] Their first son, Marie Joseph Louise Emilie Routier, was born in 1765, but did not survive childhood. Two more sons, Augustin and Cyrille, were born in Île de France in 1766 and 1767. Ricquebourg, *Dictionnaire,* 3, pp. 2589–90.

[17] Ève, *Le corps des esclaves de l'île Bourbon*, 70. Garnier and Desarthe have determined that at least one of these cyclones was a Level Five by today's classification. "Cyclones and Societies in the Mascarene Islands," 8–9. An Île Bourbon ordinance of 10 April 1771 ordered landowners to plant 100 feet of manioc per slave declared in the census and required that slaves receive at least a meal of manioc per day; these rations were to comprise one half of the slaves' rations, the other have being the same amount of maize. Ève, *Le corps des esclaves de l'île Bourbon*, 63.

Table 6.1 Environmental events on Réunion

Year	*Month*	*Event*	*Impact*
1717		Coffee introduced	Increased importation of slaves
1718		Cyclone	Sand and pebble deposit left by storm surge caused estuary to close at "River of Pebbles"; coffee harvest nearly destroyed
1770		Cyclone	
1771		Cyclone	Masters ordered to plant 100 feet of manioc per slave
1772		Cyclone	
1777–1778		Drought	Famine, increased importation of maize, rice
1779	January	Windstorm, drought	Disrupted food production in south
1779	November	Cyclone	
1780		Cyclone	Disrupted maize and rice production Impacted fertility or child mortality on Routier plantation
1781	January	Cyclone	Disrupted maize and rice production
1781	February	Cyclone	Disrupted maize and rice production
1784			Evidence of increased fertility and survival of Routier's slaves
1785		Drought	Disrupted wheat production; slaves insufficiently fed, prompting escape, theft, violence
1786		Drought	
1787		Drought	Routiers: older slaves replaced by births and purchases of Creole, Cafre, Malagasy, and Indian slaves
1788		Drought	
1789			Inflation in wheat prices
ca. 1790		Cloves, cotton introduced	
1791	4 July	Volcanic eruption	Inflation in grain prices, February–March 1792; slaves inadequately fed
1795	January	Cyclone	Disrupted maize production
1795	13 March	Cyclone	Affected both Réunion and Mauritius
1795	9–10 April	Cyclone	
1806	21 February	Cyclone	Together with 11 March 1806 cyclone, destroyed two-thirds of harvested crops, especially in the eastern districts

(*continued*)

Table 6.1 (continued)

Year	*Month*	*Event*	*Impact*
1806	11 March	Cyclone	
1806	December	Cyclone	Washed away topsoil
1807	January	Cyclone	Famine
1807	March	Cyclone	
1807	March	Drought	Maize and manioc depleted; inflated maize prices; wheat used to feed slaves; exportation of food banned; coffee replaced with maize in western districts; increased slave mortality
ca. 1810		Sugar introduced	

Sources: Prosper Ève, *Le corps des esclaves de l'île Bourbon*, 64–75; E. Garnier and J. Desarthe, "Cyclones and Societies in the Mascarene Islands 17th–20th Centuries," *American Journal of Climate Change* 2 (2013), no. 1: 3.

travel to France.[18] While in Lorient, Charles arranged for the acquisition of a 13-year-old slave, Madeleine; she had only recently arrived from her native Bengal, where periodic famines, such as that of 1760, precipitated peasant families to pawn or sell their children into bondage, thus making them vulnerable to resale through regional or transoceanic trading networks.[19] One such event probably prompted Madeleine's family to sell her in 1761 to Portuguese slave traders, who sold her to a French woman, who brought her to France, where Madeleine was given to the Routier family with instructions to return her to the Indies and free her there.[20]

[18] "Personnel colonial ancien: Routier, Charles, major des milices à Bourbon 1746/1779," ANOM COL E 358 bis. Charles Routier's mother-in-law, Marie Panon Desblottières, had died in 1769, leaving her plantation and town house to Charles's wife. The boys were at an appropriate age to begin their education in France, following a tradition upheld by the colonies' elite families.

[19] In 1760, "a fearful famine swept away a third of the population of Bengal and about a third of the cultivated area of land relapsed into jungle." Romesh Chunder Dutt, *The Peasantry of Bengal* (Calcutta: Thacker, Spink & Co., 1874), 196; Richard B. Allen, "A Traffic Repugnant to Humanity: Children, the Mascarene Slave Trade and British Abolitionism," *Slavery & Abolition* 27, no. 2 (August 2006): 219–36; See also Gwyn Campbell, "Introduction: Bondage and the Environment in the Indian Ocean World," Sravani Biswas and Subho Basu, "Environmental Disaster in Eastern Bengal: Colonial Capitalism and Rural Labour Force Formation in the Late Nineteenth Century," and Steven Serels, "Famine and Slavery in Africa's Red Sea World, 1887–1914," Chaps. 1, 10, and 11 of this volume.

[20] R. Datta, "Crises and Survival: Ecology, Subsistence and Coping in Eighteenth-Century Bengal," *Calcutta Historical Journal* 18, no. 1 (1996): 5.

Table 6.2 The Routiers' slaves according to the censuses of 1776, 1780, 1784, and 1787

	Children 0–14		*Adults*		*Aged*		*Total*		*Total*
	Male	*Female*	*Male*	*Female*	*Male*	*Female*	*Male*	*Female*	
1776	**18**	7	**56**	**32**	4	7	**78**	**46**	**124**
1780	**15**	6	**55**	**27**	5	7	**75**	**40**	**115**
Net loss/gain	−3	−1	−1	−5	+1	0	−3	−6	−9
Net loss/gain	−4		−6		+1		−9		
1784	**13**	6	**52**	**30**	5	**8**	**70**	**44**	**114**
Net loss/gain	−2	0	−3	+3	0	+1	−5	+4	−1
Net loss/gain	−2		0		+1		−1		
1787	**23**	**10**	**55**	**31**	5	7	**83**	**48**	**131**
Net loss/gain	+10	+4	+3	+1	0	−1	+13	+4	+17
Net loss/gain	+14		+4		−1		+17		

Source: Recensement, île Bourbon, 1776. ANOM COL G/1/479. "Grand Recensement, R. Recensement des Sieurs Routier père, et fils, habitants de la paroisse Sainte-Marie pour l'année 1780," Archives Départementales de la Réunion 71C.

During the long ocean voyage from France to Île Bourbon in 1773, Madame Routier gave birth to their first daughter, Eugénie, no doubt attended in her labour by the 14-year-old Madeleine.[21] A second daughter, Euphémie, was born the following year.

The Île Bourbon census of 1776 indicates that the Routiers, now one of the largest landowning families in the region, planted acreage in both staples and coffee: 120,000 feet were planted in maize, 75,000 in wheat, 50,000 in manioc, 12,000 in vegetables, 5000 in rice, and 10,000 in coffee. The food crops sustained the Routiers and their 124 slaves, and the surplus was exported and traded to Île de France, transoceanic ships, and areas along the African coast.[22] The Routiers' workforce consisted of 88 adult men and women, 11 elderly people, and 25 children under the age of 14 (see Table 6.2).[23] Madeleine, about 17 years old in 1776, was almost

[21] *Etat nominative des passagers embarqués pour les colonies, pendant les années 1769–1780, Port Lorient*, ANOM COL F/5B/51. Marie Charles Eugénie Routier was born on 27 February 1773 "au bord du *Brunoy*." Ricquebourg, *Dictionnaire*, 3, p. 2590.

[22] Evidence from the early nineteenth century shows that rice grown in Île Bourbon was traded for slaves in Zanzibar, for example.

[23] Recensement, Ile Bourbon, 1776, ANOM COL G/1/479. Routier also owned 64 head of beef, 1 sheep, 40 goats, 80 pigs, and 7 horses.

certainly counted among these slaves, as was her first child, a son named Maurice.[24]

The Routiers' last child, Élie, was born in May 1777, just as a devastating drought took hold of Île Bourbon, lasting into 1778.[25] The north-eastern parish of Sainte Suzanne ran out of food, necessitating the importation of maize and rice to feed slaves, poor whites, and free people of colour resident there. In January 1779, a wind storm lashed Saint Paul, reputedly the most violent seen there since 1749. This was followed by an extreme drought in the south that drove down food production to dangerously low levels. In November 1779, another major cyclone whipped the island. The powerful hurricane of 1780 further devastated half the maize plantations and one-quarter of the rice harvest. Cyclones hit the island yet again in January and February 1781, lowering maize and rice production still further.[26] Thanks to care by an anonymous enslaved wet nurse, Élie survived these calamities and survived into adulthood.[27]

The 1780 census shows that the Routier family had lost the equivalent of nine slaves in the previous four years, yielding a total of only 115 slaves (Table 6.2).[28] Yet the reality is more complicated than this. The number of children had been reduced by four, and the adults by six (mostly women), while the number of aged persons increased by one.[29] Since none of the enslaved children in 1780 were under the age of three, it seems likely that the poor harvests lowered fertility and/or increased child mortality on the Routier plantation.

[24] Recensement, Ile Bourbon, 1776, A.N. COL G/1/479. This record totals the number of slaves but does not itemize individuals by name.

[25] Ricquebourg, *Dictionnaire*, 3, p. 2590.

[26] Ève, *Le corps des esclaves de l'île Bourbon*, p. 70.

[27] George D. Sussman, *Selling Mother's Milk: The Wet-Nursing Business in France, 1715–1914* (Urbana: University of Illinois Press, 1982), 19. I deduce the Routier family's dependence on enslaved wet nurses from the widespread practice among elite contemporaries in France and the rapid succession of Madame Routier's many births (eight babies in 11 years).

[28] ADR 71C, "Grand Recensement, R. Recensement des Sieurs Routier Père, et Fils, habitants de la paroisse Sainte Marie pour l'année 1780."

[29] Since the 1776 census did not itemize slaves by name, origin, and age, as did that of 1780, it is impossible to tell whether the reduction in that category of children was due to mortality or simply aging out of that category.

By 1784, the Routiers had eliminated (probably through sale) eight male slaves.[30] At the same time, they added four creole (i.e. island-born) girls (through birth or purchase), while the aging of other girls into puberty helped to balance the workforce sex ratio on the plantation. It is difficult to know whether the Routiers replaced male slaves with females as an economizing measure or with an eye to improving the fertility of their slave population; perhaps they liquidated the more valuable males to raise cash and purchased a pubescent girl, Blandine, with the hope of eventually increasing their workforce through reproduction. The year 1784 was important for the Indian slave Madeleine as well: she gave birth to her second child, a girl named Constance, probably the illegitimate child of Charles Routier or one of his older sons: Augustin, aged 18, or Cyrille, 17.

From 1785 to 1788, the island's southern districts suffered a long period of drought that devastated wheat production.[31] In 1785, an anonymous writer complained that "two thirds of slaves are not fed by their masters," resulting in slave escapes, thefts, and even violent crimes. As Prosper Ève notes, this humanitarian writer nevertheless neglected to point out that masters' negligence resulted in slave malnourishment and lower productivity through weakness.[32] During this period, on 7 October 1786, Madeleine's third and last child, a son, was christened "Fursi."[33] Although the Routier family paid less attention to Furcy's baptism than to Constance's, subsequent developments suggest that he too had a Frenchman for a father.[34] Only a year later, the Routier family hit difficult

[30] ADR 71C, "Recensement des Sieurs Routier Père, habitants de la paroisse [Saint Marie] pour l'année 1784." There is some damage at the extreme margins of this document, making it impossible to tally the ages of individuals, but the total numbers of slaves by age group and ethnicity are tallied at the end.

[31] Ève, *Le corps des esclaves de l'île Bourbon*, p. 70.

[32] Ève, *Le corps des esclaves de l'île Bourbon*, p. 66.

[33] ADR, Registre Paroissial de Saint-Denis, 1786 (2 MIEC11 [J38]). The baptismal record spells his name "Fursi" and the 1787 census was to spell it "Furci," but all later records spell it "Furcy."

[34] Letter from Furcy, Port Louis, Ile Maurice, to Louis Gilbert Boucher, place unknown, 15 May 1826, ADR 1Jp2007-1, no. 71. "Fursi" is the original baptismal spelling; all subsequent references spell his name "Furcy." Furcy's baptismal record describes him as "fils naturel [illegitimate son] of Magdelene" (2 MIEC11 [J38]). In 1826, he would write to a French magistrate and ally, "je suis né Colon Français et je suis fils d'un Français de naissance."

times when Charles died at the age of 53.[35] The 1787 census records a marked increase in the family's slaveholding: 131 slaves, an increase of 17 individuals, or almost 15 per cent over the previous 1784 census.[36] Careful comparison of the names, ages, and origins of these slaves also suggests a dramatic shift in their demographic make-up. Ten men and one woman, most of them rather old, disappeared from the Routier roster: whether through marronage, sale, or death is impossible to tell.[37] Meanwhile, nine new creole children under the age of three (including Furcy), undoubtedly the sons and daughters of the Routiers' slaves, joined the rolls; indeed, the Routiers' pro-natal policy seems to have worked.[38] In addition, the family purchased nine boys imported via the flourishing transoceanic slave trade: five from the mainland of Africa, three from Madagascar, and one from India. Likewise, three Malagasy women joined the Routier plantation community, though their mature ages suggest that they were purchased from fellow colonists rather than the overseas trade.[39] Although this census return does not itemize the crops cultivated on the Routier property, maize remained the dominant crop in Sainte Marie (more than 50 per cent), followed by wheat and smaller proportions of coffee and rice.[40]

The early years of the French Revolution do not seem to have produced any major weather events. Yet a dramatic increase in the cost of wheat

[35] Ricquebourg, *Dictionnaire*, 3, p. 2589; ADR 1Mi28 (B23–27).

[36] ADR74C, "Recensement de dame Vve Routier et des héritiers de feu Sr. [Routier] pour l'année 1787."

[37] Like the environment around Cape Colony discussed by Nigel Worden in Chap. 5 of this volume ("The Environment and Slave Resistance in the Cape Colony"), Réunion's mountainous interior encouraged hardy slaves to escape. By contrast, Réunionnais maroons did not have to contend with sometimes hostile indigenous societies. On the other hand, they could not tap into the resources of established communities, other than to remain hidden on other plantations for short-term relief.

[38] The four creole boys—Adrien, "Furci," Ciayle (spelling?), and Charles—were all aged one or less, while the five creole girls all seem to have born around 1785: Sidaincie (spelling?),aged three; Joséphine, three; Vitoire, three; Adélaide, three; the fifth girl's name and age are illegible, owing to damage to the paper.

[39] Five male children catalogued as *cafre* had apparently been recently purchased from the African mainland trade: Ratema (spelling?), aged 16; Léveillé, nine; Azore, nine; Jougna, eight; and a child whose name is indecipherable, eight. The three new *malgache* boys included Pinimime (spelling?), ten; Fortuné, eight; and Fidelle (*sic*), nine. A boy from India, Lamond (spelling?), 12, had also joined the Routier plantation. Three adult women from Madagascar also joined the rolls (Barbe, 49; Sace, 19; Élevie, 21).

[40] Claude Wanquet, *Histoire d'une révolution: La Réunion 1789–1803*, vol. 1 (Marseille: Jeanne Lafitte, 1981), 35. Interestingly, Wanquet's chart does not include manioc cultivation.

(though not maize) in December 1789 strained white colonists' budgets.[41] Meanwhile, the widow Routier set about managing affairs and distributing the family property in the wake of her husband's death. She filed manumission papers for Madeleine on 6 July 1789, though Madeleine continued to live with the widow until the latter's death in 1808. Likewise, the widow arranged for the marriages of her two eldest sons, Augustin and Cyrille, to two French sisters born in Pondichéry, Anne Adélaïde and Thérèse Agnès Rathier-Duvergé.[42] This was probably when she liquidated about 40 per cent of her slaves in order to distribute her husband's property equally among his heirs. Augustin was gifted the family plantation in Sainte Marie, while Cyrille received land in Sainte Suzanne. Each son received a few slaves from his mother (their names and ages can be traced in the censuses) and purchased more to fill out the plantations' workforces.[43]

Curiously, Furcy's name disappears from all Routier censuses after 1787. Unlike his mother and sister, dutifully reported by the widow in 1796, and brother Maurice, claimed as Cyrille's slave in 1805, Furcy simply vanishes from the archival record for about 20 years. My best explanation is that Furcy, a toddler when his mother was freed, enjoyed a "paperless" informal emancipation during his youth. The common practice of informally freeing a slave (rendering the individual *libre de fait* as opposed to *libre de droit*) allowed the widow Routier to evade paying the lifetime pension on all manumitted slaves that was mandated by the colonial government.[44]

[41] Wanquet, *Histoire d'une révolution*, 1, p. 699.

[42] Ricquebourg, *Dictionnaire*, 3, pp. 2364–65, 2590. The second son, Cyrille, married the older sister, Anne Adélaïde Rathier-Duvergé (aged 16), on 27 April 1790, followed several months later by his elder brother, Augustin, who married Thérèse-Agnès Rathier-Duvergé (aged 15) on 21 June 1790. ADR L 163–164, "Isle Bourbon, Année 1813, Quartier [blank], Paroisse, Saint Denis, Recensement de Monsieur Cyrille Routier"; ADR L 166/2, "Isle Bourbon, Année 1814, Quartier de Saint Denis, Recensement de Monsieur Cyrille Routier, habitant de Sainte Suzanne."

[43] ADR L 168, "An 13e [1804], Isle de La Réunion, Quartier Ste Marie, Déclaration de Routier, habitant"; ADR L183.2, "An [1805]. Isle de La Réunion, Quartier de Ste Suzanne, Déclaration de Cyrille Routier." The widow gave Augustin at least six slaves, and Cyrille received Madeleine's son, Maurice, and the aforementioned Blandine, probably as a gift to his wife.

[44] To avoid payment of the requisite (and expensive) manumission fees, which would make made them *libres de droit*, masters sometimes gave slaves tacit freedom, making them *libres de fait*. These *libres de fait*, also known in Martinique as *libres de savane*, sometimes continued to live on the master's property, freed of labour obligations.

On 4 July 1791, there was a major eruption of La Fournaise,[45] an 8294-foot-high volcano, which may have precipitated an economic crisis that resulted in elevated grain prices from February to March 1792.[46] The Colonial Assembly passed a law on 14 June 1792 chastising slaveholders for neglecting to feed their slaves adequately, categorically prohibiting masters from "unburdening themselves of this obligation by permitting slaves to work for themselves several days per week for their own sustenance" in garden plots or provision grounds.[47] A maize crisis in January 1793 further interrupted the food supply, especially for slaves. The Assembly renewed its directives towards feeding slaves on 14 June 1793 and 18 August 1794, imposing severe penalties for non-compliance.[48] This crisis does not seem to have been prompted by a major weather event in either Mauritius or Réunion.[49] Rather, an increased demand resulting from the rapid conversion of arable land to sugar cultivation in Île de France and the need to provision French naval ships stationed in the Mascarenes to stave off British invasion drove up the cost of grains in Île Bourbon.[50]

In the midst of this political and economic turmoil, an energetic young merchant from Île de France, Joseph Lory, arrived in Île Bourbon in 1793.[51] He must have caught the eye of the widow Routier, for she arranged for him to marry her eldest daughter, Eugénie, on 7 April 1794. These must have been difficult years for Augustin Routier and his wife, Thérèse. Although she gave birth to four children, none seem to have

[45] J. Lambert, "Base de données SisFrance 'Océan Indien' et site Internet associé," BRGM/RP 53711-FR (March 2005), 25, http://www.planseisme.fr/IMG/pdf/rapport_SisFrance_ ocean_indien.pdf

[46] Wanquet, *Histoire d'une révolution,* 1, p. 699. Unlike the 1765 eruption of Mount Macaturin discussed by Warren in Chap. 4 of this volume ("Volcanoes, Refugees, and Raiders: The 1765 Macaturin Eruption and the Rise of the Iranun"), the La Fournaise eruption did not directly cause loss of life or destruction of cropland.

[47] ADR L 12, quoted in Ève, *Le corps des esclaves de l'île Bourbon*, 68–69.

[48] Ève, *Le corps des esclaves de l'île Bourbon*, 66–69.

[49] Garnier and Desarthe, "Cyclones and Societies," 5. Garnier and Desarthe list no cyclones for either Mauritius or Réunion between the late 1780s and 1806.

[50] Charles Grant de Vaux, *The History of Mauritius: Or the Isle of France, and the Neighbouring Islands; from their First Discovery to the Present Time* (London: W. Bulmer and Company, 1801), 568.

[51] ADR L 156/2, "Isle Bonaparte, Année 1810, Quartier Saint Denis, Paroisse, idem, Recensement de Joseph Lory." The name of the ship is listed in ADR L 165, "Isle Bourbon, Année 1814, Quartier de Saint Denis, Déclaration de M. Joseph Lory."

survived to adulthood, and the couple took advantage of the new revolutionary capacity to divorce in 1795.[52] Renewed bad weather hit in 1795, including a severe cyclone at the end of January 1795, which ruined much of the maize harvest. This was followed by hurricanes on 13 March and 9–10 April that completely ravaged both Réunion and Île de France. These cyclones were followed by drought that devastated maize production and rice in some quarters.[53] Meanwhile, Eugénie and Joseph Lory established themselves in the rural Saint Marie parish, near Augustin.[54]

In 1793, colonial commissioners in Saint Domingue (Haiti) abolished slavery, hoping to attract the revolted slaves to fight for the French republic against their enemies: the British, the Spanish, and their royalist allies; when news of this abolition reached Paris, the Jacobins of the Constituent Assembly proclaimed the general emancipation of slaves throughout the French Empire with the decree of 16 Pluviôse An 2 (4 February 1794). However, when Parisian officials bearing orders to implement the decree in the Mascarenes arrived in June 1796, an armed mob drove them from Mauritius—they never reached Réunion—and slavery persisted in both islands throughout the revolutionary era. When Madame Routier filed her 1796 census, just months after the abortive emancipation mission, she was living in the city of Saint Denis with only a few domestic servants (some enslaved, some free), including Furcy's mother and sister and a white woman—an orphan taken in by the widow, perhaps as a companion.[55]

[52] A third of these children died at the age of 16 months; the rest are not chronicled further in Ricquebourg, *Dictionnaire*. The National Assembly passed revolutionary legislation permitting divorce for the first time in France on 20 September 1792; the capacity to divorce continued under the Code Civil (1804), though under more circumscribed conditions, and was suppressed altogether in 1816 under the Restoration. See Elaine Marie Kruse, Divorce in Paris, 1792–1804: Window on a Society in Crisis (PhD diss., University of Iowa, 1983); Theresa McBride, "Public Authority and Private Lives: Divorce after the French Revolution," *French Historical Studies* 17, no. 2 (1992): 749–51. Ricquebourg, *Dictionnaire*, 3, pp. 2363–64, 2590–91. The Routier divorce was registered on 27 May 1795 in the Sainte Marie parish. Ricquebourg, *Dictionnaire*, 2, pp. 1515–16, and 3, p. 2590. Augustin took a new wife on 10 June 1800; she, too, would bear him four children, none of whom apparently survived to adulthood.

[53] Ève, *Le corps des esclaves de l'île Bourbon*, 70–71.

[54] ADR L 169. "An 12e, Isle de La Réunion, Quartier Ste Marie, Déclaration de M. Joseph Lory."

[55] ADR L 142/1, "Saint Denis, Recensement de la Citoyenne Veuve Routier, pour l'année 4me."

In 1799, the year of Napoleon's coup over the Directory, the extended Routier family suffered two losses: in January, Eugénie and Joseph Lory's four-year-old daughter died of unknown causes,[56] and within three weeks the family attended the funeral of the daughter's maiden aunt Euphémie (aged 25).[57]

The extreme weather of 1806–1807 impacted Réunion and Mauritius in ways heretofore unimagined.[58] Two successive cyclones of 21 February and 11 March 1806 together destroyed two-thirds of the harvested crops, hitting the eastern districts of Réunion especially hard. This prompted General Decaen's injunction on 16 March 1806 to plant sufficient food to feed slaves. Nevertheless, the subsequent months of drought created a food crisis of near-famine proportions, while speculators continued to sell their grain to Île de France at exorbitant prices. Two extraordinarily destructive back-to-back storms, each lasting more than ten days, flooded the island in December 1806 and January 1807, eroding so much topsoil that the sea turned yellow.[59] Famine conditions ensued, prompting the *agent général de police* (police superintendent) to propose a tax on foodstuffs, while the city of Saint Paul discussed extreme measures to prohibit the exportation of grain, to collect a reserve of seed grain, and to encourage wheat production. In March 1807, the mayor of Sainte Suzanne lamented:

> … the 600 individuals of the white and free population and 1500 of the blacks are reduced to famine. This lack of food causes daily desertions and dilapidations, the deaths of many blacks killed by guards and the difficulties of service in a time when it is most needed and the despair of landholders.[60]

A new cyclone hit the island in March 1807, and the succeeding dry season plunged the island into a new pattern of drought. A measure of the extremity of the crisis is that wheat, normally reserved solely for whites, was now used to feed slaves since maize and manioc were completely lost. By June, the situation was so extreme, with the price of maize having risen

[56] Date of burial: 16 January 1799. Ricquebourg, *Dictionnaire*, 2, p. 1753.

[57] Date of burial: 7 February 1799. Ricquebourg, *Dictionnaire*, 3, p. 2590.

[58] Garnier and Desarthe, "Cyclones and Societies," 8. Mauritius saw devastating storms between 1806 and 1808 as well.

[59] Ève, *Le corps des esclaves de l'île Bourbon*, 70.

[60] ADR, L 109, "Lettre du commandant du quartier de Sainte-Suzanne au gouverneur du 6 mars 1807," quoted in Ève, *Le corps des esclaves de l'île Bourbon*, 72.

sixfold in three years, that the governor prohibited the export of all foodstuffs from the colony. Farmers on the western slopes replaced some coffee plantations, many of which had been completely destroyed by recent natural disasters, with maize.[61]

Prosper Ève has attempted to calculate the impact of the 1806–1807 crisis on slave mortality, though the poor survival of birth and death records before 1805 makes this difficult to achieve with precision. Nevertheless, it is clear that certain districts, including those where the Routier and Lory families resided, suffered a significant spike in slave mortality. Two of the western districts (Saint Louis and Saint Paul), where maize constituted 70–80 per cent of the agricultural production, experienced a much smaller increase in slave mortality than the north-eastern districts, which were more dependent on maize.[62]

On 21 December 1807, Joseph Lory completed the 1808 census on behalf of his mother-in-law, the widow Routier, just ten months before her death.[63] She was living in Saint Denis with all four of her surviving adult children, but although each of them was married, none of their spouses were listed in the census. Also living in the household were Madeleine (aged 50), Constance (22), and Constance's two sons, Auguste (six) and Eugêne (one and a half). Five domestic slaves, including Madeleine's two sons—Maurice, 29, and Furcy, 20—now reappeared in the roster, listed again as the widow's slaves.[64]

If I am correct that Furcy enjoyed a paperless freedom during his youth, it seems likely that in the wake of the unprecedented natural disaster of 1806–1807, Joseph Lory spied an opportunity to reclaim him in bondage. Lory took advantage of the 1808 census report to re-enslave Furcy, whether through negotiation or, as I suspect, inveiglement.

[61] Ève, *Le corps des esclaves de l'île Bourbon*, 67, 70–72; Cheke and Hume, *Lost Land of the Dodo*, 109–10, 138; Garnier and Desarthe, "Cyclones and Societies," 8.

[62] Ève, *Le corps des esclaves de l'île Bourbon*, 72. The hardest-hit districts were Saint Denis, Saint Pierre, Saint Joseph, Sainte Marie, Sainte Suzanne, and Saint Benoît. See also Prosper Ève, *Naître et mourir à Bourbon à l'époque de l'esclavage* (Paris, Saint Denis, and Bagneux: L'Harmattan, Université de La Réunion, and Numilog, 2000), 140–63. Wanquet, *Histoire d'une révolution*, 1, p. 699. I am basing the correlation to maize production on Wanquet's map/graph for 1788, which admittedly is two decades out of date.

[63] ADR L 152/2, "Isle Bonaparte, Année 1808, Quartier Saint Denis, Paroisse [Saint-Denis], Recensement de la Dame Ursule Desblottières Ve Routier."

[64] Louison and Sophie also appeared in the 1787 census; Caroline was born in the intervening years.

In Ile Bourbon, the cyclones of 1806–1807 led directly to the introduction of sugar, which was believed to be more resilient than the coffee trees that were destroyed in the storms.[65] There were political-economic factors as well: the loss of Saint Domingue in the Haitian Revolution meant that France was looking for new sources of sugar.[66] The British occupation of 1810–1815 also probably contributed to the acceleration of sugar production, as had occurred in Guadeloupe during the Seven Years War (1756–1763).[67] Joseph Lory was one of 18 colonists in north-eastern Île Bourbon to seize the opportunity of British occupation to begin to invest in sugar cultivation and production; Lory, with his prior experience on Île de France, along with Jean-Baptiste Azéma du Tilleul and Charles Desbassayns provided the necessary technical knowledge.[68]

I do not have the space in this chapter to detail the effects of Lory's ascending power, fuelled in part by the sugar revolution, on the lives of Madeleine's children. Suffice it to say that after re-enslaving Furcy, Lory cheated Madeleine out of the accumulated pension owed her upon the widow's death, a significant sum of money, worth at least two skilled adult male slaves, with which Madeleine hoped to purchase Furcy's and Maurice's freedom. Maurice died of unknown causes, possibly resisting the British invasion in 1810, and Madeleine died two years later in 1812. Furcy, with Constance's help, attempted to free himself from Lory in judicial proceedings in 1817, but the attempt backfired. Governor Philippe Panon-Desbassayns, Lory, and their cronies effectively manipulated the ensuing battle so as to fend off the crown's muted effort at judicial reform, while Furcy wasted away in prison for a year. Eventually, Lory transported Furcy to his sister's plantation in Mauritius, from which he lobbied for justice for another quarter century. Meanwhile, Lory and the two Routier brothers, Cyrille and Augustin, who had also invested in a *moulin horizontal en fer* (sugar mill) before 1818,[69] continued to prosper in business until the latter two died in 1824 and 1828 respectively.

[65] Jean-François Géraud, "Des habitations sucreries aux usines sucrières, la 'mise en sucre' de l'île Bourbon, 1783–1848" (PhD diss., Université de la Réunion, 2002), 9; Garnier and Desarthe, "Cyclones and Societies," 9; Cheke and Hume, *Lost Land of the Dodo*, 110.

[66] Cheke and Hume, *Lost Land of the Dodo*, 137.

[67] Laurent Dubois, *A Colony of Citizens: Revolution and Slave Emancipation in the French Caribbean, 1787–1804* (Chapel Hill: University of North Carolina Press, 2004), 35–36.

[68] Géraud, "Des habitations sucreries," 9, 183. Joseph Lory established a sugar refinery in Bras Panon, to the east of Saint André, around 1815–1816. Géraud, "Des habitations sucreries," 32.

[69] Géraud, "Des habitations sucreries," 190.

But in less than a year, Lory found himself implicated in yet another colonial crisis. Despite the British ban on the slave trade that was foisted upon the French government in treaties at the conclusion of the Napoleonic wars and reiterated by French royal legislation in 1817, Île Bourbon planters demanded more slaves, especially for expanding sugar cultivation and processing on the island.[70] In October 1820, the ship *Succès*, newly arrived from Nantes on its maiden voyage, clandestinely unloaded 215 slaves upon the isolated beach of Les Cascades, at the south-eastern tip of Île Bourbon. Planters eagerly bought them up, though 44 of them, "maigre et malade" (thin and ill), ultimately succumbed to smallpox. French authorities arrived too late to stop the sale, but embargoed the ship in Saint Paul while subjecting its officers to judicial proceedings, eventually letting them off with a slap on the wrist. Lory proceeded to finance a second slave smuggling voyage to Zanzibar, but the ship was ultimately captured by the British navy off the Seychelles and tried in Mauritius.[71] The average height of the 205 males aged over 18 years on board was only five feet (152 cm), suggesting that they had grown up under famine conditions in the region now governed by Malawi and Tanzania.

Conclusion

French settlers replaced the indigenous coastal forests of Réunion, which were well adapted to storms, with plantation agriculture. Its nutrient-rich soil transformed the island into a major producer of grain for internal consumption and exportation. Colonists quickly began to import enslaved men, women, and children from around the Indian Ocean rim, first from Madagascar and India, but increasingly from the East African mainland. Severe cyclones and droughts periodically stressed the island's capacity to produce sufficient food to sustain the rapidly growing population, and the colonial government repeatedly issued injunctions against masters to feed

[70] Hubert Gerbeau, "Quelques aspects de la traite illégale des esclaves à l'île Bourbon au XIXe siècle," in *Mouvements de populations dans l'Océan Indien* (Paris: Champion, 1980), 276.

[71] Deryck Scarr, *Slaving and Slavery in the Indian Ocean* (New York: Saint Martin's Press, 1998), 113–18; R. B. Allen, "Licentious and Unbridled Proceedings: The Illegal Slave Trade to Mauritius and the Seychelles during the Early Nineteenth Century," *Journal of African History* 42 (2001): 102–03. My analysis here is also based on the many documents generated from the British confiscation of the *Succès* held in the National Archives of the United Kingdom, Kew, CO167/92.

their slaves. Early in the nineteenth century, a conjuncture of bad weather, economics, and politics prompted some of the wealthier landowners of Île Bourbon to introduce sugar cultivation and refineries. Yet the sugar revolution in Île Bourbon coincided with British determination to ban the transoceanic slave trade, forcing French colonists to adopt clandestine measures, which almost certainly worsened conditions upon the smugglers' boats.[72]

The Routier family and their slaves were touched by all of these developments, many of which were influenced by weather extremes of cyclones and droughts. Madeleine's initial enslavement in Bengal probably began when her parents pawned or sold her in the face of famine, a typical experience for the bonded household servants of the Indian Ocean world. Drawn into the Routiers' slave plantation in Île Bourbon, Madeleine and her children—like their masters, the Routiers and the Lorys—were affected by physical environment, the periods of famine and fertility, disease contagion, and the wider currents of Indian Ocean history.

[72] Marina Carter and Hubert Gerbeau, "Covert Slaves and Coveted Coolies in the Early 19th Century Mascareignes," *Slavery & Abolition* 9, no. 3 (1988): 194–208.

CHAPTER 7

The Impact of Cyclones on Nineteenth-Century Réunion

Alessandro Stanziani

The monsoon regime has shaped the agricultural production in which the vast majority of the Indian Ocean world population was, and still is, engaged. It largely dictates precipitation patterns on the Asian continent, in both coastal and inland regions, thus critically influencing production patterns and social relations. From April to September, the Asian continent heats up and hot air rises, producing a vacuum that sucks in air from the ocean, thereby creating the south-west monsoon. During the alternate "winter" months, the opposite occurs, creating the north-east monsoon.[1]

[1] Gwyn Campbell, "Africa and the Early Indian Ocean World Exchange System," in *Early Exchange between Africa and the Wider Indian Ocean World*, ed. Gwyn Campbell, forthcoming; see also J. L. Abu-Lughod, *Before European Hegemony. The World System A.D. 1250–1350* (New York: Oxford University Press, 1989); K. N. Chaudhuri, *Asia before Europe: Economy and Civilisation of the Indian Ocean from the Rise of Islam to 1750* (Cambridge: Cambridge University Press, 1992); K. N. Chaudhuri, *Trade and Civilisation in the Indian Ocean: An Economic History from the Rise of Islam to 1750* (Cambridge: Cambridge University Press, 1985).

A. Stanziani (✉)
EHESS and CNRS, Paris, France

G. Campbell (ed.), *Bondage and the Environment in the Indian Ocean World*, Palgrave Series in Indian Ocean World Studies,
https://doi.org/10.1007/978-3-319-70028-1_7

Studies indicate that major disturbances to this system could be life-threatening and drive people into human bondage.[2] It is also possible to identify a link between environment, debt bondage, and emigration.[3] This chapter discusses the interplay between environment and dependence in the nineteenth century on the Indian Ocean island of Réunion, the final destination for many emigrants. It begins by establishing what is known about cyclones and storms in the Indian Ocean in the period under investigation. It then details the impact of climatic disturbances on Réunion and on local planters and colonial elites, and their subsequent attempt to pass the cost of such disturbances on to indentured immigrants. Our focus on Réunion requires some clarification. There are relatively few academic studies of the environment on the island in comparison to the northern and eastern parts of the Indian Ocean (affected by storms and droughts and by typhoons and volcanoes, respectively), although analyses of storms and cyclones on Réunion and the neighbouring island of Mauritius (together known as the Mascarenes) improved over time. One major peculiarity of the Mascarenes is that they were affected less by monsoons, heavy rains, and/or drought than by violent storms. Small islands like Réunion and Mauritius are particularly exposed to phenomena such as changes in sea level and variations in rainfall regimes and winds. In a context where food crops were marginal and the inhabitants were dependent on a sugar monoculture, the islands were also vulnerable to external market shocks and food insecurity. However, this does not suffice to explain the highly differentiated impact of cyclones depending on the food crop (rice or maize) and the socio-economic group (planters or indentured immigrants) concerned. To this end, we need to carefully consider the relationships between the environment, production choices and strategies, and social hierarchies.

[2] J. Breman, I. Guérin, and A. Prakash, eds., *India's Unfree Workorce: Old and New Practices of Labour Bondage* (New Delhi: Oxford University Press, 2009); J. L Dull, ed., *Han Social Structure: Ch'ü T'ung-tsu* (Seattle, WA: University of Washington Press, 1972); G. Campbell, ed., *Abolition and its Aftermath in Indian Ocean Africa and Asia* (London: Routledge, 2005).

[3] R. B. Allen, "The Constant Demand of the French: The Mascarene Slave Trade and the Worlds of the Indian Ocean and Atlantic during the Eighteenth and Nineteenth Centuries," *Journal of African History* 49, no. 1 (2008): 43–72.

Weather and Cyclones: From Travel Narratives to Forecasting

Located in the intertropical zone, the south-west Indian Ocean is one of the world's seven cyclogenesis basins. Tropical cyclones make their appearance at the beginning of the austral summer, around 15 November. During the cyclone season, which generally lasts until 15 April, winds can reach up to 300 km per hour.[4] The north-eastern and extreme southern zones are the most affected.[5] On Réunion, a cyclone can be referred to as a "coup de vent" (or *koudvan*), "mauvais temps" (or *mov45 tan*), or "ouragan" (sometimes spelled "houragan" in the nineteenth century). Sailors referred to cyclones as "gros temps," "bourrasque," "trombe," "avalasse," or "avalaison"—though the last two terms relate to heavy rainfall more than cyclones. In one nineteenth-century book, cyclones are referred to as a "typhon."[6] The number of different terms is not simply a question of the way people spoke in the nineteenth century. Today, weather stations name and classify severe storms of cyclonic nature in different ways.[7] The scientific identification of major tropical storms was slow to progress. In the eighteenth century and the first half of the nineteenth century, most information concerning such storms derived from travel narratives and nautical logs. At that time, meteorologists in both France and England were rarely part of scientific circles, which considered meteorology akin to popular prophecy or, at best, the lore of sailors and planters.[8] Only during the second half of the nineteenth century did meteorology start to gain credence, thanks initially to growing interest in the weather by public administrations and colonial elites that wished to limit the effects of poor harvests

[4] J. Ecormier, *Cyclones tropicaux du Sud-ouest de l'Océan Indien: le cas de l'île de la Réunion* (La Réunion: SMR (Station Météo Réunion), 1992), 7.

[5] P. Eve, *Île à peur: la peur redoutée ou récupérée à La Réunion des origines à nos jours* (Saint André, La Réunion: Graphica, 1992).

[6] E. Pajot, *Simples renseignements sur l'île Bourbon* (Paris: Challanel Ainé, Librairie Coloniale, 1887), 307.

[7] Centre Canadien de Prévision d'Ouragan, *Comment catégorise-t-on les ouragans?*, http://www.ec.gc.ca/ouragans-hurricanes/default.asp?lang=Fr&n=AB062B74-1; Bureau of Meteorology, *Guide to Tropical Cyclone Forecasting*, http://www.bom.gov.au/cyclone/about/names.shtml

[8] K. Anderson, *Predicting the Weather: Victorians and the Science of Meteorology* (Chicago: University of Chicago Press, 2005); Fabien Locher, *Le savant et la tempête: étudier l'atmosphère et prévoir le temps au XIXe siècle* (Rennes: Presses Universitaires de Rennes, 2008).

and famine, respectively. Meteorology also benefited from improvements in measuring and forecasting instruments, from progress in probability calculation (Bayesian statistics), and from the introduction of telegraphy, which made it possible to communicate the arrival of poor weather conditions.

Improvements in meteorological analysis were slower to occur in the colonies, because of both the lack of knowledge about the local environment and the high cost of obtaining such knowledge. In the Indian Ocean, the British and French navies gradually compiled charts for currents, tides, and winds, notably for the northern sector; the southern sector remained relatively neglected. This meant that progress in the understanding of regional cyclones was slow and mainly dependent on information noted in the ships logs of naval officers and seafarers.

The first serious research into Indian Ocean cyclonic activity was carried out in 1838 by Colonel Sir William Reid. He used newspaper articles and logbooks that he collected from areas to which he had been posted: Réunion, Mauritius, England, and Barbados. Reid produced detailed descriptions of cyclones, how they formed, their routes, direction, and strength, and associated changes in atmospheric pressure and rainfall. He described in great detail the main cyclonic storms to have hit Barbados, Mauritius, Réunion, and certain parts of India in the early nineteenth century. Further, he proposed what he referred to as a "Law of Storms,"[9] which comprised a set of general principles based on the cyclones noted in his work. Reactions to his publication were mixed: some readers felt the work to be totally unsuitable and unscientific, while others considered it to be a real step forward in meteorological knowledge. Among the latter was Henry Piddington (1797–1858), an English naval captain who had served in Calcutta and the Brahmaputra estuary—regions regularly devastated by cyclones. Piddington had immediately grasped the link between tropical storms at sea and devastation on land, which was a highly innovative proposal for that period. Indeed, it was he who invented the term "cyclone" to describe the whirlwind characteristics of these major storms, which he compared to a snake winding itself into a circle—"kyklos" in Greek. He correctly distinguished between the anti-clockwise rotation of cyclones in the northern hemisphere and the clockwise rotation of those

[9] Sir W. Reid, *An Attempt to Develop the Law of Storms* (London: John Weale, 1838).

in the southern hemisphere.[10] Piddington's work, published in 1848, remained for sailors in the tropics the chief reference on cyclones for half a century.

In France, the works of Reid and Piddington were well known, but the study of tropical cyclones was chiefly based on research carried out on Réunion. The first major study there was conducted by a local botanist and naturalist, Joseph Hubert (1745–1825), famous for having introduced cinnamon and cloves to the island, who in 1818 suggested that tropical storms did not move in straight lines but turned.[11] He proposed a simultaneous movement of wind rotation (turning around an axis) and translation (moving from one place to another), which was confirmed 15 years later by the German meteorologist Heinrich Dove (1803–1879).[12] This concept was further developed by Gabriel Hilaire Bridet (1818–1894), a harbourmaster on Réunion who, in his *Étude sur les ouragans de l'hémisphère austral* (*Study of Hurricanes in the Austral Hemisphere*), noted the wind directions and their variations that sailors and landsmen should use to establish the centre and course of a cyclone.[13] Bridet used Hubert's notion of circular rather than convergent winds (Meldrum's theory) to explain the parabolic course of cyclones along first a west-south-west and then a south-south-east trajectory as they moved gradually into the southern hemisphere. Indeed, Bridet, who was harbourmaster of Réunion from 1850 to 1860, encouraged naval officers, administrative authorities, and meteorologists to liaise in the observation, analysis, and constitution of extensive archives on cyclones and the damage they caused. It is now time to take a closer look at these documents.

[10] H. Piddington, *The Sailor's Horn-Book for the Law of Storms* (London: Smith/Elder, 1848); A. K. Sen Sarma, "Henry Piddington (1797–1858): A Bicentennial Tribute," *Weather* 52, no. 6 (2012): 187–93.

[11] G. Gerard, *Un grand créole, Joseph Hubert, 1747–1825* (Saint-Denis la Réunion: Éditions Azalée, 2006).

[12] H. Dove, *The Law of Storms Considered in Connection with the Ordinary Movement of the Atmosphere* (London: Longman, 1886).

[13] H. G. Bridet, ed., *Étude sur les ouragans de l'hémisphère austral: manœuvres à faire pour s'en éloigner et se soustraire aux avaries qu'ils peuvent occasionner* (Saint-Denis la Réunion: Rambosson, 1861); J. Cox, *Storm Watchers: The Turbulent History of Weather Prediction from Franklin's Kite to El Niño* (Hoboken, New Jersey: John Wiley and Sons, 2002).

The Impact of Cyclones on Crops

In recent years, meteorologists, assisted by archaeologists, have reconstructed past climatic events in impressive detail. In particular, in the northern Indian Ocean, it is now possible to identify from 1860 onwards an increase in the temperature of the water, a reduction in rainfall, and weaker summer monsoons, which led to drought, poor rice harvests, and often famine.[14] Weaker monsoons in the northern sector of the Indian Ocean were accompanied by more frequent and stronger cyclones in the southern sector, which frequently had a major impact on the Mascarenes.

This is confirmed by evidence from historical archives that mention that 89 cyclones hit the Mascarene Islands between 1656 and 2007. Evidence for the nineteenth century indicates that the cyclones of 1844, 1863, and 1868 were particularly severe and damaging, although the number of cyclones peaked thereafter, between 1870 and 1879.[15] Changes in sea level, variations in rainfall regimes and winds, and external market shocks strongly affect small islands such as Réunion and Mauritius. Thus, Réunion was particularly vulnerable to the impact of hurricanes and cyclones because of its small size, open economy, limited natural resources, relative isolation, great distance from major markets, limited funds, and poor infrastructure.[16] However, we need to examine in closer detail why food crops were more affected than sugar cane and how other economic and social factors influenced the impact of cyclones.

Reconstituted data for the period show that the cyclones damaged food crops more than they did sugar cane, which has a remarkable capacity to recover from even the most severe storms.[17] This is due primarily to the morphological characteristics of sugar cane, notably to its root base, which is generally very firmly anchored to the soil.[18] By contrast, food crops such

[14] H. Xu, Y. Hong, and B. Hong, "Decreasing Asian Summer Monsoon Intensity after 1860 AD in the Global Warming Epoch," *Climatic Dynamics* 39 (2012): 2079–88; C. Fu and J. Fletcher, "Large Signals of Climatic Variation over the Ocean in the Asian Monsoon Region," *Advances in Atmospheric Sciences* 5, no. 4 (1988): 389–404.

[15] ANOM, FM SG/Reu c 124 d 932–48.

[16] N. Mimura et al., "Small Islands," in *Climate Change: Impacts, Adaptation and Vulnerability. Contribution of Working Group II to the Fourth Assessment Report of the Intergovernmental Panel on Climate Change*, ed. M. L. Parry et al. (Cambridge: Cambridge University Press, 2007), 687–716.

[17] E. Garnier and J. Desarthe, "Cyclones and Societies in the Mascarene Islands, 17th–20th Centuries," *American Journal of Climate Change* 2 (2013): 1–13, fig. 5.

[18] A. Courteau, "La canne à sucre et l'environnement à La Réunion—revue bibliographique," *CIRAD—Pôle canne à sucre* (2005), http://reunionmayotte.cirad.fr/index.php/dr/content

Table 7.1 Réunion Island rice imports, special trade, in francs

Year	*Francs*
1847	468,555
1848	245,542
1850	120,097
1851	96,180
1852	264,709
1853	335,712
1854	601,518
1855	895,000
1856	427,000
Average 1847–1856	337,026
Average 1837–1846	100,961

Source: Ministère du Commerce Extérieur, *Tableau général du commerce, année 1856* (Brussels: Imprimerie Nationale, 1857), 59

as maize and manioc are easily uprooted, while rice is vulnerable to flooding and winds. Thus archival records systematically mention significant damage to food crops but not sugar cane. However, between 1850 and 1860, the area of cultivated land devoted to monoculture sugar plantations increased from 50 per cent to 68 per cent, which, with a corresponding decline in the land devoted to food crops, considerably amplified the risk of famine following cyclones and increased reliance on food imports (Table 7.1).

With the influx of waves of immigrant indenture workers following the abolition of slavery, the island's population grew—from 100,000 to 178,000 in the 1850s alone. In the following decade, rice production decreased by 40–50 per cent. However, the growing amount of land dedicated to sugar cultivation meant that the quantities of rice, manioc, and maize produced per capita in 1860 were much smaller than in 1847. Moreover, while on average rice imports increased, they varied considerably from year to year and never compensated for the overall fall in local rice production. In addition, unsatisfied demand and growing imports led to rising rice prices, which in turn exacerbated local tension, not only between whites and immigrants of colour but also between landowners and *petits blancs* (small white estate owners). Landowners, especially the smaller ones, looked to reduce their workers' food rations or even to keep their workers' wages for themselves.

Economic Policies and Cyclones

An analysis of relevant historical archives on Réunion reveals the impact of cyclones, the impressions they made on people, and the reasons why such details were recorded. Two cyclones in close succession in February and March 1806 almost entirely destroyed the island's coffee plantations. In a letter to Governor-General Decaen dated 27 February, General Des Bruslys attempted to describe the extent of the catastrophe:

> There is not a single inhabitant who has not suffered damage, either to buildings or plantations. In the report that the General Agent of Police Bédier has sent to Sub-Prefect Marchant, he lists the damage that the colony has incurred. Inhabitants may reasonably expect to recover only a third of the maize and half of the coffee harvest. The clove trees have been almost completely destroyed in the eastern part of the island ... The countryside has been totally devastated ... All the black people's dwellings and the stables adjoining domestic properties have been blown away.[19]

These were major setbacks for coffee producers, especially the smaller ones. With the 1807 British embargo on the slave trade, which focused on the transatlantic traffic in slaves, sugar prices soared worldwide. Producers on Réunion, wishing to profit from the situation, abandoned coffee for sugar cane, a change that resulted in a significant increase in the importation of slaves. It is estimated that 48,900–66,400 slaves were imported between 1811 and 1848, some 45,000 alone entering the island between 1817 and 1835.[20]

After relative prosperity in the 1820s, Réunion was again hit by severe cyclones in 1830 and 1832. Although most ships were spared, the cyclones caused extensive damage to infrastructure and farms depending on the region and crop. Assessments by experts, gendarmes, and the botanist from the island's department of internal affairs estimated that between 11

[19] Papier Doyen, 2 October 1806, the Mauritius National Archives, Archives de la Marine, Île de France, 16105–07.

[20] M. Schuler, "The Recruitment of African Indentured Labourers for European Colonies in the 19th Century," in *Colonialism and Migration: Indentured Labour Before and After Slavery*, ed. P. Emmer (Dordrecht, Boston, and Lancaster: Martinus Nijhoff, 1986), 125–61; H. Gerbeau, "Quelques aspects de la traite illégale des esclaves à Bourbon au XIXe siècle," in *Mouvements de populations dans l'Océan indien* (Paris: Imprimerie Champion, 1979), 273–96; R. Allen, "The Mascarene Slave Trade and Labour Migration in the Indian Ocean during the Eighteenth and Nineteenth Centuries," in *The Structure of Slavery in Indian Ocean Africa and Asia*, ed. G. Campbell (London: Frank Cass, 2004), 37–38.

and 12.5 per cent of the manioc, coffee, and sugar harvests were lost along with 33 per cent of the maize harvest, 75 per cent of the rice harvest, and almost all fruit and cloves.[21]

Yet in a later letter to the minister of the colonies, the governor wrote that according to the agronomists, "association-owned plantations, those finding themselves in a state of collapse or bankruptcy, have exaggerated their damage and losses; damage assessment was lower in the more robust plantations."[22] During the weeks and months that followed, the governor introduced incentives and exemptions from duties on imported foodstuffs along with exemptions from duties for cloves "on the condition that they be converted into American grains."[23] Although food imports, especially from India, were on the rise, they proved insufficient to meet demand. This resulted in a deterioration in conditions for slaves on several plantations and protests from the first Indian and Chinese coolies employed during the 1830s over poor living and working conditions, food, and housing as well as mistreatment by the planters.[24]

This scenario was repeated in an even more dramatic fashion in 1844–1845. An extremely violent cyclone on 21 December 1844 was followed by floods on 4 January 1845 and another cyclone from 8 to 9 March 1845.[25] These caused severe damage to roads and plantations. In his report to the minister of the colonies, the governor of Réunion noted that, as the colonial coffers were empty, 339,000 francs had to be taken from the reserve fund in order to repair the damage inflicted by these natural disasters. He asked for special aid from Paris and for the law on the colony's tax allocations to be modified.[26]

As in 1832, food crops were more affected than sugar. Damage to the rice crop resulted in a sharp rise in the price of rice from 16–17 to 25 francs per 75 kg. Five ships were due to arrive from India with approximately 20,000 75 kg sacks of rice, but these were scheduled to meet the island's usual chronic deficit in rice and would be insufficient to meet crisis

[21] "Coup de vent du 5 mars 1832," Rapport au ministre du 13 mars 1832, Centre des Archives d'Outre Mer, Aix en Province (hereafter ANOM), FM SG/Reu c 124 d 931.

[22] ANOM, FM SG/Reu c 380 d 3288, c 370 d 3180.

[23] ANOM, FM SG/Reu c 380 d 3288, c 370 d 3180.

[24] ANOM, FM SG/Reu c 380 d 3288, c 370 d 3180.

[25] "Inondation du 4 janvier 1844," Rapport du gouverneur du 6 janvier 1845, ANOM, FM SG/Reu c 124 d 932.

[26] "Coup de vent du 8 et 9 mars 1845," Rapport au ministre du 31 mars 1845, ANOM, FM SG/Reu c 124 d 934.

food requirements.[27] As a result, conditions on the plantations deteriorated, notably for slaves and "free" immigrant (i.e. indentured) workers, comprising Indians and newly arrived Chinese, who complained bitterly about the lack of provisions.[28] Reconstruction was far from complete by 1847 when another "wind" (i.e. storm) hit the island. The archives show that this time the different authorities were divided as to the degree of damage. While the governor spoke of a "ravaged island" in his report to the minister of the colonies,[29] the director of internal affairs considered that it was essentially foodstuffs that had been damaged rather than sugar cane and that the affected areas had long been in economic difficulty. However, factories and sugar refineries were more severely affected.[30]

The abolition of slavery in 1848 accentuated the island's problems. In order to compensate for the increased costs caused by the move from slavery to indentured labour, planters once again increased the amount of land for sugar cane cultivation and looked for ways to reduce food rations for their workforce. In 1847, there were 6508 Indian, Chinese, African, and Creole *engagés* (indentured workers) on the island.[31] As occurred in the British Empire in the 1830s and 1840s, the abolition of slavery in French colonies in 1848 was followed by a call for additional recruits. Réunion looked particularly to cheaper regional sources of labour, notably Africa, which had until then supplied cheap slaves. Between 1851 and 1854, an annual average of approximately 4000 African workers were recruited, rising to 10,008 in 1858 and falling to 5027 the following year.[32] India proved a more reliable source, supplying 43,958 workers between 1849 and 1859.[33] This growing population increased the pressure on food resources precisely at a time when the amount of land devoted to the production of maize, manioc, and rice was being reduced in favour of sugar cane.

[27] "Coup de vent du 8 et 9 mars 1845," mars 1845: lettre de la direction de l'intérieur de La Réunion au gouverneur, ANOM, FM SG/Reu c 124 d 934.

[28] S. Fuma, *L'esclavage à la Réunion, 1794–1848* (Paris: L'Harmattan, 1992), 129.

[29] "Coup de vent du 25 janvier 1847," ANOM, FM SG/Reu c 124 d 935.

[30] "Coup de vent du 25 janvier 1847," ANOM, FM SG/Reu c 124 d 935.

[31] "*L'indicateur colonial,*" 12 April 1845.

[32] "Réunion, tableau de l'immigration africaine à la Réunion de 1848 à 1869," ANOM, FM SG/Reu c 454, d 5042–5074; see also Ho, Hai Huang, *Histoire économique de l'île de la Réunion, 1849–1881: engagisme, croissance et crise* (Paris: Lavoisier, 2004).

[33] J. Weber, "L'émigration indienne à la Réunion: 'contraire à la morale' ou 'utile à l'humanité'," in *Esclavage et abolition dans l'Océan Indien, 1723–1869*, ed. E. Maestri (Paris: L'Harmattan, 2002), 309–28.

It was within this context of latent shortages and virtual sugar monoculture that a new series of cyclones hit the island at the end of January 1850, and on 28 February, five sugar refineries were destroyed and the roofs of many dwellings were blown away, notably those of "this interesting segment of the population which, recently emancipated, is devoting its freedom to work."[34] Again, unlike the maize and rice crops and food stores, which were almost entirely destroyed, the sugar cane crop generally survived. In its deliberations on 28 March, the island's council asked the governor for 10,530 francs and 2480 sacks of rice in aid, but because of lack of resources, the governor agreed to only half of the requested money and only 2000 sacks of rice.[35]

After a tidal wave hit the island in July 1853,[36] a new hurricane struck in 1858 causing considerable damage to the port, crops, and food stocks; afterwards, the governor reported that "Almost all food reserves have been ruined; the maize harvest is almost entirely lost. A large number of recruits' dwellings have been destroyed."[37] The damage varied, however, from one plantation to another.[38] A few months later, during the first few days of November 1858, Sainte Rose, one of Réunion's volcanoes, erupted, the lava flowing down to the sea and cutting off the island's ring roads. The eruption continued until the end of the year. A variety of natural events hit Réunion in 1860. At the start of February, heavy rainfall caused flooding in Saint Paul and, from 9 February, an exceptional rise in water levels throughout the island. Worse followed as two cyclones hit on 11 and 26 February 1860, the first of which was accompanied by a tidal wave and flooding. Such events prompted the governor to call for the building of a proper meteorological observatory.[39] On 19 March 1860, another volcanic eruption caused further destruction through a phenomenon that is relatively rare on the island: two thick clouds that rained ash, one heading north-east and the other south-east, affecting 60,000 hectares, or one-fifth of the island's area.[40]

[34] "Ouragan du premier mars 1850," ANOM, FM SG/Reu 124 d 937.

[35] "Ouragan du premier mars 1850," ANOM, FM SG/Reu 124 d 937.

[36] "Raz de marée de juillet 1853," ANOM, FM SG/Reu c 124 d 938.

[37] "Ouragan du 16 au 17 janvier 1858," Le gouverneur au ministre, 3 février 1858, ANOM, FM SG/Reu c 124 d 939.

[38] "Ouragan du 16 au 17 janvier 1858," Le gouverneur au ministre, 3 février 1858, ANOM, FM SG/Reu c 124 d 939.

[39] "Cyclones des 11 et 26 février 1860, inondation du 11 février," ANOM, FM SG/Reu c 124 d 941.

[40] "Eruption volcanique du 19 mars 1860," Rapport de Hugoulin, pharmacien de la Marine, au ministre des colonies, ANOM, FM SG/Reu c 124 d 940.

It seemed as though this tragic series of events would never end. Although a hurricane recorded the following year was relatively weak,[41] "one of the most powerful" cyclones (according to the head meteorologist and other observers) struck the island in 1863.[42] Official reports provided an unusually detailed region-by-region, village-by-village account of damage to plantations, infrastructure, industries, transport, and communications.[43] "Everything is in ruins,"[44] wrote the governor to the minister of the colonies. The sea overran coastal land, rivers and streams overflowed their banks, and roads were blocked. The food depots and maize and manioc crops were all but destroyed, and "sugar cane suffered, though not as much as we feared."[45] According to assessments, expenditure of one million francs was needed to repair the damage. Given the customs deficit that was forecast owing to the reduction in sugar exports, the governor decided that it would be necessary to borrow from Paris.[46] The governor also renewed his request to Paris to provide a proper meteorological station.[47] The report on the cyclone that struck the island in 1868 was more detailed than that of 1863, assessing the physical damage and monetary loss in terms of bridges, roads, ports, buildings, plantations, crops, and depots for every municipality. Total losses were assessed at 10.7 million francs.[48]

Such constant cyclone damage caused significant tension between authorities in the colony and the central government and between factions within the colony. Of course, the scarcer the available resources, the greater the tensions. French colonies such as Réunion received little aid from the central government, which expected colonies to be self-financing. However, local revenues were limited. The main sources of revenue on Réunion were a head tax and customs duties until the abolition of slavery in 1848, and thereafter a registration duty on recruits (paid by hirers), taxes on wages, and customs duties, though in the 1860s such duties fell, notably on sugar exports. Following each natural disaster, there were heated debates in the general council about the distribution of aid. In the

[41] "Ouragan du 27 février 1861," ANOM, FM SG/Reu c 124 d 942.
[42] ANOM, FM SG/Reu c 124 d 942.
[43] ANOM, FM SG/Reu c 124 d 942.
[44] ANOM, FM SG/Reu c 124 d 943.
[45] ANOM, FM SG/Reu c 124 d 943.
[46] "Ouragans de 1863," ANOM, FM SG/Reu c 124 d 943.
[47] ANOM, FM SG/Reu c 124 d 942.
[48] "Cyclone du 12–13 mars 1868," ANOM, FM SG/Reu c 124 d 945.

French Caribbean, most such aid went to those hardest hit. On Réunion, local representatives—small landowners in particular—were quick to respond that the situation in Guadeloupe and Martinique was not comparable, and that if aid was allocated to those who declared themselves to have suffered the greatest losses from a cyclone, it would favour the biggest landowners affected. Some pointed out that the heavy rainfall that followed cyclones allowed many plantations to recover, and that therefore damage assessments made immediately after a cyclone had hit should be discounted. Others concluded that the most affected domains were those already in difficulty owing to incompetent owners, who should not be allowed to claim full cyclone damage. Some argued that the council should allocate two-fifths of available financial aid to the individuals most affected, two-fifths to the colonial treasury, and one-fifth to the municipalities. Yet others considered that aid should prioritize the municipalities, which would use the money to repair the infrastructure, thus benefiting everyone, and that such aid should be distributed according to the damage each municipality suffered.[49]

As the figures in Table 7.2 show, the 1860 cyclone had little effect in contrast to that of 1863, the damage from which, combined with the spread of sugar cane disease in the early 1860s, led to a decline in the quantity and value of sugar exports. Sugar prices fell in the early 1860s, stabilized towards the middle of the decade, and then collapsed again during the years that followed. The governor released 500,000 francs from the treasury to aid sugar cane planters and authorized a "temporary withdrawal" of 125,000 francs from the *Caisse d'immigration* (immigration fund). This fund had been created in 1860–1861, owing to pressure from

Table 7.2 Réunion Island exports of sugar

Year	*Millions of tons*	*Millions of francs*
1860	5.6	36
1861	6.9	31.5
1862	7.6	31.2
1863	7.6	20.8
1864	5.8	22.1
1865	5.7	20.7
1866	5.5	23.7

Source: Cyclone du 12–13 mars 1868, ANOM, FM SG/Reu c 124 d 945.

[49] "Cyclone du 12–13 mars 1868," ANOM, FM SG/Reu c 124 d 945.

the French minster, for the colonies and the British consul to protect immigrants from mistreatment by the planters, who often retained their wages under the pretext that they would be paid in full at the termination of their contracts. Employers were obliged to deposit workers' wages in the fund after deducting subsistence and accommodation costs and whatever form of guarantee they deemed to be suitable. The fund was also intended to be used to pay the medical expenses of immigrant workers and fund their passage back home. However, it became another source of injustice: planters did not pay their contributions, medical care was not provided, and workers' wages were appropriated by planters, who justified their actions through claims that, for example, workers had broken their contracts or were guilty of stealing.[50]

The minister for the colonies in Paris, under pressure from the British consul, ordered the governor to rapidly reimburse the immigrant fund.[51] Nevertheless, abuse of workers continued. Against this backdrop of mounting social tension, a major cyclone hit the island in 1872. Inhabitants of Réunion hoped that, with the end of the Empire and the start of the Third Republic, the new regime in Paris would furnish them with greater financial assistance. As with previous cyclones, a list was made of the damage incurred in each municipality. Two hypotheses were put forward in council discussions to explain why the southern part of the island, which had previously gone virtually unscathed, was badly hit. One group blamed climate change and demanded greater aid from France, while a second group argued that monoculture sugar cane growing was responsible, a theory disputed by certain planters and by the island's botanist, who pointed out that sugar cane was more resistant than other crops to hurricanes.[52] Aid nevertheless remained limited, and the island fell deeper into crisis.

When another cyclone struck in March 1873, Paris recommended a new system to assist not only direct victims of such disasters but also the islanders as a whole. To this aim a new Public Assistance Administration was established in 1869 and completed on 16 May 1873 with a decree creating a charitable office (a branch of the Public Assistance Administration) in each of the island's municipalities. This system divided the population into three categories: "natives whom the hurricane has left without

[50] ANOM, FM SG/Reu c 319 d 3317; c 319 d 3211; c 319 d 3317.

[51] "Cyclone du 12–13 mars 1868," ANOM, FM SG/Reu c 124 d 945.

[52] "Sinistre de 1872 cyclone du 17 février," ANOM, FM SG/Reu c.124 d. 946.

resources and unable to work," whose needs were estimated at 521,106 francs; "needy persons to whom aid may be granted with a view to helping them rebuild their small crops, small businesses and workshops, and reconstruct their dwellings," whose needs were estimated at 879,508 francs; and "landowners not asking for any aid," whose total losses were assessed at 4.7 million francs.[53]

This proved to be a turning point, as both immigrants and the *petits blancs* (small white estate owners) were categorized as "natives." In reality, numerous small properties disappeared as a result of the cumulative effect of cyclones, falling sugar prices, and mismanagement. While the largest properties merged, small white estate owners ended up selling their land and moving to the high plains, and an increasing number of former immigrants, chiefly Indians, bought land from former white landowners.

To sum up, the archives document an increase over time in the number of cyclones, and possibly of their strength, particularly from the 1860s. The reports show that food crops were always harder hit by cyclones than sugar cane was. The lack of food and the inappropriate use of the colonial fund reflected attempts by planters and colonial authorities to pass the cost of lower sugar prices on to the workers. However, the question remains: to what extent was this strategy effective?

Immigrants' Rights and Inequalities

The routes supplying *engagés* (indentured labourers) and immigrants partly reproduced those of the slave trade and commercial networks already in place. Indian labourers were brought to the island with the help of Arab and Indian middlemen, who were often in competition with each other. Taking advantage of the Pax Britannica and permission to sail under the British flag, Indian merchants, often from Bombay and the Malabar Coast, gained a considerable competitive advantage in transporting indentured labourers and goods in the Indian Ocean.[54] In the 1840s and 1850s, roughly half of the indentured labourers on Réunion were new immigrants, the other half comprising local freed slaves. In 1854, 34,650 male and 4709 female immigrants joined the 22,650 male and 9022 female indentured labourers already on the island. Only a tiny minority of workers, comprising 3763 men and 3116 women, engaged under agreements

[53] "Cyclone du 7 et 8 janvier 1873," ANOM, FM SG/Reu c 124 d 947.
[54] The National Archives, Kew, FO84/174.

other than indenture.[55] At the time, the law required long-term contracts of one-year or longer, and only 1683 workers had engagements of shorter duration. However, as the seasonal variations in the workforce demonstrate, planters did employ short-term workers. Thus in the first half of 1855, 61,191 labourers out of a total workforce of 86,028 were actually working, including 37,062 new male and 5049 new female immigrants as well as 9545 local males and 4785 local females. In the second half of 1855, the number of agricultural labourers jumped to 91,276, including 41,155 new immigrants, 33,228 members of the "local population," and 9214 with no contract.[56]

Whereas in the early 1850s the government in France adopted measures increasingly favourable to workers, such as a law prohibiting child labour and the legalization of worker organizations, the Second Empire enforced an engagement contract that imposed tighter restrictions on emancipated slaves and *engagés* in the colonies.[57] In principle, recruits had a legal recourse in cases of mistreatment and injustice. In practice, however, it was extremely difficult to make use of these rules, mainly because the local elite controlled the colonial law courts. In the late 1850s, following protests by Indian immigrants and British consuls, a body for the protection of immigrants was created with the authority to inspect estates and provide legal protection for immigrants. However, at least until the late 1860s, few inspections occurred and little legal assistance was offered to immigrants. Despite protests by immigrants and the British consul, local courts interpreted regulations conservatively and rejected workers' claims of unequal treatment.[58]

Legal disputes chiefly concerned issues such as healthcare, food rations, contractual performance, and physical violence. In particular, many planters reacted to rising rice prices by reducing contractual food rations and/or substituting manioc and maize for rice. This triggered an outcry from immigrant workers followed by official protests from the British consul.[59] However, it was only at the end of the 1860s that the governor of Réunion admitted that the substitution of manioc and maize for rice was legitimate

[55] Archives Départementales de la Réunion (hereafter ADR), 10M9, statistiques du travail.

[56] ADR, 10M9, statistiques du travail.

[57] *Le moniteur de la Réunion*, 3 July 1852.

[58] ANOM, FM SG/Reu c 384 d 3361.

[59] "Rapport des syndics des immigrés du 15 juin 1858, après visite de 107 établissements, situation sur le premier trimestre 1858," ANOM, FM SG/Reu, c 384 d 3348; c382 d 3321; "Protestations anglaises pour les traitements des coolies. Mai 1866," c 319 d 3317.

only in "exceptional years" and could not become the rule.[60] Yet the attitude of planters remained unchanged, and mistreatment of immigrants continued. Food rations were often below the legal minimum and of poor quality, wages were often seized, and healthcare was almost non-existent. Indeed, time taken off work due to illness was considered absenteeism and punished as such.[61] Planters also seized immigrants' wages and applied strong penalties for misdemeanours, notably "laziness" and failure to accomplish assigned tasks within a given time.[62] At the same time, employers sued several hundred workers a year for breach of contract. Sentences were usually favourable to the plaintiffs, and those found guilty faced severe fines, which were often repaid in labour. Immigrant workers were also frequently accused of robbery, for which sentences were severe—for example, five years of forced labour for stealing a chicken.[63] Aside from contracts and wages, corporal punishment and violence inflicted by owners were the most common accusations brought before magistrates by plantation workers. In the late 1860s and 1870s, special investigative commissions were set up, most often in response to British diplomatic pressure. Their archives testify not only to widespread corporal punishment but also to the resistance by members of the commission to acknowledging its existence. In most cases, abuses were described as "exceptional" despite being commonplace, and even when employers were found culpable of unfair punishment, they received the slightest of sentences: in one case where physical punishment resulted in the death of a worker, the employer was sentenced to one month in prison.[64]

Only from the 1860s, under pressure from British diplomats and metropolitan French authorities, did violent employers risk receiving more severe sentences.[65] Nevertheless, only a few dozen cases of contractual abuse and illegal wage retention reached the courts annually, and lawsuits often dragged on for years. Even when the court ruled in favour of the complainant, employers were simply forced to pay workers their due wages, with no compensation being offered or interest paid. On the other

[60] ANOM, FM SG/Reu c 319 d 3317.

[61] "Enquête réalisée sur un domaine 'ravine des figues," ANOM, FM SG/Reu c 379 d 3211; "Etablissement du baril," c 379 d 3210; "Commission d'enquête de 1877," c 379 d 3213.

[62] ANOM, FM SG/Reu c 379 d 3211.

[63] ANOM, FM SG Reu c 385 d 3367.

[64] ANOM, FM SM SG/Reu c 382 d 3323.

[65] ANOM, FM SG/Reu c 382 d 3324, 3310, 3311, 3318.

hand, many immigrant workers were granted permission to terminate illegal contracts without penalty.[66]

While estate owners retained a portion of wages to pay the tax on *engagés*, the tax offices noted that the taxes were unpaid.[67] There was thus disagreement between the various offices involved. For example, the tax office allowed the estate owner to collect and hand over the tax payment, whereas the immigration department considered this procedure illegitimate and argued that the tax should be paid directly by the immigrants once they had received their full wages.[68] In 1901, the French Council of State approved the second option and called for the *engagé* tax to be returned to the immigrants concerned. However, as the tax officials considered that these sums represented a tax, not wages, only those registered on the tax rolls could claim the amounts due. It thus proved impossible for the immigrants to recover their money.[69]

Conclusion

In the nineteenth century, cyclones, the frequency and intensity of which increased in the second half of the century, had a major impact on social and economic relations in Réunion inasmuch as they affected worker (first slaves and later immigrant recruits) access to food and the labour strategies employed by planters and the colonial elite. This was particularly the case with access to food; sugar cane is more resistant to strong winds than maize and manioc, while the price of imported Indian rice, the main food staple, increased owing to the weakening of summer monsoons in India during the very period when it was most in demand in Réunion. This accentuated a scarcity of local provisions caused by the growing conversion of plantations to monoculture sugar cane growing and insufficient and poorly protected food stocks.

Planters, especially smaller landowners, attempted to pass the cost of bad weather and poor markets on to their workers by increasing their working hours and retaining wages.[70] However, they were not necessarily

[66] ANOM, FM SM SG/Reu c 379 d 3217, 3210.

[67] ADR 12M69, Cessions d'engagés, impôts.

[68] ADR 12M69, Cessions d'engagés, impôts.

[69] ADR 12M69, Transferts d'engagés, taxes.

[70] D. Northrup, *Indentured Labor in the Age of Imperialism, 1834–1922* (Cambridge: Cambridge University Press, 1995), 31; R. Allen, *Slaves, Freedmen, and Indentured Laborers in Colonial Mauritius* (Cambridge: Cambridge University Press, 1999), 23.

successful. As a result of pressure from the British and subsequently from Paris, workers gained increasingly successful legal recourse against planters from the 1870s.[71] This was resisted most notably by small planters with few resources, who suffered more than big planters from the impact of cyclones and the falling price of sugar. Consequently, they faced increasing costs and decreasing income, and many sold their properties and moved to the highlands.[72]

[71] A. Stanziani, "Beyond Colonialism: Servants, Wage Earners and Indentured Migrants in Rural France and Reunion Island, c. 1750–1900," *Labour History* 54, no. 1 (2013): 64–87; A. Stanziani, "Local Bondage in Global Economies: Servants, Wage Earners and Indentured Migrants in 19th Century France, Great Britain and the Mascarene Islands," *Modern Asian Studies* 47, no. 4 (2013): 1218–51.

[72] See ANOM, FM SG/Reu c 400 d 3688 and c 514 d 5970. See also A. Bourquin, C. Prudhomme, and H. Gerbeau, *Histoire des petits-blancs à la Réunion* (Paris: Karthala, 2005).

CHAPTER 8

Egypt's Slaving Frontier: Environment, Enslavement, Social Transformations, and the Local Use of Slaves in the Sudan, 1780–1880

George Michael La Rue

Introduction

Recent research on enslaved Sudanese has uncovered new and overlooked narratives, travellers' accounts, and other sources that provide insight into the lives of Sudanese slaves, including their capture, their processing as slaves, their service to their captors, and their export to Egyptian and other markets.[1] This research also provides a new understanding of local social transformations and the impact of environmental factors. Environmental changes had an impact on enslavement processes, the composition and direction of the enslaving forces, and the geographic loci

[1] For a sample of this, see G. M. La Rue, "'My Ninth Master was a European': Enslaved Blacks in European Households in Egypt, 1798–1848," in *Race and Slavery*, ed. Terence Walz and Kenneth M. Cuno (Cairo and New York: The American University in Cairo Press), 99–124.

G.M. La Rue (✉)
Clarion University, Clarion, PA, USA

G. Campbell (ed.), *Bondage and the Environment in the Indian Ocean World*, Palgrave Series in Indian Ocean World Studies,
https://doi.org/10.1007/978-3-319-70028-1_8

of their operations and, in turn, significant effects on local social structures and the lives of slaves within them.[2] This chapter will analyse and discuss those effects based on accounts of enslaving techniques, biographical material on a few enslaved individuals, and contemporary descriptions of the impact of enslavement on social structures. Geographically focused on Darfur and Kordofan (Kurdufan), the source materials are primarily Muhammad ibn Umar al-Tunisi (1789–1857) for information relating to Darfur's sultanic slave raids, Léon de Laborde (1807–1869) for the Turco-Egyptian slave raids, and Georg Schweinfurth (1836–1925) for the operations of the *zariba* (forts and trading posts) system.[3] The biographical and environmental sources are more varied.

The Environmental Context

Environmental factors, notably droughts and epidemic disease, had a major impact on the enslavement processes in the Sudan in the period 1780–1880. The best evidence for this relates to a regional drought that affected the entire area from Wadai to Egypt, including the Sudan.[4] First felt in Wadai and northern Darfur—possibly as early as 1828 but certainly by 1829[5]—the drought influenced the relations between Wadai and Darfur, relations between nomads and Darfur's rulers, and the market for slaves in Egypt. The drought exacerbated tensions between the neighbouring sultanates, causing encroachments by the nomads on settled populations on better-watered lands in Darfur's heartland, and affected the price, supply, and profitability of using agricultural slaves in Egypt, as will be shown below.[6] Drought may also have rendered Egypt's economic

[2] J. Ewald, *Soldiers, Traders, and Slavers* (Madison, WI: University of Wisconsin Press, 1990). See also G. M. La Rue, "Land and Social Stratification in Dar Fur, 1785–1875: The Hakura System," in *The State and the Market*, ed. C. Dewey (New Delhi: Manohar Press, 1987), 24–44.

[3] Muhammad ibn Umar al-Tunisi, *Voyage au Ouaday* (Paris: Benjamin Duprat, 1851); Léon de Laborde, "Chasse aux Nègres," *Revue française* 9 (1838): 319–33; G. Schweinfurth, *The Heart of Africa*, vol. 2 (New York: Harper and Bros., 1874).

[4] Muhammad ibn Umar al-Tunisi (Le Chaykh Mohammed Ebn Omar El-Tounsy), *Voyage au Darfour*, trans. Dr Perron (Paris: Benjamin Duprat, 1845), 347.

[5] Sharon Nicholson, "A Climate Chronology for Africa: Synthesis of Geological, Historical and Meteorological Information and Data" (PhD diss., University of Wisconsin-Madison, 1976), 147.

[6] H. Barth, *Travels and Discoveries in North and Central Africa*, vol. 2 (1890; repr., London: Frank Cass, 1965), 646; G. Nachtigal, *Sahara and Sudan*, vol. 4, *Wadai and Darfur*, trans. Allen G. S. Fisher and Humphrey J. Fisher (London: C. Hurst, 1971), 304;

oppression in the northern Sudan more unbearable. Darfur's invasion of Wadai occurred in late 1834 or early 1835 and was caused by internal unrest and nomadic raids on western Darfur: "During the famine which had ravaged Wadai during the last years of Sultan Abd al-Aziz, the eastern tribes of that country had in their distress made frequent raids in Dar Fur."[7] Darfur invaded Wadai and replaced the incumbent sultan with a rival local prince, Muhammad al-Sharif, who opened up a new trans-Saharan trade route from Wadai to Benghazi.[8]

Disease, afflicting humans and animals, also had a major impact. Muslims from all parts of the IOW made the pilgrimage to Mecca and were potentially exposed to contagious diseases, such as cholera and the plague. Moreover, drought could accentuate the impact of disease. Following the great drought of the late 1820s and early 1830s, a source from the riverain Sudan reported: "The years 1251, 1252 and 1253 [1835–1838] were years of famine; in 1252 the cholera morbus spread through the country striking down a population already weakened by hunger."[9] The plague struck at the same time. Beginning in the port of Alexandria in the summer of 1834, the plague claimed over 4000 victims, more than a third of the town's civilian population. By June of 1835, Cairo similarly lost a third of its 250,000 inhabitants. For the country as a whole, mortality from the plague was about 10 per cent, or up to an estimated half a million out of a total of five million inhabitants.[10] Many observers, including John Bowring (1792–1872) and Antoine Barthelemy

M. J. Tubiana, "Un document inédit sur les sultans du Wadday," *Cahiers d'études africaines* 2 (1960): 78–79. For the consequences of drought induced famine in a slightly later period, see S. Serels, "Famine and Slavery in Africa's Red Sea World, 1887–1914," Chap. 11 in this volume.

[7] G. Nachtigal, *Sahara and Sudan*, vol. 4, pp. 219–20.

[8] D. Cordell, "Eastern Libya, Wadai and the Sanusiya: A Tariqa and a Trade Route," *Journal of African History* 18, no. 1 (1977): 21–36.

[9] R. Hill, *On the Frontiers of Islam: Two Manuscripts Concerning the Sudan under Turco-Egyptian Rule, 1822–1845* (Oxford: Clarendon Press, 1970), 53.

[10] D. Panzac, *La peste dans l'empire Ottoman, 1700–1850* (Leuven: Editions Peeters, 1985), 365; Afaf Lutfi al-Sayyid Marsot, "Muhammad Ali's Internal Politics," in *L'Égypte au XIXe siècle*, ed. R. Mantran (Paris: Éditions du Centre National de la Recherche Scientifique, 1982), 164; Panzac and Marsot follow J. McCarthy, "Nineteenth-Century Egyptian Population," *Middle Eastern Studies* 12, no. 3 (1976): 15. This figure is challenged by L. Kuhnke, *Lives at Risk: Public Health in Nineteenth-Century Egypt* (Berkeley: University of California Press, 1990), 204, footnote 59. For a later statement of Panzac's views, see "The Population of Egypt in the Nineteenth Century," *Asian and African Studies* 21 (1987): 18–19.

Clot (1793–1868)—known locally as Clot-Bey—reported particularly high levels of mortality among slaves in Egypt. For example, Bowring observed, "The mortality among the black slaves in Egypt is frightful. When the epidemic plague visits the country they are swept away in immense multitudes, and they are the earliest victims of almost every other domineering disease,"[11] while Clot noted that in the Cairo slave market, "Negroes died en masse."[12] Another European observer, Félix Mengin (1772–1862), specified that "In Cairo, of the blacks of both sexes who are employed as domestics, fifteen thousand individuals have perished, as well as Nubians and other strangers."[13] Again, in 1863, a devastating cattle murrain in Lower Egypt killed between 600,000 and 900,000 head of cattle, amounting to most of the cattle stock, and was followed shortly in 1865 by an outbreak of rinderpest that caused high mortality among Egyptian cattle.[14]

Such events had a huge impact on the labour market. In the 1830s, for example, high mortality due to famine and disease led to a significant increase in demand for servile labour from the Sudan, as Egyptian slave owners sought to replace slaves who had died.[15] Such demand was accentuated by animal diseases that decimated draught animals traditionally used in agriculture. The difficulties may have been felt first by nomads near the Nile: "The nomads of the prairies, the cattle-owning tribes, generally paid [their taxes] in cattle, though in 1835, a lean year, the Viceroy permitted Khurshid to accept negroes in place of the customary cattle."[16] These reports are confirmed for Egypt and the Sudan by numerous sources.[17]

[11] J. Bowring, "Report on Egypt and Candia," in British Parliamentary Papers, Reports from Commissioners, vol. 11 (1840), 92.

[12] A.-B. Clot-Bey, *De la peste observée en Egypte: recherches et considerations sur cette maladie* (Paris: Fortin, Masson, 1840), 111–12, footnote 2. My translation from the French here, and throughout.

[13] F. Mengin, *Histoire sommaire de l'Égypte* (Paris: Librarie de Firmin Didot, 1839), 472.

[14] R. Owen, *Cotton and the Egyptian Economy, 1820–1914: A Study in Trade and Development* (Oxford: Clarendon Press, 1969), 98–100.

[15] G. M. La Rue, "African Slave Women in Egypt, from ca. 1820 to the Plague Epidemic of 1834–1835," in *Women and Slavery*, vol. 1, *Africa and the Western Indian Ocean Islands*, ed. G. Campbell, S. Miers, and J. C. Miller (Athens, Ohio: Ohio University Press, 2007), 168–89; G. M. La Rue, "Treating Black Deaths in Egypt: Clot-Bey, African Slaves and the Plague Epidemic of 1834–35," in *Histories of Medicine and healing in the Indian Ocean World*, ed. A. Winterbottom and F. Tesfaye (New York: Palgrave Macmillan, 2015).

[16] R. Hill, *Egypt in the Sudan, 1820–1881* (London: Oxford University Press, 1959), 40.

[17] For more details, see G. M. La Rue, "The Hakura System: Land and Social Stratification in the Social and Economic History of the Sultanate of Dar Fur (Sudan), ca. 1785–1875" (PhD diss., Boston University, 1989), 116–18.

The problem was highlighted again during the American Civil War (1861–1865), when traditional supplies of cotton to world markets from the American South were cut off. This led to a search for alternative sources and a fivefold increase in cotton production in Egypt.[18] However, disease decimated local cattle stocks, leading to a sharp growth in demand for Sudanese slaves to work the *shadufs* (devices used to raise water, comprising a long suspended rod with a bucket at one end and a weight at the other) and *sāqiyas* (devices designed to lift water from the Nile with containers attached to a rope or chain flung over a wheel, traditionally pulled by draught animals) that irrigated Egyptian cotton fields. The French traveller Raoul Lacour, visiting Egypt in the late 1860s, saw that blacks were often manning *shadufs* along the Nile.[19] In 1880, Count della Sala noted that male slaves fit to work *shadufs* were readily available in Upper Egypt. He observed, "On my arrival at Darawi I saw a slave in irons who was working a shaduf ..."[20] A British report stated that "As late as 1884, nine-tenths of the men who worked the water-pumps in Isna province were slaves."[21] Given the cattle murrain and the drought of 1863, which required additional pumping of water from a low Nile to grow cotton (and other crops), cultivators must have turned to slaves to work the *shadufs*. A similar situation prevailed in the northern Sudan.

In 1860, slaves were already being smuggled into Egypt despite a ban on slave imports. Petherick reported, "The Cairo market ... is indebted for the greater number of its slaves purchased in l'Obeid, in Kordofan, emanating from Darfour, and the Baggara, nomade Arabes, inhabiting the southern districts of the province."[22] Although the conventional wisdom is that slaves were not used as agricultural labourers in Egypt, there is growing evidence that they were. Slave imports, notably from the Sudan, spiked during the cotton boom for markets in the Egyptian areas most

[18] Owen, *Cotton and the Egyptian Economy*.

[19] R. Lacour, *L'Égypte d'Alexandrie à la seconde cataracte* (Paris: Hachette, 1871), 8.

[20] G. M. La Rue, "The Capture of a Slave Caravan: The Incident at Asyut (Egypt) in 1880," *African Economic History* 30 (2002): 102.

[21] G. Baer, *Studies in the Social History of Modern Egypt* (Chicago: University of Chicago Press, 1969), 165.

[22] Colquhoun to Russel, Alexandria, 8 June 1860, and enclosure: "Memo by Petherick." (quoted) Great Britain. Irish University Press Series of British Parliamentary Papers. Slave Trade. Shannon: Irish University Press, 1968. vol. XLVII, (1861), no. 163.

affected.[23] They were in demand both to help irrigate the expanded acreage of cotton and, as incomes increased for cotton producers, as "superior goods"—especially servants and concubines. Nineteenth century sources hint at this: "... village sheikhs, who after the increase wealth consequent on the development of cotton culture during the American Civil War, in a few instances bought slaves to help in field labor."[24] French archival sources are more emphatic: "... every small farmer knows that in a certain street in Cairo he could buy a black slave to cultivate his land. On the Nile, convoys frequently arrive of black slaves who are sold secretly ... near Cairo ..." Those slaves arrived from Kordofan and Darfur by caravan to Egypt before embarking on the Nile.[25]

Structures of Slave Raiding

Under Muhammad Ali (r. 1805–1849), the Sudan was Egypt's major source of slaves for internal labour, for the new Egyptian army, and for export to the Middle East to serve in harems, protect the holy places of Islam, and perform other essential labour. After 1824, Egypt used Sudanese slave troops to fight in Greece, the Hijaz, and Ethiopia, and in doing so gained control of key ports and commodities that were traded along the Red Sea in and out of the IOW. Muhammad Ali had designs on two potential sources of slaves: the upper Nile and Darfur.[26] The drought, famine, and loss of slaves from the plague spurred renewed Egyptian interest in sources of Sudanese slaves.

By 1837, when Muhammad Abu Madyan (*ca.* 1803–1847), a pretender to Darfur's throne, came to Muhammad Ali's attention, a plan was quickly hatched to install him as sultan in return for a generous tribute, including slaves. Diplomatic preparations for war are suggested by the

[23] Baer, *Studies*, 165; R. Austen, "The Trans-Saharan Slave Trade: A Tentative Census," in *The Uncommon Market: Essays in the Economic History of the Atlantic Slave Trade*, ed. H. A. Gemery and J. S. Hogendorn (New York: Academic Press, 1979), 34; K. Cuno, "African Slaves in Nineteenth-Century Rural Egypt: A Preliminary Assessment," in *Race and Slavery*, 77–98.

[24] J. C. McCoan, *Egypt As It Is* (New York: 1882), 306n.

[25] "Consul au Ministère d'Affaires Étrangères", Le Caire, 16 July 1863; see also Consul de France au Caire à M. Drouyn de Lhuys, 27 July 1863. France: Archives du Ministere des Affaires Etrangeres, Correspondance commerciales, Alexandrie, vol. 30.

[26] P. Holt and M. Daly, *The History of the Sudan: From the Coming of Islam to the Present Day* (Boulder, Colo.: Westview Press, 1979), 67.

issuance of a firman by the Ottoman Sultan for Muhammad Ali including Darfur as one of the provinces of Egypt.[27]

Slave raiding, which was an additional source of income in hard times when crops failed or when herds decreased because of poor rainfall, occurred primarily through the kidnapping of individuals by nomads, retail purchases by small-scale traders, and the capture of entire villages during large-scale slave raids and wars. The enslavement of individuals changed relatively little, but the large-scale capture of slaves varied over time and occurred in three different periods. The first, in early nineteenth-century Darfur under sultanic rule, represents the patterns of the Sudanic kingdoms. During the second, in the late 1820s, Turco-Egyptian slave raids were carried out by the new Egyptian army (the *nizam al-jadid*), staffed largely by enslaved Sudanese captured during the Egyptian invasion of the Sudan in 1820–1821 or in subsequent slave raids.[28] Soldiers were trained initially by European military instructors—"Turks" typically recruited from the Ottoman Empire—or sometimes by Europeans officers. The third period, in the 1860s, was characterized by patterns of enslavement typical of the *zariba* system reported by Schweinfurth—an elaborate slave-raiding and trading network south-west of Kordofan and Darfur.

Darfur

For Darfur, the analysis of the slaving frontier is based on al-Tunisi's descriptions of Dar Fartit and Darfur in the early nineteenth century. The slave raids depicted by de Laborde in the Nuba hills of southern Kordofan between 1828 and 1838 were carried out by Turco-Egyptian forces based in El Obeid. Schweinfurth's description of the *zariba* system emphasizes raids carried out in the late 1860s in the Bahr al-Ghazal area of the

[27] Hill, *On the Frontiers*, 173; F. Werne, *African Wanderings; or, An Expedition from Sennaar to Taka, Basa, and Beni-Amer, with a Particular Glance at the Races of Bellad Sudan* (London: Longman, Brown, Green and Longmans, 1852), 5–6; Hill, *Egypt in the Sudan*, 75–76; Khosrew Effendi to Col. Barnett, Kafr Mugar, 6 December 1842, Great Britain: Public Records Office (Kew). Foreign Office. FO 141:8, unfoliated; al-Tunisi, *Darfour*, 370–96; F. Fresnel, "Mémoire sur le Waday," *Bulletin de la Société de Géographie de Paris* 14 (1850): 153–92.

[28] For the *nizam al-jadid*, see E. A. Helal, "Muhammad Ali's First Army: The Experiment in Building an Entirely Slave Army," in *Race and Slavery*, 17–42; for Sudanese in the Egyptian military, see R. Hill and P. Hogg, *A Black Corps d'Elite* (East Lansing, MI: Michigan State University Press, 1995).

north-western south Sudan, where, however, raiders' *zaribas* had been established from about 1843. These processes of enslavement overlapped in time, competed with each other, and drew upon the experiences of earlier raiders.

The enslaving frontiers slowly moved south-west from the Sultanates of Darfur and Sinnar and reflected shifts in power, as when Darfur engaged in slave raids following its invasion of Kordofan in 1785 or when Egypt invaded the Sudan in 1820–1821, took the northern riparian Sudan, toppled Sinnar, and captured the urban centres of Kordofan from the Darfur garrison in 1820–1821. Similarly, the creation of the *zariba* system supplanted the earlier raids of the Fur Sultans in the Bahr al-Ghazal area, pushing them to raid further to the west and south-west. These major military forces also interacted with lesser polities that paid tribute in slaves and with nomadic Arab groups—often from areas that were most sensitive to drought—who served as auxiliaries or independent suppliers for the slave-trading networks supplying markets such as Kubayh in Darfur, El Obeid in Kordofan, or Shakka in southern Darfur, and through them also supplying markets in Egypt.

In the early nineteenth century, the organization of slave raids was taken on the initiative of enterprising individuals rather than the Sultan of Darfur.[29] These men had to be of sufficient wealth and status to be able to make a present—often a fully outfitted warhorse with a slave attendant—to the sultan in order to receive his written permission and the symbolic broad-bladed spear (*salatiya*) to lead a *ghazwa* (armed expedition).[30] Once permission was granted, preparations would begin and the raid leader would announce his commission and plans in Darfur's capital in order to attract financial support and armed followers.

Merchants supplied raid leaders with trade goods, particularly imported cloth, expecting to receive a portion of the captured slaves. The terms varied according to where the merchants obtained the slaves: nearer the point of capture, slaves were cheaper (so merchants could afford up to three times as many for a given quantity of trade goods), while in al-Fashir, the same quality of slaves cost more. This difference reflects the cost of

[29] This discussion is based on al-Tunisi, *Voyage au Ouaday*, 466–95; R. S. O'Fahey, "Slavery and the Slave Trade in Dar Fur," *Journal of African History* 14, no. 3 (1973): 29–43; G. M. La Rue, *Land and Social Stratification in Dar Fur, 1785–1875: The Hakura System* (Boston: Boston University, 1984).

[30] Al-Tunisi, *Voyage au Ouaday*, 468–69.

transporting slaves from the slaving frontier and the fact that a portion of them would die en route. Merchants accompanying the raiding party assumed these costs and risks, thus enabling raiders to dedicate themselves to enslaving yet more victims. A leader with a good reputation could obtain goods worth the equivalent of 500 to 600 average slaves. The raid leader gathered notables with military experience and commoners seeking their fortunes and then arranged for ten or more squads of raiders to rendezvous annually, usually well before the rains began in June, near the sultanate's southern borders. He announced his mission and the proposed division of spoils en route to the rendezvous. Once the raiders had assembled, he took the title of "sultan" of the raid and appointed squad leaders with titles reflecting those of the sultanic court.

The distribution of slaves was multi-tiered. Those paid in tribute or captured without resistance belonged to the raiders' "sultan." Periodically, a *jabaya* (distribution ceremony) was held, wherein all captives were brought to a central location for inspection by the "sultan," who took his share according to the announced terms and his own status. Highly regarded leaders took up to half, while others got as few as a quarter of such slaves. Sometimes the terms varied so that the raid leader took a third of the first *jabaya* but only a quarter of the second. After the ceremony, the "sultan" repaid the accompanying merchants who had advanced goods to the raiders by delivering slaves to them. Upon returning to al-Fashir, the raid leader gave a portion of the slaves to the sultan, presented slaves to high officials who had supported him, and, finally, gave slaves to the merchants who had stayed in the capital as payment for the trade goods they had advanced in support of the raiders.

The areas raided varied over time. Umm Busa, Sultan Muhammad Fadl's mother, was a Beigo slave woman, and when Fadl began his rule in 1799, he forbade the enslavement of her people. Thereafter, slave raiders had to go further south beyond Dar Beigo for slaves. Generally, the raids targeted "pagans" who lived in Dar Fartit, south-west of Darfur. By the early nineteenth century, the leaders of some polities threatened by *ghazwas*, including "Runga, Fongoro, Bandala, Binga and Shala," transformed the area under their control into tributaries of Darfur in order to avoid slave raids. Local terminology for the enslaved suggests other aspects of the process: those known as *denguyeh* had not resisted, while those captured in the mountains were called *fekk el-djebal*.[31]

[31] O'Fahey, "Slavery," 36.

The captives were divided into three categories. The first, acquired by trans-Saharan merchants in Kubayh for shipment to Asyut in Upper Egypt, on the west bank of the Nile, and sale further north, notably in Cairo and Alexandria, pass out of our consideration here.[32] The second category was retained within Darfur for productive purposes, and the third supplied the sultanic court.

Absorption of the retained slaves gradually transformed societies at the slaving frontier, and their statuses are again reflected in local terminology. The *foutyr* were those not yet acclimated to life in Darfur and thus less valuable. By contrast, *mougeddeh* were fully acculturated slaves. The distinction here between the two terms may indicate a process of "seasoning"—the practice of acculturating some new slaves for subsequent resale at higher prices. Some captives were *hamil*, stray slaves, who had run away from their masters. That they had run away makes a clear statement about their conditions of enslavement in Darfur.[33]

Broadly speaking, all "free" groups in Darfur held slaves. Nomads used slaves in herding and raiding as well as to perform domestic tasks, and, particularly after famines characterized by high mortality, they launched slave raids to rebuild their populations. Farmers employed slaves to work their fields, while merchants used them as domestic servants, to provide hospitality in their households to both local and long-distance merchants, and to cultivate their land. For example, the *khabir* (caravan leader) Ali employed over 80 slaves growing wheat on irrigated fields at Naburu, his *hakura* (estate) in Kubayh.[34] Faqihs who ran quranic schools also received slaves occasionally as donations and used them to grow crops.

Many slaves were retained in the sultanic court and in the large households of other notables. Most performed domestic tasks or grew food on the *hakuras* of the royal family. Quite a number were assigned to the military as ordinary soldiers or as members of elite corps, such as the *korkwa* (spear men), *saringa* (swordsmen), and *koriat* (grooms). Some slave women served as concubines or wives to high officials, as in the case of Umm Busa, a rather plain woman who began as the future Sultan Abd

[32] G. M. La Rue, "The Export Trade of Dar Fur, ca. 1785–1875," in *Figuring African Trade*, ed. G. Liesegang, H. Pasch, and A. Jones (Berlin: Reimer, 1986), 636–38.

[33] O'Fahey, "Slavery," 36.

[34] G. M. La Rue, "Khabir ʿAli at Home in Kubayh: A Brief Biography of a Dar Fur Caravan Leader," *African Economic History* 13 (1984): 56–83; G. M. La Rue, "Land Documents in Dār Fūr Sultanate (Sudan, 1785–1875): Between Memory and Archives," *Afriques* (2016), http://afriques.revues. org/1896

al-Rahman's sole slave, accompanied him to war in Kordofan, and gave birth to the next sultan. At court, there were page boys from the *soming dogala* (a group of young boys, both slave and free, who were being trained for service at court) and about 20 titled eunuchs, who held official government positions as provincial governors, generals, and bureaucrats. The most successful eunuchs became guardians of the sultan's harem, his most trusted officials and top generals. One eunuch, Muhammad Kurra, Governor of Kordofan, effectively became a kingmaker, ensuring the succession of Sultans Abd al-Rahman and Muhammad Fadl and serving as Governor of Kordofan. Through their loyalty to the court rather than to local sedentary or nomadic lineages, such slaves enabled the sultans to rise above the ordinary population of Darfur and to dominate them. Royal slaves helped to collect agricultural taxes in kind, led the sultan's most loyal troops, and supported the royal household in increasing their social and economic wealth at the expense of the broader population. Their presence fostered and signalled a significant degree of social stratification at the slaving frontier. Challenges to sultanic power were generally greatest during droughts and the subsequent famines, as although the sultanic state collected agricultural taxes in kind and redistributed grain during famine years, it was often difficult to provide for all those who expected help. The sultan's court, including royal slaves, always ate well, and the members of the court used their control of the sultanic granaries as yet another weapon to control challenges from royal rivals and powerful ethnic groups.

Egypt

Following the Egyptian invasion of the Sudan in 1820–1821, slave raids targeted the Nuba Hills in south Kordofan, in the process changing the social organization of the societies on the slaving frontier. De Laborde described a particular raid launched from El Obeid, the provincial capital, in about 1828. The raiding party was a mixed group directed by a "Turkish" (that is, of Ottoman mamluk, Albanian, European, or non-Sudanese origin) commandant and "consisting of four hundred Egyptian troops and their Turkish officers, all armed in the French style, wearing Egyptian uniforms, exercising according to French standards, and marching to drums and Napoleonic-era marches."[35] The troops often comprised

[35] De Laborde, "Chasse," 324, my translation.

Sudanese soldiers captured in the 1820–1821 invasion or in raids. The commandant took along his *aides-de-camp*, a comptroller, and a European doctor and two surgeons to treat the wounded. Although the black Egyptian soldiers had fairly high mortality, the "Turkish" officers and their staff did not stay long in the Sudan. While there, however, they often established households in El Obeid, where slave and free residents laboured to supply them with provisions and other services.

Also accompanying the main force were 100 light cavalry supplied by their local Arab allies, the Baggara, and a dozen chiefs from villages near El Obeid, each with about twenty or more followers and camel convoys carrying provisions. Nomadic allies were most available during famine years. They tended the captives, helping to reduce the incidence of slave rebellion and increase the survival rate among the wounded. The raiders' guides were "local blacks," often herders and "traders in hides" who knew the area and occasionally visited El Obeid and who provided advance intelligence on the defence and demographics of target villages in return for payment.[36]

The *zariba* system, which combined ivory gathering with a third type of slave raiding, developed across the Bahr al-Ghazal area south of Kordofan. It peaked during the 1860s, a decade marked by significant droughts and outbreaks of cattle disease.[37] Individual *zaribas* were usually headed by Bahhari (northern Sudanese from the Dongola area) and supported by an elaborate structure of soldiers, military slaves, local allies, and subordinate populations. Some of the social transformations were comparable to those that occurred in the regions that supplied slaves and ivory to Zanzibar to be distributed throughout the IOW.[38] Companies of traders and *jallabas* (merchants) from Darfur, Kordofan, and the northern Sudan linked the *zaribas*. Slave raids provided ample funds and personnel to create new societies within the raided areas. The nature of the enslavement process can be deduced from the description of the new societies that were created.

[36] De Laborde, "Chasse," 324–25.

[37] On the mid-1860s drought, see G. Douin, *Histoire du règne du Khédive Ismail*, tome III *L'Empire Africain* (Cairo: Société Royale de Géographie d'Égypte, 1938), 1, 174; Nachtigal, *Sahara and Sudan*, vol. 1, p. 44; A. Bjorkelo, *Prelude to the Mahdiyya* (Cambridge: Cambridge University Press, 1989), 93.

[38] For a classic introduction to this literature, see Abdul Sheriff, *Slaves, Spices and Ivory in Zanzibar: Integration of an East African Commercial Empire into the World Economy 1770–1873* (London: James Currey, 1987).

Processes of Enslavement

The raiding party generally advanced to within two days' journey of the target, often a village on one of the isolated and forested small mountains that rose from the plain. Because the mountains were discontinuous, forested, and natural fortresses, the attackers travelled more easily along the plains, so they ventured into the mountains only for raids. To prepare the attack, they moved into position at night, using cavalry and infantry to surround the targeted hill village, but waited until first light to attack, as they were clambering over rough terrain up to the village. They fired cannons at dawn to rouse and frighten the villagers; caught off guard, the villagers strained to see their attackers, and women hurriedly gathered up the children and led the infirm to safety. The attackers' goal was to capture the entire population, so four squads set off at once up the mountain after the cannon-fire, bayonets at the ready, while the rest of the attackers provided covering fire and a general distraction.

There is a vivid account of a "slave hunt" from Almas, a 12-year-old boy from the Nuba Hills, captured in about 1836, presented personally to Dr Arthur T. Holroyd (1806–1887), a traveller in the Sudan:

> ... the pasha's troops made the attack during the night, whilst the negroes were sleeping; that they fired repeatedly upon the district with cannon and muskets, both loaded with shot; and that they burnt the straw huts of the negroes. As they escaped from their burning huts they were seized by the troops: many, especially the children, were burnt to death, and many were killed. Those who ran away and were pursued by the soldiers defended themselves with stones, spears and trombashes; the latter, an iron weapon in common use among the natives of these mountains.
>
> The negroes retreated to the caves in the sides of the mountains, from whence they were eventually obliged to come forth, from fear of suffocation from the fires made at the entrances, or from want of food and water ...

Once they were in the hands of their captors, the newly enslaved were roughly treated:

> Pronged stakes were fastened round the throats of the men, and their hands were fixed in blocks of wood nailed together. Boys, of twelve or fourteen years, had their hands only manacled, and the young children and women were without any incumbrance. Two or three times Almas saw a stubborn

> slave drawn (his expression) like a carriage, by a horse across the rocks, until he was dead. He cannot say how many were killed in the attack; he thinks 500 were taken along with him from Korgo, but many of these died of thirst, hunger, and fatigue on their march to Kordofan.

Almas's own family was divided by the attacking allies:

> Almas's father and brother were captured along with him … They are both soldiers at Sobeyet [El Obeid]. His mother was seized by the sultan of Baggarah, who makes expeditions continually against the inhabitants of Gebel Noobah.[39]

De Laborde wrote:

> The blacks have huts and cabins on the mountaintop where they live in ordinary times, but for defensive purposes they have dug holes, real rabbit warrens where they hide everything precious, including their women and children. They are courageous in the defence of these, and they throw poisoned spears with one hand, while using the other to hold a shield. But before the enemy is within reach of their weapons, they have been hit by the attackers' bullets. They are tough and often take four or five bullets before they succumb, rubbing the entrance of their wounds with a bit of dirt, perceiving them as mere scratches, and fighting until exhausted by the loss of blood they fall, dead. As long as the head of the family defends himself, his wife and children stay with him to encourage him with their shouts, to help him by throwing stones at the enemy, but once he is killed they surrender.[40]

The less brave fled to the mountaintop or hid in caves and tunnels. The attackers forced them into the open by shooting a gun loaded with pepper into the tunnels or smoking them out. As the villagers emerged coughing and eyes stinging, they fell into nets deployed by the attackers and were then bound in chains by the raiders. If no one emerged, the attackers knew that the villagers had committed suicide—mothers smothering children, husbands killing wives and themselves—to avoid capture.[41] De Laborde continues:

[39] T. F. Buxton, *The African Slave Trade and its Remedy* (London: Cass, 1967), 96–97.
[40] De Laborde, "Chasse," 326.
[41] De Laborde, "Chasse," 327.

> As for their prizes, they drag them along. But resistance continues: some refuse to walk, others grab hold of a tree and refuse to let go, another clutches at his family members, forming a knot that can only be broken by iron ...
>
> Some of the unwilling are tied by the legs to a horse and dragged through the thorns and over the rocks to the bottom of the mountain, where they arrive, bleeding, scratched, disfigured but without letting go. They are killed, as nothing else can be done. If they give in, the punishment stops, they are enchained and around their necks is attached a long heavy forked branch which they have to lift to take a step.[42]

The raiding soldiers brought their prisoners to their commander's tent. A Turkish clerk made a list of each hour's catch of slaves, and the village chiefs who would guard and watch over them:

> These chiefs with their peasants know the country, the habits of the blacks, their language and how to calm their anger. They manage to control those who were captured. Formerly the Egyptian soldiers would guard the captives, but the captives used their own language to coordinate their sudden attacks on them, knocking them out with their chains or with the forked branches around their own necks. Such actions required a military response, and put the expedition at risk.[43]

The squads took turns raiding and bringing captives down the mountain to be recorded on the clerk's list and guarded by local allied forces as long as there remained villagers on the mountain. When these basic tactics failed against larger villages, raiding generals besieged them. Water sources were typically located below the village, so attackers surrounded the wells and waited until thirst forced villagers to submit. Locals tried to "chew on the bark of trees to slake their thirst, but with the burning sun and without food the roof of the mouth dries out, and the suffering begins."[44] Some held out for a week, but for all the stark choice was between death and enslavement.

[42] De Laborde, "Chasse," 327.
[43] De Laborde, "Chasse," 327.
[44] De Laborde, "Chasse," 328.

Following the capture of a village, and after checking the clerk's list which enumerated which raiding unit had captured each group of slaves, soldiers were assigned to guard each group and the raiders moved off:

> From this moment, a remarkable change occurs: hostility is transformed into touching compassion. An old man who is so bent over from age that he can't go on is put on a stretcher or a camel, and made comfortable with a fortifying drink. Women are given a bit of time to give birth or to nurse their children, while the wounded are allowed to be treated, and everyone is given abundant food. But don't be fooled—it is the care of the butcher, the pity of the executioner. The goal is to bring the victims alive to the capital, because they must be answered for. But this humanity only lasts for the week-long trip between the mountain and the capital, a short respite from the misfortune of a lifetime.[45]

At the capital, another division of the slaves was made. Each allied chief brought forward the slaves in his charge and sales began. The governor enquired to see if the "king of the mountain" and the religious chief of each attacked village would step forward. If they were present among the living captives, they were permitted to select 20 male and female relatives from among the captives from their village. The selected group was then given "two camels, some provisions, some trade goods to start a business with, and is sent back to their mountain in the hope that they will repopulate it within a few years, so that it can be raided again in the future."[46]

Of those captives who remained behind, the oldest, weakest, maimed, and disfigured were given to the nomadic Arabs who had accompanied the raid as payment for their help, presumably because they were less valuable in the long-distance trade but could eventually recuperate and be of local service. The best captives were reserved for the officers and for trans-Saharan export. Officers and soldiers chose the slaves they wanted, as in-kind payment of their salaries, with the value of slaves varying according to age and strength for men and beauty for women. The story of Selim, captured in about 1835 in Taqali, gives insight into the use of slaves by the Egyptian military in El Obeid and their subsequent fate:

> During our stay here, Medina and I were taken to the camp of the Turks, not far away from the village, where we were put through different exercises. The first thing we were desired to do was to show our tongues, and then our

[45] De Laborde, "Chasse," 329.
[46] De Laborde, "Chasse," 329.

teeth. The rest of our limbs underwent a serious examination also. Having undergone this examination, we were taken back to our lodgings again. The next day our master joined the Turks, who were returning to Kordofan, and by that means ensured our fate of never returning to our native country ... My time while with the first three masters was employed doing nothing. The Turkish gentleman found work for everybody; and all the testimony I can bear to his good character is that he was one of the cruelest men in existence.

Being an officer of the rank of an *aga*, his men suffered many harsh cruelties under him. On one occasion a soldier having been brought to his house for a small offence, he took the office of corporal; and commanding four men to hold him down, beat the poor man, till the blood was running from his cheeks. The keeper of his camels often suffered in a similar way. My office was what might be called a general house-servant. The duties of waiting the table, washing dishes, making coffee, and waiting for orders, were allotted to me as my share of the work. Medina was made assistant cook for a short time, but I had the disagreeable misfortune to see her sold to another Turk; thus I was left to suffer alone.

Some six months, however, relieved me of my hardships. To mention all the cruelties I suffered at that time, would be quite needless. I will only notify a few of them. My master, on whom I had continually to attend, punished every small fault with great severity. If he called, he said I ought to hear him at whatever distance I might be. At one time, being sent from home by my mistress, my master interrogated me on my return with where have you been, and began to thrash me. Self-justification was of no use. No moderate blows did with him, for while he struck one side of my head, he met it at the other side also. I became almost insensible, while the blood was running out of my ears. At another time, having made some coffee by his own orders, I happened to make a few cups more than was required. He said nothing at the time; but after I was in bed, he got hold of a horse whip, and coming upon me unawares, thrashed me till I was quite speechless. I am persuaded he would have killed me had not one of the domestics heard my cries, and come to my rescue. Here I may mention that a very small child can stop a Mahomedan from revenging himself to too great an extent, by taking the whip, or whatever he uses, from him. One of the slaves was the means of preventing my master from whipping me any longer. In Kordofan the houses are all of one storey high. The part in which I lived was chiefly occupied by officers in the Pacha's service. My master was married, had two children, two female slaves, two males, and myself. The other two, being grown-up men, were taken out to exercise along with the rest of the soldiers. When coming home from exercise, my master was sure to be heard

> crying my name a quarter of a mile's distance from the house, at which I had to run out to meet him and carry his sword home. These, and other sufferings of the like nature, prepared me for my subsequent career, and fitted me for the journey on the desert.[47]

Indeed, the "… bulk of the captives are sent in a caravan to Dongola on the Nile, from whence they descend to Cairo by boat. There, the relevant administrators working with customs agents collect the import duties and register the slave in the slave market, where the sales begin."[48] No allowance was made for family connections, as parents were separated from their children, and spouses from each other.[49] Selim, for example, was trained in basic domestic duties and then re-sold into the trans-Saharan trade, ending up in Alexandria by 1836 as part of a wave of slaves replacing those lost in the plague epidemic.[50]

Impact on Target Societies

The raids had major demographic, social, economic, cultural, and environmental impacts on the targeted settlements, which in turn were most vulnerable when their water supplies and granaries were depleted. The mortality rate among societies subject to slave raiding was high because of both the violence of the initial attack and capture and the subsequent enslavement processes. *Ghazwas* had a major demographic impact: "For six thousand slaves gathered, there are two thousand cadavers."[51] A quarter of the captives died before reaching the raiders' base in El Obeid, and most survivors faced exportation rather than retention by local raiders. In addition to raids in Kordofan, which in the years 1821–1825 alone captured between 50,000 and 80,000 individuals,[52] four raiding parties from Sinnar annually collected between 5000 and 6000 slaves in the 1830s. More slaves were brought from Ethiopia to Sinnar for sale to local elites

[47] S. Aga, *Incidents Connected with the Life of Selim Aga, a Native of Central Africa: Electronic Edition*, 20–23, http://docsouth.unc.edu/neh/aga/aga.html

[48] De Laborde, "Chasse," 329.

[49] De Laborde, "Chasse," 330.

[50] For his later life, see J. McCarthy, *Selim Aga: A Slave's Odyssey* (Edinburgh: Luath Press, 2006), 78–193.

[51] De Laborde, "Chasse," 330.

[52] J. Michaud and B. Poujoulat, *Correspondence d'Orient*, vol. 7 (Paris: Ducollet, 1835), 223.

and for further sale to Egypt. According to de Laborde, on the basis of information from the 1830s, the Sultan of Darfur exported 8000–9000 slaves annually to Dongola and Asyut in Upper Egypt, with a quarter dying on that trans-Saharan route. Caravans crossing the Sahara carried food only for a minimal transit time, which meant that the caravan had to keep a brisk pace in order to arrive before the provisions were depleted. Human bones along the route showed the fate of those abandoned by earlier caravans.[53] De Laborde's information suggests that one-quarter of the captives died en route to the raiders' base camp at El Obeid and that, at least for the slaves exported from Darfur, another quarter died in the trans-Saharan journey. These estimates do not include the loss of life during the sieges and attacks on the targeted villages. Clearly, not even half of the original population of the targeted villages survived to reach the slave markets of Egypt.

Repeated slave raiding transformed Nuba hill societies. As families were split up after capture and family members often sold separately, the possibility of reconstituting the target society at some point in the future was remote.[54] Villagers who escaped capture retreated to defensive positions or prepared elaborate underground hideouts. They chose safety over close proximity to water or the ease of agriculture on fields below the hills; only larger polities and armies dared to settle on the plains.[55] Moreover, no target society was immune from the widespread social, economic, and political disturbances caused by the *zariba* system. Schweinfurth noted these changes when describing the Nubians, meaning the people of the northern riparian Sudan who were the dominant free residents of the *zariba* system. This system required the constant acquisition of slaves through violence and their local acculturation. Slaves were a constant topic of conversation and worry in the *zaribas*. Disputes broke out over ownership and failure to pay for slaves. Slaves often ran away and were hunted; if they escaped, hunger often drove them to another *zariba*, where they were viewed as lucky finds to be seized or sold to the north. Anyone later claiming to be a runaway's owner provoked a violent dispute.

The trading companies (whose members lived in Khartoum) left the running of the *zaribas* to their agents, typically "slaves reared in their master's house." These agents had the opportunity to cheat their masters in

[53] De Laborde, "Chasse," 330–31.

[54] De Laborde, "Chasse," 329–30.

[55] For Tagali, one such kingdom, see Ewald, *Soldiers, Traders, and Slavers.*

trade, sell off the slaves, and retire to Darfur. In the lesser *zaribas*, things were even more corrupt, and merchants often sought out slaves. Agents there might view all slaves as merchandise, including the boys and girls who were designated future cultivators in the larger *zaribas*.[56]

On average, every free resident owned three slaves; in total there were 50,000–60,000 privately owned slaves, excluding those captured for immediate sale to the north. Schweinfurth described four categories of slaves. The first consisted of boys between seven and ten years old who carried guns and ammunition for each Nubian soldier. As they aged, they joined the second category, which included most of the adults in *zaribas*. Called Farookh, Narakeek, or Bazingir, they carried guns and accompanied the locals whether for war or trade. Comprising half the armed men in the *zaribas*, they played an essential part in all combat. They collected grain from local villages, gathered porters, and kept local villagers in line. These slave soldiers had wives, children, and land in the *zaribas*, and the older ones even had their own young slave gun-bearers. After each raid, their numbers grew, as captured young men were tempted by the prospect of obtaining a cotton shirt, a gun, and better food. Slave women were the third category. Every soldier had at least one, and some had more. Usually one slave woman became his favourite, and the rest then did menial tasks. They were treated as currency and passed from soldier to soldier, which helped spread venereal diseases. Following Muslim law, the children of these slave women were treated as free. Women slaves had to do all the domestic work for the soldiers: hauling water, washing, laboriously grinding millet by hand, cooking, and carrying everything on journeys. In larger households, the tasks were more specialized. Schweinfurth estimated that it took one slave woman working full time to grind millet for five or six men. Agricultural slaves of both sexes were the fourth category. Only wealthy and powerful people in the *zaribas* actually owned land and herds. Poor people did a little gardening, older women did the weeding, and even slave soldiers joined in at harvest time.

Schweinfurth did not enumerate a fifth category of slaves who passed into supervisory positions after having been raised in their masters' houses. While detailed biographies of such men are rare, one example has been found. Katherine Petherick reported that "some Neam Neams had accompanied our men from their station at the village of Mundo."[57] Ringa (a.k.a.

[56] Schweinfurth, *Heart of Africa*, vol. 2, pp. 420–26.

[57] J. Petherick and K. Petherick, *Travels in Central Africa*, vol. 1 (London: Tinsley Brothers, 1869), 62–63.

Ringio), described as "son of Goria, sheikh of Beringi," was soon waiting on the Pethericks at table and became a loyal and useful servant. The traveller Junker told Ringa's story as chief of the Idios, more generally known as the Makarakas:

> Ringio, the principal chief of the Bombehs and Makarakas, and the head Dragoman of the province ... was captured in his boyhood and taken as a slave to Khartum. Here he entered the household of John Petherick, the English vice-consul, in whose service he remained many years. Having heard from a Makaraka woman, a fellow slave, of the new territory acquired by the Idios, he resolved to join them; Petherick gave his consent and sent him to his zeriba, Neangara. Ringio was well received by the Makarakas and Bombehs and was acknowledged as Ngerria's son and their chief. At his suggestion, the late Fadl Allah came as governor of Petherick's zeribas to Makaraka Land, and was followed by Ahmed Agha Atrush and Ahmed Agha Akhuan. The independence of the negroes quickly came to an end, and the heavy yoke of the Khartum traders depopulated the previously thickly inhabited land.[58]

Ringio is well known in southern Sudan history, particularly among the Azande.[59] Similar figures who rose to positions of power included Rabih ibn Fadlallah, a former protégé of al-Zubayr, who took over his operations in the *zaribas* and passed over the continental divide to raid and conquer Dar Runga, Bagirmi, and Bornu, before he was finally defeated.[60]

Conclusion

In all, drought and disease influenced both the suppliers of slaves and the demand for slaves in societies from the Sudan to Egypt and, from the Egyptian commercial perspective, helped stimulate and maintain the Sudanese slave trade.[61] They also affected the supply of commodities such as ivory that were vital to IOW trade. In the classic desert-side "demo-

[58] W. Junker, *Travels in Africa during the Years 1875[–1886]*, trans. A. H. Keane (London: Chapman and Halle, 1890) vol. 1, 481.

[59] R. Gray, *A History of the Southern Sudan, 1839–1889* (London: Oxford University Press, 1961), 147–48, 160.

[60] D. Cordell, *Dar al-Kuti and the Last Years of the Trans-Saharan Slave Trade* (Madison, WI: University of Wisconsin Press, 1977), 53–65.

[61] R. Austen, "The Trans-Saharan Slave Trade," 23–76. See also G. M. La Rue, "The Export Trade of Dar Fur," 636–38; and La Rue, "African Slave Women in Egypt," 168–89.

graphic pump" described for the central Sudan, populations that expanded along the desert's edge in wet years were forced to move in arid years. The most mobile and aristocratic residents could plan for this and build up a network of dependants and slaves extending from the desert into the savannah.[62] But in nineteenth-century Egypt and the Sudan, three sets of large-scale enslavement processes overlapped each other and overlaid localized versions of the demographic pump. The local patterns meant that in drought years, nomadic populations used slave and cattle raids to "recruit" slaves and cattle to replace those they had lost. But those nomadic raiders also had the option of selling human and bovine captives to other markets. Initially, these were local sultanates, such as Darfur, Sinnar, and Wadai, which each used slaves locally and hosted trans-Saharan traders with connections to Egypt and beyond.

Over time, those affected by the local demographic pumps included not only the nomadic populations but also residents of the sultanates. These residents augmented lower incomes from poor harvests with regional trading, seeking grain from better-watered regions to the south or from local highlands. For example, residents of the Nuba Hills (such as Almas and Selim and their families) became targets of slave raids by nomads or sultanic armies with nomadic or other local allies. Their choices were stark: to be captured, to flee uphill to less fertile and drier defensive perches, or to ally with the raiders. In dry years, they were more likely to be raided and were possibly more vulnerable, as they had to turn to the best-watered land in their local environment.

After 1821, the Egyptian army used local allies in Kordofan much as the sultanic forces had done earlier. By the 1840s, the expanding *zariba* system was making the temporary alliances more permanent, using more modern weapons, and adding ivory into the mix of exports. By the 1860s, the *zariba* system was the principal supplier of slaves across the Sahara to Egypt. Throughout this time, the successive methods of enslavement created local demand for slaves as soldiers, as farmers to feed the new army, as servants to officials, and as slave-concubines to meet the soldiers' domestic needs.

Epidemic diseases and economic factors, such as the Egyptian cotton boom, also affected these patterns by disrupting the ordinary flow of enslaved populations in the regional economy. For humans, the clearest

[62] P. Lovejoy and S. Baier, "The Desert-Side Economy of the Central Sudan," *International Journal of African Historical Studies* 8 (1975): 551–81.

case is the plague epidemic of 1834–1835, which reduced the free and servile population of Egypt, which in turn led to a surge in slave exports from the Sudan to Egypt from 1838 to 1840. It also led to increased attempts by Egypt to expand control over its slaving frontier in the Sudan.

The period 1861–1865 similarly saw an increase in the supply of enslaved Sudanese to Egypt. The cotton boom was a factor, as Egyptian farmers increased acreage under cotton and benefited from higher cotton prices. Male slaves were purchased as farmhands to help with irrigation, and slave women were sometimes obtained with the profits from the boom. The regional drought of 1863, the ensuing famine, and the cattle epidemic of 1865 also had major impacts. Egyptian farmers, already under demand-side pressure to produce more cotton, had to find alternatives to the cattle-powered *sāqiyas*. Mechanical alternatives were unfamiliar, costly, and sometimes unreliable, but the human-powered *shadufs* could be worked by slaves imported from the Sudan.

In the Sudan during the same period, the classic demographic pump led to increased cattle and slave-raids by nomadic populations and to a search for alternative economic activities by sedentary populations. The *zariba* system was quick to respond to increased demand for slaves and cattle for export to Egypt and interacted with other suppliers of slaves and cattle to facilitate the movement of these commodities towards northern markets. Schweinfurth depicted the complexity of *zariba* societies immediately after the expansion of the 1860s. The *zaribas* had scaled up their activities by incorporating enslaved populations. This ultimately led to the advancement of a few slaves such as Ringio and Rabih ibn Fadlallah to positions of power.

The interactions between environmental factors, changing enslavement processes, social transformation, and the local uses of slaves are complex. A full Braudelian analysis might emphasize annual cycles of life, larger patterns of wet and dry years, and cultural factors such as religion and the rise and fall of polities. Here, the emphasis has been on a period of change where older patterns, such as the demographic pump, have been overlaid with new developments, such as the expansion of the sultanates, the Egyptian invasion of the Sudan, and the growth of the *zariba* system. Economic factors and epidemic disease disrupted ordinary markets and demographic expectations. Changes in the processes of enslavement led to local social transformations that could be observed not only in reports of slave raids but in the lives of enslaved individuals—local transformations that, as other chapters in this volume demonstrate, had parallels and ramifications elsewhere in the IOW.

CHAPTER 9

Environmental Knowledge and Resistance by Slave Transporters in the Nineteenth-Century Western Indian Ocean

Hideaki Suzuki

Introduction

Focusing on the environmental knowledge of slave traffickers, this chapter examines the resistance by those slave traffickers to the British naval campaign against the slave trade in the nineteenth-century western Indian Ocean. It addresses the subject from two challenging points of view: first, it places slave traffickers as its central concern while maintaining a wide perspective on maritime trade in general, and second, it examines the traffickers' resistance in the context of their knowledge and exploitation of their marine environment.

In contrast to the rapid growth of studies on other aspects of slavery in the Indian Ocean, research on the actual trade in slaves has not advanced much. If the nineteenth-century slave trade in the western Indian Ocean

H. Suzuki (✉)
Nagasaki University, Nagasaki, Japan

G. Campbell (ed.), *Bondage and the Environment in the Indian Ocean World*, Palgrave Series in Indian Ocean World Studies,
https://doi.org/10.1007/978-3-319-70028-1_9

consisted of three elements—the slaves themselves, British attempts to suppress the slave trade, and the slave traffickers—the first two have received far more scholarly attention than the last.[1] Of course, slave traders were both deeply involved in the fate of the slaves and the main target of the British campaign of suppression. For example, Raymond W. Beachey's classic study of the anti-slave-trade campaign contains little detailed information on the traffickers.[2] A critical problem with Beachey's work and of most studies of the western Indian Ocean slave trade is that they treat the slave trade as a business distinct from other commercial activities. This tendency leads to a great deal of misunderstanding in terms of the nature of slave traffickers and thwarts the huge potential of studying them as a means to explore much larger questions of maritime transport in the Indian Ocean region.

This chapter focuses on slave traders from an environmental perspective—a novelty in Indian Ocean world (IOW) slave trade studies. The confrontations between slave traffickers and those involved in suppressing the slave trade were characterized by a mixture of diplomacy and violence. The diplomacy involved negotiations and the making of treaties between local rulers and British forces. Needless to say, slave traffickers took no part in such negotiations, although they had to deal with the resultant policies and regulations. Most of the violence involved clashes between British naval forces and dhows suspected of carrying slaves. Such encounters have been graphically portrayed in the memoirs of naval officers. However, actual physical combat between the opposing sides was not so common. Although ships in the western Indian Ocean tended to be armed regardless of whether or not they had slaves on board,[3] their crews were disinclined to be aggressive towards the generally larger patrol vessels and would rarely offer resistance if challenged. By contrast, British warships

[1] One rare exception is Mandana E. Limbert's 2013 work, which uses documents concerning the *Yasmeen* captured by HMS *Vulture* in 1872. This work is devoted to an in-depth consideration of documents, focusing on particular details that were included and omitted while also highlighting the details of slave traders. "'If You Catch Me Again at It, Put Me to Death': Slave Trading, Paper Trails, and British Bureaucracy in the Indian Ocean," in *Indian Ocean Slavery in the Age of Abolition*, ed. Robert Harms, Bernard L. Freamon, and David W. Blight (New Haven and London: Yale University Press, 2013), 120–40. In addition, see also, Hideaki Suzuki, *Slave Trade Profiteers in the Western Indian Ocean: Suppression and Resistance in the Nineteenth Century*, New York: Palgrave, 2017.

[2] Raymond W. Beachey, *The Slave Trade of Eastern Africa* (New York: Barnes and Noble, 1976).

[3] Philip H. Colomb, *Slave-Catching in the Indian Ocean: A Record of Naval Experiences* (New York: Longmans, Green, and Co., 1870), 39–40.

were often belligerent and would more readily open fire with ships' cannons and firearms. Indeed, careful observation of the illustrations in British naval memoirs reveals that it was often the naval ships that initiated conflict. The foremost weapon the traffickers possessed against patrol ships was not force but their extensive seamanship and knowledge of the local environment—coastal topography, currents, and winds—that they had developed over many years and kept a closely guarded secret.

The British tried to develop their knowledge of the IOW environment in the nineteenth century when British naval officers conducted a series of detailed investigations of the coastal geography. Treaties intended to suppress the movement of slaves gave such studies added importance because they divided the ocean geographically into zones where slave movement was either legal or illegal. The marine environment was thus a hotly contested zone. The advanced environmental knowledge of the local sailors became highly effective in their resistance to the British when it was combined with their seamanship and their understanding of the treaties for suppressing the slave trade. This made it difficult for British naval forces to apply anti-slave-trade policies and obliged them to adopt more complex responses. It is such dynamics that are examined here.

Slave Traffickers and Maritime Trade in the Western Indian Ocean

There are several reasons why slave traffickers are an important subject for understanding maritime transport in general in the nineteenth-century western Indian Ocean. Our examination of resistance by slave traffickers is not limited in scope to the context of the simple contrast between the slave traffickers and the naval campaign against them; slave transport was inseparable from other forms of transport in the western Indian Ocean. Herbert F. Disbrowe, an officer of the Indian Navy engaged in the anti-slave-trade campaign around the Gulf of Oman and the Persian Gulf in the 1850s, stated that "the slaves that may be on board constitute but a minimum part of the cargo in the vessel ...."[4] Other commodities on board, such as salt, salted fish, grain, and mangrove poles indicate a much wider range of transoceanic exchange—one that transforms an otherwise limited

[4] "Report on the Slave Trade in the Persian Gulf extending from January 1 1852 to June 30 1858, compiled by H. Disbrowe," OIOC (Oriental and India Office Collection), British Library, London, IOR/R/15/1/171/23.

concept of the western Indian Ocean into a more meaningful historical unit known as the western IOW. A large proportion of the East African commodities brought into the Gulf of Oman and the Persian Gulf, such as grain and mangrove poles, were unavailable locally or could not be produced in sufficient quantity to satisfy local demand. That demand was the engine of the creation of a durable and mutually advantageous trans-oceanic exchange system.[5] Salt and salted fish brought by ships from the Gulf of Oman and the Persian Gulf turned out to be important foodstuffs for coastal-dwelling and inland East Africans, while grain from East Africa played the same role in the Gulf of Oman and the Persian Gulf, where coastal people also lived in houses built with East African mangrove poles.[6] One of the consequences of this inseparability between the nature of slave transport and other forms of transport was that all local ships were suspected by the colonial regime of involvement in the slave trade.[7]

An analysis of slave trafficking networks has enormous potential in terms of allowing researchers to explore almost all aspects of local maritime transport in the western Indian Ocean. Viewing the maritime trade in slaves as separate from other aspects of maritime trade, as is done in the Atlantic world, would in the IOW present a definite obstacle to scholarly discovery.[8] Conventional Atlantic models of slave transportation can rarely be applied to the western Indian Ocean context. For example, the term "slaver," which obviously emerged from the Atlantic experience of slave trading and its suppression is, as Disbrowe noted, largely inapplicable to the Indian Ocean:

> The term "Slaver" is scarcely applicable to vessels that engage in slave trade between Zanzibar & the Persian Gulf. No such thing as a slaver, that is, a ship specially rigged for, or solely occupied in, the transport of human flesh is to be found in these waters.[9]

[5] Hideaki Suzuki, "Indo-you nishi-kaiiki sekai no kanousei: kaiiki-shi kara sekai-shi he," *Rekishigaku-kenkyuu* 911 (2013): 178–85.

[6] Hideaki Suzuki, "Arabia-hantou Omān-wan, Persia-wan enganbu to Mangrōbu: Indo-you nishi-kaiiki ni okeru koukan ga sasaeru Arabu shakai no jyuu kuukan," in *Mangrōbu: Arabu no nariwai seitaikei*, vol. 3, ed. Ryou Nakamura and Hiroshi Nawata (Kyoto: Rinkawa Shobou, 2013), 151–68.

[7] Erik Gilbert, *Dhows and the Colonial Economy of Zanzibar, 1860–1970* (Oxford: James Currey, 2004), 59–83.

[8] Gwyn Campbell, "Slave Trades and the Indian Ocean World," in *India in Africa, Africa in India: Indian Ocean Cosmopolitanisms*, ed. John C. Hawley (Bloomington: Indiana University Press, 2008), 20–25.

[9] "Report on the Slave Trade in the Persian Gulf extending from January 1 1852 to June 30 1858, compiled by H. Disbrowe, OIOC," IOR/R/15/1/171/22–23.

Similarly, the Royal Navy's Captain Christopher Colomb, who engaged in anti-slave-trade operations in the Indian Ocean in the 1860s, stated,

> Between the two trades [i.e. the Atlantic and the Indian Ocean slave trades] there are many differences. Ships for the Atlantic trade are—or were—large well found vessels, specially fitted with slaves carrying appliances, extra water tanks, extra cooking places, special decks for slaves, and other things unknown in the Arabian Sea ... his [the Arab's] slave carrying vessels are fitted just as any other—that is, not fitted at all.[10]

As this implies, there was also a large disparity in the number of slaves carried by transatlantic and western Indian Ocean vessels. While the average number of slaves on transatlantic slaving ships is generally estimated at over 250, the number of slaves carried by local ships in the western Indian Ocean was comparatively small. Indeed, as Fig. 9.1 shows, of the 326 ships noted that were captured with slaves on board between 1837 and 1892 by either the Royal Navy or the Indian Navy, 47 per cent were carrying fewer than ten slaves.

Further, in contrast to the Atlantic, there are few reliable sources available on the maritime slave trade in the Indian Ocean. Researchers are thus obliged to rely chiefly on the reports of British naval officers and consular staff, who were keen to monitor the traffic in slaves with the aim, of course, of suppressing it. Given the inseparable nature of the maritime transport of slaves and other commodities, the most effective way of analysing the actual conditions of general local transport in the nineteenth-century western Indian Ocean is to explore such official British reports.

Dividing the Ocean: Progress in the Suppression of the Slave Trade

During the nineteenth century, a series of treaties were concluded between Britain and various local potentates around the western Indian Ocean with the aim of abolishing the slave trade.[11] They were not only the first treaties concerned with the suppression of slave trading in the region but also the

[10] Colomb, *Slave-Catching*, 38–43. See also Jones to Anderson, Bushire, 28 August 1856, OIOC, IOR/R/15/1/157/208.

[11] For the treaties concluded between Britain and the polities along the Persian Gulf, see John G. Lorimer, *Gazetteer of the Persian Gulf, Oman and Central Arabia*, vol. 1 (London: Archive Editions, 1986), 2476–84.

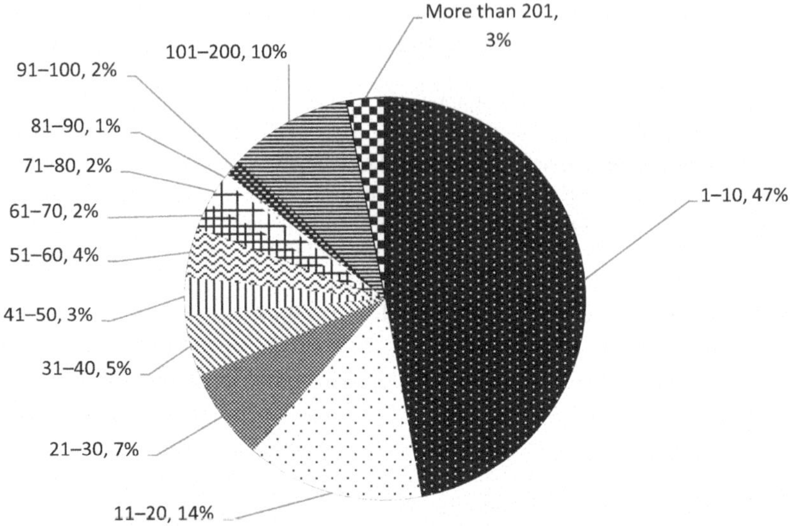

Fig. 9.1 Number of slaves on board local ships captured by either the Indian Navy or the Royal Navy, 1837–1880 (the number of slaves, percentage). Sources: *BPPST*, vol. 24, Class D, pp. 28, 33, 69, 380; *BPPST*, vol. 25 Class A, pp. 363–64, 380–81, 383–84, 386, 388, 423–24; *BPPST*, vol. 39, Class B, pp. 198, 203–05; *BPPST*, vol. 40, Class B, pp. 367, 371–75; *BPPST*, vol. 41, Class B, pp. 292–94, 304; *BPPST*, vol. 46, Class A, pp. 85–88; *BPPST*, vol. 48, Class A, pp. 12–13, 29–31, 164; *BPPST*, vol. 49, Class A, pp. 42–43, 188, Class B, p. 73, Class C, pp. 64–65, 68, 180, Class D, pp. 120, 125; *BPPST*, vol. 50, Class A, pp. 15, 17–18, 91, 99–100, 102, 104–05; *BPPST*, vol. 51, Class A, pp. 10, 68, 71; *BPPST*, vol. 52, Class A, p. 86; *BPPST*, vol. 91, pp. 138, 187–89, 198–99, 203, 488; MAHA PD/1837/78/854/393–400; MAHA PD/1855/1457/93, 115; MAHA PD/1856/93/335/203; MAHA PD/1864/54/704/279; MAHA PD/1864/54/942/14; MAHA PD/1865/52/780/13; MAHA PD/1877/149/689/66; NAUK ADM123/179/n.d.; NAUK FO84/1090/86–98, 105; NAUK FO84/1224/205; NAUK FO84/1245/189; NAUK FO84/1325/166; NAUK FO84/1344/139; NAUK FO800/234/66, 93; NAUK FO881/1703/3, 6; NAUK FO881/3342/5–7; OIOC IOR/L/P&S/9/42/349–51; OIOC IOR/L/P&S/18/B84/65, 72, 81; OIOC IOR/R/15/1/123/13, 14, 17, 22, 25, 26, 27, 29–30; OIOC IOR/R/15/1/127/3, 22, 25, 27; OIOC IOR/R/15/1/134/1, 5; OIOC IOR/R/15/1/143/306–14; OIOC IOR R/15/1/157/226; OIOC IOR R/15/1/168/119; OIOC IOR R/15/1/177/11; OIOC IOR R/20/A1A/255/9; OIOC IOR R/20/A1A/255/26; OIOC IOR/R/20/A1A/285/77; OIOC IOR R/20/A1A/318/90–1, 143; ZZBA AA12/2/9–10; ZZBA AA12/29/32; Jerome A. Saldanha, "Précis on Slave Trade in the Gulf of Oman and the Persian Gulf, 1873–1905 (With a Retrospect into Previous History from 1852)," in *The Persian Gulf Précis*, vol. 3, Gerrards Cross, Archive Editions, 1986, ed. Jerome A. Saldanha (1st. in 18 vols., Calcutta and Simla, 1903–1938), 90

first legal attempt in the history of the region to restrict maritime transport between countries. What is remarkable here is that, in some cases, the restriction was closely associated with the idea of dividing up the ocean. Among European countries, especially since the end of the fifteenth century, this idea of dividing the ocean and thereby claiming sovereignty over maritime space had gained ground following repeated disputes over commercial transport, maritime resources, and the discovery of new lands.[12] The most well-known and striking case must be the Treaty of Tordesillas, forged between Spain and Portugal, which, after concluding in 1494, resulted in a line being drawn along 46° 37′W. Newly discovered lands to the west of the line would become Spanish possessions, and any to the east would belong to the Portuguese. Some time later, in the seventeenth century, legal debates over the sovereignty of the sea were resurrected, triggered by Hugo Grotius's *Mare liberum* in 1609.

On the other hand, there are few known cases of territorial claims over the western Indian Ocean staked by local powers. What is generally known as the Moresby Treaty seems to have been the first case in which the concept of dividing up the maritime sphere was realized in the region. The Moresby Treaty was aimed at suppressing the slave trade and was concluded in 1822 between Fairfax Moresby, the senior British officer in Mauritius, and Sa'id b. Sultan, ruler of Oman and Zanzibar.[13] In order to avoid fierce resistance from local powers and the traffickers and because of a lack of sufficient naval forces for the purpose, British anti-slave-trade policy did not demand an immediate prohibition but instead proposed the division of the Indian Ocean into zones where slave trading would either be outlawed or remain legal. The Moresby Treaty established a line running from Cape Delgado, the southern limit of the acknowledged territory of Sa'id b. Sultan, to Diu Head on the west coast of the Indian subcontinent via a point 60 miles east of Socotra Island. Beyond that line to the east, slave transport was made illegal, and British naval ships were given the authority to confiscate any vessels found with slaves on board. The treaty was amended in 1839 to reduce the area available for legal slave

[12] For the details, see Chiyuki Mizukami, "Kaiyou jiyuu no keisei (1)," *Hiroshima Hougaku* 28, no. 1 (2004): 1–22; Chiyuki Mizukami, "Kaiyou jiyuu no keisei (2)," *Hiroshima Hougaku* 28, no. 2 (2004): 1–25.

[13] Its official English title is "Treaty Concluded with the Imam of Muscat for the Suppression of Slavery." For the articles of this treaty, see Charles U. Aitchison, *A Collection of Treaties, Engagements and Sanads: Relating to India and Neighbouring Countries*, vol. 7 (Calcutta: Office of the Superintendent of Government Printing, India, 1909), 211–15.

trafficking.[14] In 1847, a new treaty, known as the Hamerton Treaty, reduced the space even more, limiting it to the area between 9° 2′S (near Kilwa) and 1° 57′S (near the Lamu Archipelago).[15]

However, the effect of the treaties was limited. The local potentates were not cooperative, and even Sa'id b. Sultan and his officials lost no time in explaining to the British consul at Zanzibar how important slaves were to them.[16] The British officers could not expect enough support from the local powers, which meant that their only hope was to intercept slave traffic before the slaves were landed.[17] However, the small size of the Indian Navy and its wide-raging obligations in various missions and wars elsewhere meant that the progress of naval efforts at suppression was rather slow before the 1860s.[18] Indeed, this shortage of naval force often allowed slave traffickers to escape confiscation of their cargoes.[19] Things improved following frequent requests to the government in Bombay to reinforce the naval force for the anti-slavery campaign,[20] and at the same time, further diplomatic pressure was applied. Eventually, the number of slaves released increased during the 1850s as the campaign slowly changed from what was effectively a watching brief to an effective operation with the capture

[14] Aitchison, *A Collection of Treaties*, vol. 7, p. 226.

[15] Its official English title is "Agreement between Her Majesty the Queen of the United Kingdom of Great Britain and Ireland and His Highness Syud Saeed Bin Sultan, 'the Sultan of Muskat', for the Termination of the Export of Slaves from the African Dominions of His Highness the Sultan of Muskat." For the articles of this treaty, see Aitchison, *A Collection of Treaties*, vol. 7, pp. 227–28.

[16] Rigby to Bombay Government, Zanzibar, 5 September 1861, ZZBA (Zanzibar National Archives, Zanzibar, Tanzania) AA12/2/12.

[17] For example, Kemball to Thomson, Bushire, 12 January 1854, OIOC, IOR/R/15/1/143/304.

[18] For example, John B. Kelly, *Britain and the Persian Gulf 1795–1880* (Oxford: Clarendon Press, 1968), 237–82; Charles R. Low, *History of the Indian Navy 1613–1863*, vol. 2 (New Delhi: Manas, 1985), 137.

[19] For example, Hennell to Porter, Camp near Bushire, 26 June 1851, OIOC, IOR/R/15/1/127/20; Ethersey to Jones, Bassadore, 24 July 1854, OIOC, IOR/R/15/1/157/227; Jones to Anderson, Bushire, 25 April 1857, OIOC IOR/R/15/1/168/34–35.

[20] For example, "Extract Para 4 from a Despatch from the Honorable the Court of Directors dated the 1st March No 1 of 1854," OIOC, IOR/R/15/1/143/367; Jones to Anderson, Bushire, 26 May 1856, OIOC, IOR/R/15/1/157/193–197; Jones to Jenkins, Amulgavine Roads, 30 April 1859, OIOC, IOR/R/15/1/168/46–47; "Extract paras 28–31 from a letter from Brigadier W. M. Coghlan, in charge Muscat, Zanzibar Commission, dated 1st November 1860, no. 14," OIOC, IOR/R/15/1/171/63–65.

of local ships transporting slaves and the freeing of the slaves themselves. In addition to the improvements in the Indian Navy's efforts, the Royal Navy's Cape of Good Hope squadron joined the campaign in the early 1860s,[21] pushing progress further.[22]

Although the number of rescued slaves increased rapidly once the Royal Navy had become involved, the negative effect on local trade soon became obvious. The destruction of ships after misidentification naturally caused deep resentment not only among the local sailors, merchants, and rulers but also within European and American companies in East Africa and among the Christian missionaries who were otherwise the most vociferous and determined supporters of the campaign against the slave trade.[23] It is, of course, difficult to make even an educated guess at the extent of the damage done to the entire local maritime transport network, but this quotation from a letter written by an American merchant in Zanzibar at the end of the 1860s gives some idea:

> The English cruizers are raising the devil with the trading Dows on the coast and will ruin trade if they keep on. Not less than 70 have been destroyed by them the past year and there is no doubt that a great proportion of them were harmless traders with no slaves on board except their crews, which seem to be sufficient now to cause their destruction.[24]

As the anti-slave-trade campaign progressed, local traders were victimized in ever greater numbers regardless of whether they were involved in transporting slaves or not. This fact then should not be overlooked; because of

[21] Gilbert, *Dhows and the Colonial Economy*, 61.

[22] Between 1865 and 1869, the Royal Navy rescued and sent about 3000 slaves to Aden alone. The number of slaves sent to Aden was 17 times the number of slaves released by the Indian Navy between 1852 and the end of June 1858. "Memorandum of Number of Slaves landed and liberated at Aden, and how disposed of, Aden Residency Written by W. R. Goodfellow," 13 July 1869, OIOC, IOR/L/P&S/18/B84/69. For further details of the progress of the campaign accelerated by the Royal Navy, see Hideaki Suzuki, "Indo-you nishi-kaiiki ni okeru 'dorei-sen gari': 19-seiki dorei koueki haizetsu katudou no ichidanme"' *Afurika Kenkyuu* 79 (2011): 17–18.

[23] Suzuki, "Indo-you nishi-kaiiki ni okeru 'dorei-sen gari'," 19. As the background to destroying under misidentification, the Royal Navy's prize regulations and naval officers' Atlantic experience and the reality they faced in the western Indian Ocean were crucial. For further details, see Gilbert, *Dhows and the Colonial Economy*, 60–61; Suzuki, "Indo-you nishi-kaiiki ni okeru 'dorei-sen gari'," 17–21.

[24] F. R. Webb to E. D. Ropes, Zanzibar, 13 April 1869, PPEM (Philips Library, Peabody and Essex Museum, Salem, MA, USA), MH201/Box 3.

the very nature of the way slaves were transported in the western Indian Ocean, the campaign against the slave trade proved to be a serious threat to the maritime transport of all other goods in the region.

The Parallel Progress of the Campaign Against the Slave Trade and Its Resistance by Traffickers

As the campaign against them progressed, the slave traffickers in the western Indian Ocean developed their own methods to combat it. A frequently observed scheme was to disguise slaves as sailors or spouses of sailors.[25] In 1837, a *nakhoda* (shipmaster) and seven sailors of the Qawāsim kidnapped 233 girls and brought them to Basra (or Bandar-e Bushehr) in the guise of their wives and there sold them into slavery.[26] Moreover, it was usual to see African sailors in the western Indian Ocean, and patrolling naval officers who were unfamiliar with local sailing customs and languages found it difficult to determine whether Africans found on board the ships they intercepted were really crew members or were actually en route from East Africa to be sold as slaves.[27] They took interpreters with them on

[25] Hennell to Willoughby, Karrack, 31 August 1841, *British Parliamentary Papers: Slave Trade* [*BPPST*] (Shannon: Irish University Press, c. 1968–1969), vol. 25, Class A, p. 385; Hennell to Willoughby, Karrack, 7 September 1841, *BPPST*, vol. 25, Class A, p. 386; Hamerton to Secretary to the Bombay Government, Zanzibar, 17 September 1855, ZZBA AA12/29/136–37; Beachey, *The Slave Trade*, 54, 58; Colomb, *Slave-Catching*, 59–60; Arnold B. Kemball, "Paper Relative to the Measures Adopted by the British Government, between the Years 1820 and 1844, for Effecting the Suppression of the Slave Trade in the Persian Gulf," in *Arabian Gulf Intelligence*, ed. Robert H. Thomas (Bombay: Bombay Education Society Press, 1856), 651; George L. Sulivan, *Dhow Chasing in Zanzibar Waters* (Zanzibar: Gallery Publications, 2003), 32–35.

[26] Hennell to Willoughby, Bushire, 24 September 1837, MAHA (Mahārāshtra State Archives, Mumbai, India) PD/1837/78/854/393–400. Pretending to be married was a ruse used in the coastal region of the Gulf of Aden and the Red Sea for masking kidnapping. Playfair to Coghlan, Aden, 7 April 1860, OIOC, IOR/R/20/A1A/285/73–76.

[27] Trotter to the Secretary to the Admiralty, Simon's Bay, 5 April 1857, *BPPST*, Class A, vol. 44, p. 151; Hennell to Malet, Bushire, 30 October 1851, OIOC, IOR/R/15/1/127/31; "Extract paras 28–31 from a letter, from Brigadier W. M. Coghlan, in charge Muscat, Zanzibar Commission, dated 1st November 1860, no. 14," OIOC, IOR/15/1/171/64. On the contrary, often the people on board as passengers were recognized as slaves and protected. See, for example, Crawfold to Anderson, Bombay, 5 January 1856, MAHA, PD/1856/93/335/207–212.

patrol,[28] but even then there remained the suspicion that an interpreter would take bribes or might otherwise be hand in glove with the traffickers.[29] In addition, the inseparable nature of slave transport from all other transport forced the consulate to issue permits for departure to the ships engaged in the transporting of slaves: simply, they could not distinguish those ships carrying slaves from those did not.[30]

Another resource frequently relied upon by the traffickers was their skill at ship handling and their knowledge of seamanship; naval officers often noted the advanced skill of the local seaman.[31] In addition, the traffickers' ships were fast—able to reach as many as 11 knots when conditions allowed[32]—and able to surprise the Navy as a result.[33] The traffickers sometimes discovered new sailing routes. For example, Johann L. Krapf, who travelled from Mombasa to Cape Delgado in 1850, reported that a local slave ship heading north from Kilwa sailed along the east coast of Zanzibar Island.[34] Usually, ships sailed along its west coast, where Stone Town, the island's main port, was located. However, in the case reported by Krapf, the slave ship took the east coast route to avoid attracting the notice of other ships.

In addition, traffickers found and used places where they could minimize the danger of discovery by naval patrols and officials from the Sa'id b. Sultan In the case of the East African coast, Mkokotoni and Pemba Island were far enough from Stone Town to land and load slaves in secret.[35] For the coast of the Gulf of Oman and the Persian Gulf, traffickers also sought out places that would not be easily found by the patrolling ships and local agents of the British. After being landed, the slaves were taken to cities on

[28] For example, Frere to Granville, Mombasa, 24 March 1873, NAUK (National Archives, Kew, UK), FO84/1389/352.

[29] "Extract paras 28–31 from a letter, from Brigadier W. M. Coghlan, in charge Muscat, Zanzibar Commission, dated 1st November 1860, no. 14," OIOC, IOR/R/15/1/171/64.

[30] "Half Yearly Report on the Slave Trade (Crawfold to Walker, HMS *Sidon* at Mauritius, 30 June 1861)," NAUK, ADM123/179/n.p.

[31] Their great skill as sailors was frequently mentioned in the contemporary reports and letters of naval officers. For example, see Hamerton to Chief Secretary to the Bombay Government, 25 October 1849, ZZBA, AA12/29/85.

[32] Richard F. Burton, *Zanzibar: City, Island, and Coast*, vol. 1 (London: Tinsley Brothers, 1872), 74.

[33] Colomb, *Slave-Catching in the Indian Ocean*, 44–45.

[34] Johann L. Krapf, *Travels, Researches and Missionary Labours during an Eighteen Years' Residence in Eastern Africa* (London: Trübner and Co., 1860), 424.

[35] Rigby to Anderson, Zanzibar, 13 September 1858, OIOC, IOR/R/15/1/171/101–102. See also Beachey, *The Slave Trade*, 59.

foot or in smaller boats.[36] According to Moollah Houssein, the British agent at Muscat, small boats were used because when the patrol ships found them, the small boats could go into creeks and inlets inaccessible to the Navy's patrolling vessels.[37] For example, boats of shallow draft known as *mtepe* were used effectively along the East African coast where, according to William Cope Devereux, who was engaged in patrolling there, even if the *mtepe* were found, the British Navy could do little about it because of the shallowness of the water.[38] Furthermore, the effectiveness of the locals' advanced knowledge of coastal geography and their more suitable vessels was augmented by their skill at gathering information. Moollah Houssein reported that local ships would come in to port at Sur, located at the entrance to the Gulf of Oman, and would gather the latest information about the patrols before going on their way.[39]

After the Hamerton Treaty had been promulgated,[40] Lamu Archipelago became one of the most convenient spots for the traffickers because, under the treaty, it formed the northern border of the legal zone of slave transport. Thus ships carrying slaves could depart from Lamu legally if they declared that they were going to sail to other ports within the legal zone; those ships would not have their cargoes confiscated by the patrols as long as they remained within the legal zone.[41] Furthermore, the complex series of channels created by inlets and mangrove swamps provided endless hiding places, and another advantage of the Lamu Archipelago was its favourable currents. Referring again to the observation made by Captain Colomb

[36] Moollah Houssein to Hennell, sine loco (s.l.), 25 Rujub/6 June 1850, OIOC, IOR/R/15/1/123/19–20; Kemball to Thomson, Bushire, 12 January 1854, OIOC, IOR/R/15/1/143/304; Hajee Yacoob to Jones, s.l., 2 Moohurrum 1276, OIOC, IOR/R/15/1/171/150. Moollah Houssein to Hennell, s.l., 13 June 1851, OIOC, IOR/R/15/1/127/23. The Sultan of Ra's al-Khaymah ordered his subjects to use a land route instead of sea route for slave conveyance.

[37] For example, Taylor to Hennell, Basra, 16 September 1851, OIOC, IOR/R/15/1/127/56; Kemball to Malet, Bushire, 12 November 1852, OIOC, IOR/R/15/1/130/292; Jones to Forbes, Bushire, 2 September 1861, OIOC, IOR/R/15/1/168/109; Samuel B. Miles, *The Countries and Tribes of the Persian Gulf* (London: Harrison and Sons, 1919), 490.

[38] William Cope Devereux, *A Cruise in the "Gorgon,": or, Eighteen Months on H.M.S. "Gorgon," Engaged in the Suppression of the Slave Trade on the East Coast of Africa. Including a Trip up the Zambesi with Dr. Livingstone* (London: Bell and Daldy, 1869), 122.

[39] Moollah Houssein to Hennell, s.l., 17 May 1851, OIOC, IOR/R/15/1/127/18–19.

[40] Rigby to Anderson, Zanzibar, 28 March 1860, MAHA, PD/1860/159/830/194–195.

[41] Devereux, *A Cruise in the "Gorgon,"* 391.

of the Royal Navy, the currents and winds to the north of Lamu were generally more favourable for sailing northwards, with the south-west monsoon, than southwards, where there were less desirable sea conditions. To the north of the archipelago, a favourable current of 50 miles per day might easily be encountered, and north of Baraawe, as much as 90 miles per day has been recorded.[42] Thus, as Colomb described, "there await shipment to the northward when the wind and the absence of a British cruiser permits."[43] As an example, we have the following case: when Majid b. Sa'id, Sultan of Zanzibar, succeeded from his father, Sa'id, he prohibited any transporting of slaves by water between 1 January and 1 May,[44] and a certain number of the slave were transported by land up to Juba or Baraawe, north of the Lamu Archipelago. Meanwhile, local ships sailed from Zanzibar legally without slaves on board, and when the two parties came together at Juba or Baraawe, the ships picked up slaves and sailed northwards with the currents.[45] The campaign against the slave trade gave new significance to the geography of the western Indian Ocean, but the traffickers accommodated the changes and, with the help of their vast knowledge of their own environment, found ways of using them to their greatest benefit.

Nonetheless, those same British sources on which I have had to rely to obtain information about the traffickers' tactics reveal how, in time, the British patrols found out about the illegal trafficking and devised countermeasures.[46] The naval officers on board were keen to collect the latest infor-

[42] Colomb, *Slave-Catching*, 45.

[43] Colomb, *Slave-Catching*, 34.

[44] Majid first issued this proclamation in 1864 (Playfair to Russell, Zanzibar, 4 January 1864, OIOC, IOR/L/P&S/9/41/240; Playfair to Havelock, s.l., 1 May 1864, MAHA, PD/1864/54/942/12–13; Playfair to Russell, Zanzibar, 4 January 1864, NAUK, FO84/1224/181–182). However, even in 1859, he adopted measures including the following: a frigate investigated all the dhows passing the small channel in front of Stone Town: and dhows owned by "northern Arabs" were prohibited from anchoring as close as six miles from Stone Town, and their rudders would be confiscated to prevent them from entering the town (Rigby to Anderson, Zanzibar, 30 November 1859, MAHA, PD/1860/159/12/7; Rigby to the Bombay Government, Zanzibar, 14 May 1861, ZZBA, AA12/2/117).

[45] Colomb, *Slave-Catching*, 33–34.

[46] For example, Disbrowe's journal shows his consideration of the timing of the slave transporters' arrival in the Persian Gulf. "Extracts from Lieutenant-Colonel Disbrowe's Diary of his Journey (overland) from Muscat, Capital of the Sultan of Muscat's Dominions, to Ras-el-Khymah, Capital of the Joasmee Chief's Dominions, Extract 4," Ghallut Kulbah, 16 April 1865, *BPPST*, vol. 51, Class B, p. 125. See also Pelly to Anderson, Lingah, 5 December 1863, *BPPST*, vol. 51, Class B, p. 131.

mation from the locals and watchfully cruised along the coast looking for potential hiding spots and watering places for slave transporters. Eventually, at the end of the 1860s, "a belief … [had] sprung up in slave-catching circles" that, in order to avoid facing British patrol ships, which now obtained much information on the coast, slave traffickers had "braved the dangers of the open sea" to sail directly from Zanzibar to Ra's al-Hadd.[47] It is questionable whether such a route was actually taken by the traffickers, and even if it was, there would have been much risk involved for seafarers despite the fact that they might well have had sufficient knowledge and skill to make the voyage without their usual reliance on coastal navigation.[48] Nonetheless, the belief satisfied the Royal Navy of the value of their own efforts at improving their own effectiveness in the coastal regions.[49]

The Traffickers Outwit the Navy: The Monsoon

The most important thing governing the seasonality of trade in the Indian Ocean is the alternating system of monsoon winds. From April to September the wind blows from the south-west, while from October to March it blows from the north-east. If sailors respect this cycle, their ships can sail relatively safely and regularly.[50] Slave transports were no exception, so the anti-slave-trade patrols were planned accordingly. During the south-west monsoon, when ships sailed north from East Africa, patrols were stationed at the entrance of the Persian Gulf or along the East African coast to co-ordinate with other operations in suppressing the transports.[51] For instance, in a memorandum written on 1 September 1864, Robert L. Playfair, the British consul in Zanzibar, suggested the following method of catching slave traffickers while at sea: he considered it worthwhile for

[47] Colomb, *Slave-Catching*, 45–46.

[48] Yasuyuki Kuriyama, *Umi to tomoni aru rekishi: Iemen kaijyou kouryuu-shi no kenkyuu* (Tokyo: Cyuuou University Press, 2012), 46–50.

[49] Colomb, *Slave-Catching*, 24.

[50] However, between June and August, a quite strong south-westerly wind blows, so sailing activities are generally suspended. For further details, see Hikoichi Yajima, *Kaiiki kara mita rekishi: Indo-you to Chicyuu-kai wo musubu kouryuu-shi* (Nagoya: Nagoya University Press, 2006), 59–60; Colomb, *Slave-Catching*, 26–27; B. A. Datoo, "Misconceptions about the Use of Monsoons by Dhows in East African Waters," *East African Geographical Review* 8 (1970): 1–10.

[51] For example, Montrion to Clarke, Aden, 3 May 1854, OIOC, IOR/R/15/1/143/320–321; Playfair to Haveloch, Zanzibar, 3 February 1864, OIOC, IOR/L/P&S/9/41/245 (the same document is in MAHA, PD/1864/54/143); Rigby to Anderson, Zanzibar, 30 November 1859, MAHA, PD/1860/159/7.

the naval force to be positioned in the sea lane between Cape Guarduafi and Socotra Island from early September to the end of May, as this was where and when the most seaborne traffic between the Persian Gulf and the East coast of Africa passed.[52] The south-west monsoon season, which excluded May to August, was called "the slave season" by naval officers in their memoirs—apparently quite accurately.[53]

However, as the quotation below reveals, following the progress of the patrols, in the early 1860s some traffickers made extensive use of their understanding of the relationship between monsoon winds and seafaring (described above) as a means of outwitting the British yet again. Lewis Pelly, British Consul in Zanzibar, reported as follows:

> A is a British Indian subject, the owner of a dhow: B is an Arab who wants to run a cargo of slaves to Muscat, but who fears to fall in with a Cruizer if he run coastwise. So B comes to A and says, "I hear you want a crew to take your dhow to Bombay; you have lent me money on occasion. It would now be a pleasure to me to lend you a crew, with one of my Arab Nakhodas in command. I should not ask any remuneration, only when the dhow arrives at Bombay, just give the Nakhoda his discharge." A thus gets his dhow run for nothing. B puts a crew, slaves, to the number of fifty, on board, in charge of his Nakhoda. The dhow arrives at Bombay ...
>
> Presently the firm to whom the dhow is consigned, are about to run a bugalow to Muscat. "Very well" says B, "I must return to Zanzibar in any event. You have fed my people while in Bombay, I will run your bugalow to Muscat, en route, no charge only a 'Bukshis' for my myself, if you are content with me and my crew, on discharge at Muscat."
>
> Now observe, every body is contented with this little arrangement; no body is taxable with slave dealing. Cross question the boat owners, they have engaged a crew. Ask the Nakhoda; he has of his own free will tendered his services; ask the crew, they are all free men, and by no means willing to be interfered with. Yet, in point of fact, a cargo of slaves is run from Zanzibar to Muscat, through Bombay.[54]

[52] *BPPST*, vol. 50, Class B, p. 45. Using a naval squadron to guard the watering spots used by traffickers was also suggested in Merewhether to Secretary to the Bombay Government, Aden, 17 May 1865, OIOC, IOR/L/P&S/9/42/352.

[53] Colomb, *Slave-Catching*, 24, 32; Sulivan, *Dhow Chasing*, 444. See also Devereux, *A Cruise in the "Gorgon,"* 395.

[54] Pelly to Stewart, Zanzibar, 10 July 1862, NAUK, FO800/234/95. Part of this was also quoted in Colomb, *Slave-Catching*, 97–98.

Pelly's report can be summarized in the following two points. One is that an Arab who planned to transport slaves—without necessarily clarifying his intention to do so—could be expected to move them using the regional design of a vessel called a dhow, which would be owned by a local Indian resident. The second is that the transit route for slaves to Muscat was via Bombay, and the slaves were disguised as sailors, indeed actually working as sailors until they were sold in Muscat. Things went as follows. First of all, when a local ship departed from Zanzibar, it declared Bombay as its destination and ordinary commercial goods as its cargo. On top of that, the owner of the ship was an Indian resident who could, of course, claim that he was a British subject. The local ship could therefore obtain valid permission and so leave Zanzibar without difficulty. Heading for Bombay brought another advantage: because the East India Company had abolished the import of slaves into the Bombay Presidency in 1805, and because the Bombay government subsequently abolished slavery itself in 1843, slavery and slave trading became less conspicuous in Bombay territory.[55] In the mid-nineteenth century, ports and coastal waters around British India were less heavily policed by the Indian Navy than they had been before.[56] That meant that, even if the crew of "sailors" on board a local trading ship actually consisted of slaves on their way to being sold, the boat would still be able to enter port without much trouble.

As we have already seen, patrols and other operations against slave traffickers were planned and operated according to the south-west monsoon season, whereby ships headed usually from the East African coast to ports in the Gulf of Oman or the Persian Gulf. However, in complete contrast, the favourable sailing season in the opposite direction, from Bombay to Muscat, was the north-east monsoon season. Thus the route via Bombay to Muscat with the north-east monsoon wind was a highly effective means for slave traffickers to move their cargoes safely. Their apparent ability to turn the general view of seafaring in the Indian Ocean on its head demonstrates their innovative approach to the business of carrying slaves, available to them only because of their experience and knowledge of the environment and extensively evolved networks over the vast sphere of the

[55] Dady R. Banaji, *Slavery in British India* (Bombay: D. B. Taraporevala Sons and Co., 1933), 403; Beachey, *The Slave Trade*, 40.

[56] Colomb, *Slave-Catching*, 99–101; Kemball, "Paper Relative to the Measures," 651–52; Robert N. Lyne, *Zanzibar in Contemporary Times: A Short History of the Southern East in the Nineteenth Century* (New York: Negro University Press, 1969), 41–42.

western Indian Ocean. Such a case was reported in 1866, when at last the Bombay government investigated this slave route via Bombay.[57] However, as for the result of the investigation, the Chief of Police in Bombay judged that crewmen on board the local ships owned by Arabs or captained by Arab sailors entering Bombay were not slaves.[58]

Slave traffickers relied on their superior sailing skills and advanced knowledge of coastal geography and sea currents, tricks, such as disguising slaves as regular crew members, and their extensive networks across the whole western Indian Ocean. By combining all their skills and knowledge, especially of the monsoon winds, they found they could outsmart the British suppressors.

Conclusion

In the context of the nineteenth-century western Indian Ocean, looking at the transport of slaves yields important insights into the nature of maritime transport in general. This is true partly due to the nature of regional transport more generally in this period, the limits of available resources, and the nature of naval suppression, particularly after the 1860s when the Royal Navy began to fight alongside the Indian Navy.

The most significant advantage available to slave traffickers in their resistance to the British campaign to abolish—or at least heavily subvert—the slave trade was their skill at sailing and their knowledge of the environment: namely the coastal geography, currents, and winds. Ironically, the treaties and patrolling efforts used to combat the slave trade made the trader's methods of resistance more effective. The treaties drew imaginary lines to divide the ocean into legal and illegal zones for slave transport. Traffickers accepted the new meanings given to their familiar coastal geography, which they used to their maximum possible advantage. A striking example of this is the Lamu Archipelago, which the Hamerton Treaty designated as the northern boundary of the legalized slave transport zone.

[57] Commodore commanding the Indian Division of the Royal Navy to the Governor of Bombay, Bombay, 6 February 1867, *BPPST*, vol. 51, Class B, pp. 133–34; Bedingfeld to the Commodore commanding the Indian Division of the Royal Navy, Zanzibar, 1 December 1866, *BPPST*, vol. 51, Class B, p. 134; Wedderburn to the Commissioner of Police, Bombay and the Commissioner of Customs, Bombay, Bombay Castle, 25 February 1867, *BPPST*, vol. 51, Class B, p. 135.

[58] Souter to Secretary to the Bombay Government, Bombay, 6 March 1867, MAHA, PD/1867/126/509/55–56. The same document is in *BPPST*, vol. 51, Class B, pp. 135–36.

Traffickers combined this artificial border with the environmental advantages the archipelago provided, such as its complex inlets and changes in the currents, and continued their work of transporting slaves. For the same reasons—the position of the boundary and attendant environmental advantages—the ports of Juba and Baraawe assumed a new importance for the traffickers. Furthermore, traffickers' geographically vast networks, which extended over the entire western Indian Ocean, enabled them to collect up-to-date information on the anti-slave-trade campaign. Judging from their dynamic application of their extensive environmental knowledge of the region, we can see how the traffickers adapted to the circumstances by having recourse to develop an innovative approach to transporting slaves and resisting the anti-slave-trade campaign. In this context, which involved a dynamic combination of factors, it is no longer possible to accept a bare dichotomy between the historically developed environmental knowledge of the traffickers and the restrictive methods of the British, with their use of treaties and more modern equipment. The relationship of both parties was more interactive, and the western Indian Ocean was, in this sense, an arena of such interaction over environment.

CHAPTER 10

Environmental Disaster in Eastern Bengal: Colonial Capitalism and Rural Labour Force Formation in the Late Nineteenth Century

Sravani Biswas and Subho Basu

The eastern and southern portions of the island of Dukhin Shabazpore and the islands on the south and east of it, and the thanas of Bowful and Golachipa in the Patuakhali, have been almost entirely destroyed. It is reported that in these places there is no rice, dal, or any sort of food to be had, and that the people who have survived are in extreme distress for want of food and cloth. The exact number of people who have died by drowning has not yet been ascertained; but it is apprehended that a very large percentage of the population is gone.

E. J. Barton, Magistrate-Collector, Backergunge[1]

[1] Letter from E. J. Barton, Magistrate-Collector of Backergunge, to the Commissioner of Dacca Division, dated 6 November 1876, in *Proceedings of the Lieutenant-Governor of Bengal. Financial Department: Branch V.-Industry and Science. Head No.3.-Meteorology. Calcutta 1876*, 99, British Library (hereafter BL), London, Oriental and India Office Collection (hereafter OIOC), IOR/P/894/4/47.

S. Biswas (✉)
Earth Observatory of Singapore, Nanyang Technological University, Singapore, Singapore

S. Basu
McGill University, Montreal, QC, Canada

G. Campbell (ed.), *Bondage and the Environment in the Indian Ocean World*, Palgrave Series in Indian Ocean World Studies,
https://doi.org/10.1007/978-3-319-70028-1_10

On 31 October 1876, the southern districts of eastern Bengal (currently Bangladesh) and the numerous small remote silt islands at the confluence of the Meghna River and the Bay of Bengal were devastated by a cyclone and tidal bore. Colonial reports and newspapers termed it the "Backergunge Cyclone" or "Bengal Cyclone." The quotation above describes the impact of the cyclone on Dakshin Shabazpore, an island subdivision in the Backergunge district of eastern Bengal. Following the cyclone, a cholera epidemic decimated a large proportion of those who survived the cyclone and tidal bore. Such calamities were unexceptional in late nineteenth-century British India, where the prefix "natural" was a label superimposed on a disaster that was embedded in socio-economic and political processes. During the viceroyship of the Earl of Lytton (1876–1880) at the high noon of British imperial rule in India, when the "ornamentalism" of the Raj was expressed through the pageantry of Delhi Durbar,[2] environmental disasters, famines, and epidemics constituted critical components of the quotidian lives of the Indian rural working classes.

"Environment" is the key locution here. In the closing years of the long nineteenth century, British India experienced massive environmental transformations. No doubt, such transformations were informed by global environmental upheaval. In 1876, an increase in ocean water temperatures in the central and east-central equatorial Pacific, known in scientific parlance as Southern Oscillation, wreaked havoc in India and China. In India, its impact was further complicated by colonial policies aimed at extracting agrarian surpluses and promoting cultivation of cash crops. The principal aim of the British was to meet trade deficits with continental Europe through the extraction of Indian resources.[3] In India, this process was

[2] The term 'ornamentalism" is from David Cannadine's work. David Cannadine, *Ornamentalism: How the British Saw their Empire* (Oxford: Oxford University Press, 2001). Indeed, Delhi Durbar of 1877 was the beginning of the royal pageantry that was designed to impress India's ruling elites, namely maharajas and nawabs, with the prowess of the proclaimed empress of India. Bernard S. Cohn also refers to the Durbar as a method of representing authority. Bernard S. Cohn, "Representing Authority of Tradition in Victorian India," in *Invention of Tradition*, ed. E. J. Hobsbawm and Terrence Ranger (Cambridge: Cambridge University Press, 2000), 165–210.

[3] B. R. Tomlinson, an economic historian of the Raj, explains this economic relationship in terms of imperial commitment. The commitment, from the perspective of colonial administrators, was threefold: to provide a market for British goods, to pay interests on the sterling debt and other changes that fell due in London, and to maintain a large number of British troops from the Indian revenue. B. R. Tomlinson, "The Political Economy of the Raj: The Decline of Colonialism," *Journal of Economic History* 42, no. 1 (March 1982): 133–37. For

backed by a communication revolution, involving investment in railways, telegraphs, and steamships that, alongside colonial legal interventions, accelerated the integration of India into the world market.[4] This late nineteenth-century globalization of Indian agricultural production had an uneven impact on Indian rural society, bringing periodic economic prosperity to large landholders and farmers who produced regular surpluses but economic instability and sometimes ruin to subsistence peasants and agricultural workers, both bonded and free.

British colonial censuses, albeit imperfect, clearly indicate a stagnation in population growth for this period, which later historians have identified as an epoch characterized by famine, epidemics, and mass mortality.[5] Mike Davis, who incorporates China and Brazil into his analysis, called it the late Victorian holocaust.[6] As Sandeep Banerjee has shown, even the engaging iconic nationalist novels of the era used famines as a cultural backdrop,[7] and natural disasters have been used in recent literature to interrogate the liberal paradigm "that depicts late nineteenth century British imperialism as benign".[8]

a general account of India's role in Britain's rise as an industrial power, see Eric Hobsbawm, *Industry and Empire: The Making of Modern English Society*, vol. 2, *1750 to the present* (New York: Partheon Books, 1968); S. B. Saul, *Studies in British Overseas Trade, 1870–1914* (Liverpool: Liverpool University Press, 1960), Chap. 8; Irfan Habib, "Studying a Colonial Economy without Perceiving Colonialism," *Modern Asian Studies* 19, no. 3 (1985): 355–81. For an ecological reading of the imperial economic history, see Prasannan Parthasarathi, *Why Europe Grew Rich and Asia Did Not: Global Economic Divergence, 1600–1850* (Cambridge: Cambridge University Press, 2014).

[4] For a discussion of the spatial integration market, see Manu Goswami, *Producing India: From Colonial Economy to National Space* (Chicago: University of Chicago Press, 2004). For more on market integration, see Ritu Birla, *Stages of Capital: Law, Culture and Market Governance in India* (Durham, North Carolina: Duke University Press, 2009).

[5] Ira Klein, "When the Rains Failed: Famine Relief and Mortality in British India," *Indian Economic and Social History Review* XXI, no. 2 (April–June 1984): 91; Ira Klein, "Population Growth and Mortality, Part 1—the Climatric of Death," *Indian Economic and Social History Review* 26, no. 4 (December 1989): 387–404; Ira Klein, "Death in India," *Journal of Asian Studies* 32, no. 4 (August 1973): 639–59.

[6] Mike Davis, *Late Victorian Holocausts: El Niño Famines and the Making of the Third World* (London: Verso, 2001).

[7] Sandeep Banerjee, "Landscaping India: From Colony to Postcolony" (PhD diss., Syracuse University, 2013), 155–87.

[8] Upamanyu Pablo Mukherjee, *Natural Disasters and Victorian Empire: Famines, Fevers and the Literary Cultures of South Asia* (London: Palgrave Macmillan, 2013).

Late nineteenth-century famines, as the economist Amartya Sen asserted in a later context, were pointers to the critical failure of policies that paid little attention to food security of the rural poor.[9] Colonial policy makers, ensconced in their fundamental faith in the market mechanism and fiscal orthodoxy, believed that market rules of supply and demand would ensure food safety. Even irrigation projects, meant to increase agricultural productivity, frequently blocked earlier rainwater drainage systems, thus creating cesspools that transformed fields into malarial swamps.[10] Massive railway construction programmes initiated deforestation in the Himalayan foothills[11] and depletion of forest resources in central India.[12] Commercialization of forests, enforced through legislation and the dispossession of forest-dwelling communities, created waves of migration by impoverished workers.[13] Thus the lives of subsistence farmers, agricultural workers, and forest dwellers were sacrificed on the altar of colonial policy.

Mass migration, assisted by the burgeoning railway system, broke the earlier relative immunities of rather isolated communities. During famines, it became commonplace for rural poor to migrate in search of food, shelter, and sustainable livelihoods. The mass migration of rural poor in crammed railway compartments[14] often led to the wider circulation of diseases and enabled localized epidemics to become near-pandemics.[15]

[9] Amartya Sen, *Poverty and Famines: An Essay on Entitlement and Deprivation* (Oxford New York: Clarendon Press Oxford University Press, 1982).

[10] Sheldon Watts, "British Development Policies and Malaria in India 1897–c. 1929," *Past & Present*, no. 165 (November 1999): 141–81.

[11] Pallavi Das, "Colonialism and the Environment in India: Railways and Deforestation in 19th Century Punjab," *Journal of Asian and African Studies* 46 (February 2011): 38–53.

[12] Mahesh Rangarajan, *Fencing the Forest: Conservation and Ecological Change in India's Central Provinces 1860–1914* (Delhi: Oxford University Press, 1996).

[13] Crispin Bates, "Regional Dependence and Rural Development in Central India: The Pivotal Role of Migrant Labour," *Modern Asian Studies* 19, no. 3 (July 1985): 573–92.

[14] Manu Goswmi used the emotive phrase "mobile incarceration" to describe third-class carriages of Indian railways. Most poor Indians travelled in such third-class carriages. Manu Goswmi, *Producing India: From Colonial Economy to National Space* (Chicago: University of Chicago Press, 2004), 112.

[15] David Arnold, "Cholera and Colonialism in British India," *Past & Present* 113, no. 1 (November 1986): 118–51; Ira Klein, "Plague Policy and Popular Unrest in British India," *Modern Asian Studies* 22, no. 4 (1988): 723–55; Ian Kerr, "Reworking a Popular Religious Practice: The Effects of Railways in 19th and 20th Century South Asia," in *Railways in Modern India*, ed. Ian Kerr (Delhi: Oxford University Press, 2001), 304–27; Ravi Ahuja, "'The Bridge Builders': Some Notes on Railways Pilgrimage and 'British Civilizing Mission'

Quarantine policies directed at migrant workers not only often failed but generated significant resistance as well. Railways, the symbol of colonial "modernization" and the tool of "time-space compression"[16] in Gandhi's anti-modern romantic nationalist perspective, were the carriers of plague: an oblique reference to their role in undermining ecologic isolation and immunities.[17]

This chapter locates the micro-history of the tempest in the Bay of Bengal littoral in 1876 within the context of the macro-history of the impact of British colonial policy in late nineteenth-century India. As pointed out in Chap. 1 of this volume, debt-bondage resulted from a combination of environmental factors and human activities. In the case of eastern Bengal, the colonial administration's support for the rapacious landholding classes further impoverished the small peasants. Through an excavation of colonial archives, this chapter demonstrates how the logic of capital was embedded in the colonial state's rhetoric of progress and argues that colonial policy exacerbated the impact of "natural disasters," thus creating mass indebtedness among the Indian peasantry that further amplified disaster-induced labour migration. Colonial officials, preoccupied with securing the right conditions for investment in growing and exporting cash crops—notably jute in the case of Bengal—and with laying down a modern communications network that would serve to promote the economy and facilitate the transport of troops, were little concerned with the complex interplay of colonial policy, natural disasters, and peasant impoverishment. Ensuring the safety of capital at the expense of labour was thus deeply inscribed in the beginning of a modern developmental regime under the conditions of colonial capitalism in India. Natural disasters escalated the ongoing process of peasant oppression by landholding classes and moneylenders. The triad of *Sarkar* (government), *sahukar* (moneylenders), and *zamindar* (landlords) that came to represent colonial

in Colonial India," in *Colonialism and Civilizing Mission*, ed. Harold Fischer-Tine and Michael Mann (London: Anthem Press, 2003), 89–110; Anand Yang, *Bazar India: Markets Society and the Colonial State in Bihar* (Berkeley and Los Angeles: University of California Press, 1998), 112–61.

[16] David Harvey, "Time-Space Compression and Postmodern Condition," in *The Condition of Modernity: An Enquiry into the Origins of Cultural Change* (Oxford: Blackwell, 1989), 284–307.

[17] M. K. Gandhi, *Hind Swaraj or Indian Home Rule* (Ahmedabad: Navjivan Publishing House, 1938). In a 1910 critique of colonial civilizing mission, Gandhi wrote, "The railways, too, have spread the bubonic plague. Without them the masses could not move from place to place. They are the carriers of plague germs. Formerly we had natural segregation."

capitalism at the lowest level combined to make small peasants into debt-ridden, land-poor agricultural workers trapped within a system of extraction of surplus for foreign commodity trade.

Situating the Locale: Political Economic Rhetoric of Colonial Developmentalism

The Bakarganj district comprised an estimated area of 4891 square miles—or, excluding the Meghna estuary and the larger rivers, 3840 square miles—in which the human population was concentrated on islands, where population density varied according to the quality of the soil.[18] This geographic situation kept changing, as alluvial formation continuously added new exploitable land that the colonial administration immediately appropriated.[19] From the mid-nineteenth century, the delta in the eastern part of the district witnessed a steady increase in population due to migration from the western sector, which experienced a decline in agricultural productivity because of both declining fertility of the soil and high labour morbidity due to frequent occurrences of malaria.[20] Thus, despite natural population increase, Bakarganj was beginning, on the eve of the disaster, to be characterized by "a state of relative deficit" of labour.[21]

Colonial records for Bakarganj, which date to the seventeenth century, are a particularly rich source of information on the region. They include a description made in 1876 by Henry Beveridge, district magistrate and collector, who compared Bakarganj's resources and monotonous landscape to Manchester:

> It strikes one as being a sort of agricultural Manchester, producing breadstuffs instead of cotton cloth, but without the art culture for which Manchester is so justly famous ... Committees for International Exhibitions can get nothing from it to show in Europe as trophies of Indian skill or taste.[22]

[18] J. C. Jack, *Final Report of the Survey and Settlement Operations in the Bakarganj District 1900 to 1908* (Calcutta: Bengal Secretariat Book Depot, 1915), 1.

[19] Jack, *Final Report of the Survey and Settlement Operations*, 1.

[20] Sugata Bose, *The New Cambridge History of India, Peasant Labour and Colonial Capital: Rural Bengal since 1770*, vol. 3, part 2 (Cambridge: Cambridge University Press, 1993), 26.

[21] Bose, *The New Cambridge History of India*, vol. 3, part 2, p. 26 (see map). The Bengal census of 1881 recorded a relatively low immigration rate in the district. Migration occurred within the district especially from the northern part to the southern silt islands.

[22] H. Beveridge, *The District of Bakarganj: Its History and Statistics* (London: Trubner & Co., 1876), 5.

The colonial government gained much of its revenue from rent on land. As Beveridge wrote, "there is probably no other district in Bengal Proper where Government has so many and such valuable estates, and therefore we may say that there is no other district in which Government has so large and direct an interest."[23]

By the late nineteenth century, as world demand for jute production grew, lower Bengal became increasingly integrated into the international economy. Rice, another of the region's staple crops, was exported to the deficit areas within the empire. With an eye to enhancing revenue, the colonial state adopted a policy of aggressive reclamation of land on which it sought to promote the cultivation of cash crops.

J. C. Jack, a revenue surveyor, reported in 1915 that private companies involved in the inter-regional trade first arrived in Bakarganj in the late nineteenth century "when the northern rivers silted too much to allow steamers to reach the Brahmaputra. The Sunderban route was then opened and remained for some time the only service which passed through Bakarganj."[24] In 1884, another route to Calcutta was established via Khulna, a nearby port town, and Barisal, located in the eastern part of Bakarganj district, became a steamer terminal point and main headquarters for both inter- and intra-regional steamer networks running from Chittagong, Noakhali, and Madaripur. Local Bengali shipping companies also existed, but they were relatively small-scale and could not compete with the large Indo-British shipping companies, the India General Steam Navigation Company and the River Steam Navigation Company.[25]

Forest clearance and transport innovation projects inevitably affected the sensitive ecology of deltaic Bengal.[26] In 1868, J. E. Gastrell, a revenue surveyor, warned against reclamation drives in the Sundarbans forests and urged for the preservation of a broad belt of forest between the settlements and the bay to protect the former from the sea during storms.[27] However, his warning was unheeded, and the unabated land reclamation

[23] Beveridge, *The District of Bakarganj*, 181.

[24] Jack, *Final Report of the Survey and Settlement Operations*, 12.

[25] Jack, *Final Report of the Survey and Settlement Operations*, 12.

[26] For details on the transformation of the Bengal delta and the impact of British commercialization, see Iftekhar Iqbal, *The Bengal Delta: Ecology, State and Social Change, 1840–1943* (Basingstoke: Palgrave Macmillan, 2010).

[27] Colonel J. E. Gastrell, *Geographical and Statistical Report of the Districts of Jessore, Fureedpore and Backergunge* (Calcutta: Office of Superintendent of Government Printing, 1868), 25.

projects exposed the littoral population's vulnerability to recurring cyclones and storm-waves.

The Cyclone and the Colonial Administration

The cyclone that hit India at the end of October 1876 had two vortexes. The first struck Vizagapatnam (in present-day Andhra Pradesh, India) on the night of 30 October, and a day later the second vortex hit the southern islands at the mouth of the Meghna River in the Bakarganj, Noakhali, and Chittagong districts.[28] The establishment of the Meteorology Department in 1875 in Calcutta resulted in the collection of ship logs from which the path and nature of the cyclone were studied in order to develop warning systems for future calamities.[29] As Stanziani's contribution to this volume (Chap. 7) points out, the process of collecting cyclone data in the colonies during the early nineteenth century was generally left to the personal efforts of colonial officials like Sir William Reid and Henry Piddington. Describing the landscape after the destructive cyclone of 1876, John Eliott, the meteorological reporter, wrote,

> At the noon ... the only evidence at sea of the passage of one of the fiercest cyclones which has visited the coast of Bengal were some twenty vessels, the majority dismasted and reduced to mere wrecks, struggling feebly towards the nearest port; whilst a few scattered sheets of water on land were all that was left to indicate that the panic-stricken, houseless, starving inhabitants of one of the richest and most populous districts of Bengal had been visited by a flood more destructive in its character than history has yet recorded.[30]

[28] J. Eliott, *Report of the Vizagapatnam and Backergunj Cyclones of October 1876* (Calcutta: Bengal Secretariat Press, 1877), 181. The meteorological reporter wrote, "An area of diminishing pressure began to be formed on the 23rd [November]. The area of depression extended northwards, and on the 26th and 27th the winds in the neighborhood of the area began to show a vorticose motion and were of considerable intensity ... With the continuance of the cyclonic condition the area of diminishing pressure not only expanded, but the amount of the depression at the center increased, and on the evening of the 29th the gale of cyclonic character, which had set in some time previously md was rapidly increasing in violence, became at length a cyclone of great force."

[29] Tirthankar Roy, "'The Law of Storms': European and Indigenous Responses to Natural Disasters in Colonial India, c. 1800–1850," *Australian Economic History Review* 50, no. 1 (March 2010): 10.

[30] Eliott, *Report of the Vizagapatnam and Backergunj Cyclones*, 182.

The meteorological reporter recorded in detail the path of the cyclone from ships' logs. At 10 p.m. on 31 October, John Westcott, master of the ship *Allahabad*, travelling from Melbourne to Calcutta, wrote in his log: "This is the heaviest part of the cyclone; blowing with inconceivable fury; sails torn from their gaskets; the canvas in the mizzen-rigging blown away."[31] The log entry at 4 a.m. on 1 November for the ship *Scottish Chieftain* read,

> All hands trying to get clear of the wreck; at 10am succeeded in getting all clear forward and commenced on the main; ship still over on her beam ends ... As an inducement for the crew to extra exertions, Captain promised them one month's extra pay, providing we succeeded in getting the ship into pilot water without assistance.[32]

The surveyors sent by the Commissioner of the Dacca Division described the extent of the destruction on the islands in Noakhali district:

> The reports that had reached us had not been exaggerated. Three storm-waves of from 15 to 20 feet high have swept over the place, literally leveling it with the ground. Not a single hut and hardly a post was left standing ... It is as yet too soon to attempt to compute with anything like accuracy the loss of life which has occurred.[33]

The first-hand reports from the initial relief workers, too, corroborated the extent of the calamity. The journal of a Baptist missionary, Rev. George Kerry, commissioned by E. J Barton, the district magistrate, to assist Police Superintendent H. N. Harris in relief measures in Dakshin Shabazpore (especially Dowlutkhan), provides a more detailed account of the impact of the cyclone. In one of his earliest entries for the cyclone, written while crossing the Hilsa River, Kerry noted,

> I am writing as I go along ... I have been passing along the north-west, north, and north-east of the island of Shabazpore. The people of several villages have met us at different points, all telling the same doleful tale of the destruction of everything in their possession, and begging for food. We have given small supplies to every applicant. Their distress and suffering were

[31] Eliott, *Report of the Vizagapatnam and Backergunj Cyclones*, 103.
[32] Eliott, *Report of the Vizagapatnam and Backergunj Cyclones*, 115.
[33] Eliott, *Report of the Vizagapatnam and Backergunj Cyclones*, 165.

> beyond doubt great and terrible. The villages are at a long distance from the shore, so that I could not see them. But the whole of the vast plain has been swept by the wave, and is strewn with dead bodies of men, women and cattle ... I stopped writing just now to look upon a hundred or more cattle lying dead, with the bodies of dead men and women lying here and there among them. This terrible calamity is something frightful to think of and I only see a small part of it.[34]

Kerry also advised, "the starving people cannot have their wants supplied by sailing and rowing boats. One or two river steamers with their flats should be sent to selected centers, laden with grain, salt and cloth."[35] While colonial documents highlighted some philanthropic services rendered, they failed to recognize the logistical issues underscored by social activists like Kerry. They were concerned rather with praising the efforts of individual administrative officials, and a mass celebration was held on 1 January 1877 to proclaim Queen Victoria as Empress of India in an effort to strengthen relations with Bengal's powerful regional community leaders in the lead-up to the inaugural Delhi Durbar.

The death toll from the cyclone created an acute labour shortage at a time when demand for labourers to harvest undamaged crops and clean

[34] "Crossing the Hilsa River," demi-official from the Reverend G. Kerry to E. J. Barton, Magistrate and Collector, Backergunge, 5 November 1876, BL, OIOC, IOR/P/894/4/36. Kerry further wrote, "[5 November] 4 P.M.—Just now more than twenty men and boys gathered along the bank imploring something to eat. They were wild with the excitement of desire for food. It was painful to see them. They say that Moddopura and all the adjacent villages are gone—absolutely nothing left but starving people. For the first time I came near some growing paddy, and it seems to be just destroyed. *Sunday night*, We had to anchor in a creek near Ganeshpur on account of the bore expected during the night. I am not sorry for this, as it has enabled me to distribute relief to a large number of people here. The storm-wave suddenly passed over the chur on which the village is to the height of 14 feet and more. There must have been immense loss of human life; the people say half the inhabitants at once were swept away. They are in a most pitiable state of destitution. The plain between the village and the river is strewn with the dead—animal and human. The effluvia is sickening. The people are absolutely without food, and have been living on unripe plantains gathered from their broken down trees, and from what vegetables they can find in their ruined gardens. Close by here there were 50 cows tethered on the night of the storm; there they are now dead and putrefying. The people will do nothing to remove the dead bodies, and, as they are scattered over an immense area, a pestilence seems likely with starvation to complete what the storm left undone. I am told Ganeshpur is a khas mehal." [*sic*]

[35] "Crossing the Hilsa River," demi-official from the Reverend G. Kerry to E. J. Barton, Magistrate and Collector, Backergunge, 5 November 1876, BL, OIOC, IOR/P/894/4/36.

up and reconstruct in the wake of the disaster peaked. Writing from a Patuakhaly sub-division adjoining Dakshin Shabazpore, the assistant magistrate K. G. Gupta pointed out that

> The crops are ripening and being gathered, but there is a great paucity of reapers. In ordinary years the reaping is done by people from Furreedpore and Dacca, who flock in large numbers about this time of the year; but it is feared that the rumour of a total loss of the crops may deter these people from undertaking the annual immigration. I should request that the Magistrate of Dacca and Furreedpore be asked to make it known as widely as possible that the crops have sustained only a partial loss, and that the demand for reapers is this year greater than ever, owing to the death of a large number of able bodied men. You [E. J. Barton, Magistrate and Collector] have yourself witnessed the fearful destruction of life and property that has taken place in these islands, and *the mortality has exceeded all our calculations.* It now appears that upward of 70 per cent have perished; but women and children have suffered most...[36]

However, in the same report Gupta concluded on an upbeat note:

> there is no fear of scarcity; the crops have been injured no doubt, but an average of 8 to 12 annas can be confidently expected. The *loss of human life* is out of all *proportion to the damage sustained by the crops*; and if the latter can

[36] Letter from K. G. Gupta, Assistant Magistrate of Patuakhally, to E. J. Barton, Magistrate and Collector of Backergunge, 22 November 1876, BL, OIOC, IOR/P/894/4108, p. 199 (italics added). The assistant magistrate's estimate in Patuakhaly's char islands indicated a massive death toll to say the least. The total population of about 2305 before the cyclone was reduced to 1677. In fact he wrote, "these figures give a death rate for men 38, women 80, boys 85 and girls 90 per cent, a result which was to be expected from the varying physical strength of the different classes. It is awful to think of such a dire calamity as this, which has in one night swept away nearly 70 per cent of the total population of the islands. One of the dak manjhees [post carriers via boats] had 17 in his family, but all of them including himself are gone ..."

Commenting on the loss of cattle, the assistant magistrate wrote, "it may be said that not more than 5 per cent of the cattle have survived, but the buffaloes are strong and are used to the water and consequently a smaller number has perished; but they have been swept over vast distances, and the jungles are teeming with buffaloes ... Parties of men in small canoes are wandering about in search of their buffaloes and many are the disputes that are arising regarding them. In ordinary years the buffalo is a valuable beast, and now that the cattle have almost disappeared, it may be said to have become priceless."

now be properly reaped and gathered, not only will the *food-supply be sufficient, but an ample margin will be left for exportation.*[37]

Like most colonial officials, Gupta was more concerned with the profitability of the colonial grain trade than with providing disaster relief, which remained a secondary issue. Bakarganj was one of the surplus rice-producing districts in Bengal, but in their emphasis on an "ample margin for exportation," officials failed to appreciate the support the surviving population of the region required in order to rebuild their lives. Almost five months after the calamity, R. C. Dutt, the Sub-Divisional Officer of Dukhin Shabazpur, reported the large importation of rice through private trade into the sub-divisions to meet the needs of the population who *could* buy rice. He noted further, "Although, therefore, the statement that the greater portion of the people have no rice in their homes is, I believe, literally true, I would on no account recommend interference with private trade ..."[38] Six months after the disaster, despite such trade, a large number of impoverished people remained unable to purchase rice.[39]

It is evident from the exchange of administrative missives that the initial optimism of colonial officials that the cyclone's impact was minimal was misplaced. For example, the police administrative report for 1876–1877 pointed out that rice in the district was selling at a price much higher than normal, as most of the crop had been destroyed. Indeed, there were instances of peasants looting rice from local *mahajans* (moneylenders) who were selling the grain at exorbitant prices.[40] Not only was the

[37] Letter from K. G. Gupta, Assistant Magistrate of Patuakhally, to E. J. Barton, Magistrate and Collector of Backergunge, 22 November 1876, BL, OIOC, IOR/P/894/4/108, p. 200.

[38] Letter from R. C. Dutt, Sub-Divisional Officer of Dukhin Shabazpur, to E. J. Barton, Magistrate and Collector of Backergunge, dated 15 April 1877, BL, OIOC, IOR/P/894/4/143–44, p. 2. R. C. Dutt was the famous author of *The Bengal Peasantry* and *The Economic History of India*.

[39] Letter from R. C. Datta, Sub-Divisional Officer of Bhola to Magistrate and Collector of Backergunge, dated 30 May 1877, BL, OIOC, IOR/P/894/4/163–64, p. 2. Dutt's insistence on providing cooked rice to the distressed met with considerable resistance from the district magistrate, who wrote, "With reference to Mr. Datta's remark, I have only to state that millions of poor people in Europe and elsewhere have recourse to greater toils and to means to earn a livelihood, than those he enumerates, and do not expect the Government to relieve them of the responsibility of keeping themselves alive." Letter from E. J. Barton, Collector of Backergunge, to the Commissioner of the Dacca Division dated 1 June 1877, BL, OIOC, IOR/P/894/4/163–64, p. 2.

[40] J. Monro, *Report on the Police of the Lower Bengal Presidency for the Year 1877* (Calcutta: Bengal Secretariat Press, 1878), 70.

quantity of provisions in the district below average but also their quality was reportedly "very bad."[41]

According to Lieutenant Governor-General Sir Richard Temple, the initial estimate for deaths by drowning was about 215,000. Fortunately, this proved an over-estimation, but the cholera epidemic that followed the cyclone raised the number of deaths.[42] As Bourdillon, commissioner of the 1881 census, wrote,

> [T]he first estimate of the deaths by drowning for the Backergunge district was 105,000 persons. Fortunately this estimate was found to be too high, and careful investigation reduced the number of deaths from this cause to 73,914. In the cholera epidemic which followed the subsidence of the inundation the number of reported deaths was 41,537, giving a total of 115,451 deaths from both causes; and bearing in mind the many deaths which in such seasons are not reported, it is reasonable to assume that the mortality reached 120,000. The population of that part of the district which suffered most severely from the inundation was roughly estimated at the time upon the census figures of 1872 to be 437,000. Assuming that the population in this tract had increased at the moderate rate of 1.5 per cent per annum, it must have amounted to 25.72 per cent, or more than one-fourth of the whole population.[43]

Colonial agents also exchanged views concerning the impact of the calamity on the economic future of the sub-divisions. In a letter dated 10 July 1877, an official from the Geographical Department drew attention to the partial salinization of land due to inundation by "salt water waves."[44] Nonetheless, most official attention was focused on the more immediate impact of the cyclone on cash crops exports because, by the late nineteenth century, inter-dominion foreign trade in agricultural products was the most important revenue source for the colonial government of India.

[41] H. W. Bellew, *The History of Cholera in India from 1862–1881: Being a Descriptive and Statistical Account of the Disease as Derived from the Published Official Reports of the Several Provincial Governments during that Period and Mainly in Illustration of the Relation between Cholera Activity and Climatic Conditions Together with Original Observations on the Causes and Nature of Cholera* (London: Trübner, 1885), 287.

[42] C. C. D. Black, "Backergunj Cyclone of November 1876", *Letter from India Geographical Department* (Calcutta: Bengal Secretariat Press, 1877), 145–46.

[43] J. A. Bourdillon, *Report on the Census of Bengal, 1881*, vol. 1 (Calcutta: Bengal Secretariat Press, 1883), 46–47.

[44] Black, 'Backergunj Cyclone of November 1876', 147.

Thus a report submitted to the government by the Commissioner of the Chittagong Division lamented the detrimental effect of the cyclone on a new route being developed to transport jute from eastern Bengal to Europe:

> … all the preliminaries for a direct trade in jute between Narayangunj and Europe, via Chittagong, which had been established have now been interrupted by the gloom cast over business prospects. There was good reason to hope that all the products of Eastern Bengal which are now exported by the longer and much more expensive route by Calcutta, would have duly followed this export of jute.[45]

The report further bemoaned the loss incurred by European vessels that were destroyed by the storm-wave. For example, it noted an estimate of "damage done to the rice stored for shipment at 56,000 maunds, equivalent to about Rs. 84,000 … Besides this, of 50,000 lb of tea stated to have been stored for shipment, about 17,000 lb have been damaged, and about another 22,000 have suffered more or less from inundation."[46]

While there was relatively little official consideration of the indigenous victims of the cyclone, the loss of human life could not be entirely ignored, as high mortality among lower-level Indian personnel paralysed local government and security. For example, in Daulatkhan, the sub-division headquarters, large numbers of lower-level Indian officials and especially Bengali policemen perished in the cyclone and its aftermath. This caused the lieutenant governor to transfer the sub-division's headquarters and the *Munsif* (civil) court from Daulatkhan, which had served as the administration centre since 1845, to Bhola Sadar, on the northern part of the island.[47] Nervous about the security implications of high mortality among policemen, the lieutenant governor used relief centres that were opened in the different sub-divisions of the districts "as much for guard as for relief … for the purpose of restoring order, of preventing confusion, of keeping rustic society together …"[48] Moreover, despite stating, "Our object is to let the poor people see that the authorities are thinking, caring, feeling for them

[45] Eliott, *Report of the Vizagapatnam and Backergunj Cyclones,* 174.

[46] Eliott, *Report of the Vizagapatnam and Backergunj Cyclones,* 173–74.

[47] P. M. Basu, *Survey and Settlement of the Dakhin Shahbazpur Estates in the District of Bakergunge, 1889–1895* (Calcutta: Bengal Secretariat Press, 1896), 7.

[48] Richard Temple, "Cyclone and Storm-Wave in Backergunge and Noacolly," *Minute by the Lieutenant-Governor of Bengal, dated 21st November 1876*, 251.

in their misery,"[49] Lieutenant Governor Temple ordered official relief work to cease. Thus, in mid-November 1876, despite noting the widespread signs of devastation, pitiful condition of the cyclone survivors, and the cholera epidemic raging in the region, the Dacca commissioner claimed that "all distress has ceased," and instructed both the collector and the police superintendent to stop all forms of assistance to the victims.[50]

Cholera, Colonial Bureaucracy, and Cyclone

In the months following the cyclone, cholera became an acute problem in the districts of Bakarganj, Noakhali, and Chittagong, especially in the affected sub-divisions and their constituent villages. R. Porch, the officiating magistrate of Noakhali district, reported that many of his crew members and even police officers died of cholera. On 10 November 1876 he wrote, "[c]holera stands greatly in the way of measures undertaken, and giving medical relief is as great urgency as giving food supplies."[51] The storm-wave and the ensuing cholera apparently caused the Noakhali police to abandon their posts. In a report dated 12 November 1876, A. Smith, Commissioner of Chittagong Division, noted the rapid spread of the disease "following fast in the wake of the immediate desolation resulting from the storm-wave."[52] The cholera epidemic hindered recovery from the cyclone because of high mortality from the disease.[53] On estates in Bakarganj, the cyclone and disease caused cultivators to migrate elsewhere. The epidemic also deterred hired hands from unaffected neighbouring districts from migrating to affected areas to harvest the remaining crops, despite being offered half of the produce.[54]

[49] Temple, "Cyclone and Storm-Wave in Backergunge and Noacolly," 251.

[50] Letter from F. B. Peacock, Commissioner of Dacca Division, to the Secretary to the Govt. of Bengal, Financial Department, dated 19 November 1876, BL, OIOC, IOR/P/894/4/103, p. 4.

[51] Letter from R. Porch, Magistrate of Noakholly, to the Commissioner of Chittagong, dated 10 November 1876, BL, OIOC, IOR/P/894/4/93–94, p. 7.

[52] Letter from A. Smith, Commissioner of the Chittagong Division, to the Secretary to the Government of Bengal, Judicial Department, dated 12 November 1876, BL, OIOC, IOR/P/894/4/93–94, p. 1.

[53] Letter from F. H. Pellew on Special Duty, Sundeep and Hatya, to R. Porch, Magistrate and Collector, Noakholly, dated 9 December 1876, BL, OIOC, IOR/P/894/4/1–2, p. 10.

[54] Letter from A. Smith, Commissioner of Chittagong Division, to the Secretary to the Government of Bengal, General Department, dated 23 November 1876, BL, OIOC, IOR/P/894/4/131, p. 1.

In early December, the Dacca Commissioner was informed that cholera had killed five to six people in every village in Noakhali and that there was urgent need for pills and medicine.[55] In the following months cholera spread to Bakarganj, Noakhali, and Chittagong districts.[56] As outlined by the District Magistrate, exposure to rotten food and water and the insufficient supply of both were among the common causes for the spread of this virulent disease. Thus, on 21 January 1877, L. Cameron, the civil surgeon posted in Bakarganj, indicated that the disease was spreading because of putrefying bodies in village water tanks. He advised the administration to send police to ensure that the villagers clear the debris left behind by the storm-wave.[57] Medical personnel were assigned to the area. For example, Arthur Tomes, an Indian army surgeon, was sent to Bakarganj, eight medical attendants were dispatched to Chittagong, 19 "native" attendants were assigned to Noakhali, and 17 were sent to Backergunge.[58] In another instance, E. E. Lowis, Commissioner of the Chittagong, in consultation with Dr Coates, the surgeon, arranged for "a system of parceling out the distressed tracts under native doctors, who were to be supervised by sub-assistant surgeons ... this system has been found to work well and to afford sufficient relief."[59]

Despite persistent reports of high mortality due to cholera, those higher up in the administration started playing down the impact of the epidemic. E. J. Barton, the district magistrate, responded to claims of high mortality in Dukhin Shabazpore in January 1877 by claiming that the number of deaths from cholera in the district "cannot at present, and probably never

[55] Letter from F. H. Pellew on Special Duty, Sundeep and Hatya, to R. Porch, Magistrate and Collector, Noakholly, dated 2 December 1876, BL, OIOC, IOR/P/894/4/1–2, p. 3.

[56] Letter from L. Cameron, Civil Surgeon, Backergunge, to the Magistrate of Backergunge, dated 21 January 1877, BL, OIOC, IOR/P/894/4/37–38, p. 1.

[57] Letter from L. Cameron, Civil Surgeon, Backergunge, to the Magistrate of Backergunge, dated 21 January 1877, BL, OIOC, IOR/P/894/4/37–38, p. 1.

[58] *Statement Showing the Extra Assistant Surgeons and Native Doctors who have been Detailed for Special Duty in the Cyclone Tracts.* Chittagong (including 2 Assistant Surgeons; 5 Vernacular Licentiate Class and Apothecary Class Native Doctors; and 1 Local Native Doctor); Noakholly (including 4 Assistant Surgeons and 15 Vernacular Licentiate Class and Apothecary Class Native Doctors) and Backergunge (1 Assistant Surgeons; 7 Vernacular Licentiate Class and Apothecary Class Native Doctors; and 9 Local Native Doctors), BL, OIOC, IOR/P/894/4/17.

[59] E. E. Lowis, Commissioner of the Chittagong Division, to the Secretary to the Government of Bengal, Financial Dept, 15 January 1877, BL, OIOC, IOR/P/894/4/35, p. 28.

will, be ascertained with correctness."[60] Claims of high mortality were declared inaccurate, the result of "illiterate and unintelligent" estimates made by village *chowkeedars* (village watchmen).[61] The authorities ignored police reports from the villages that consistently emphasized the heavy toll exacted by the epidemic, such as one indicating at least 96 deaths from cholera in Bhola between 11 and 17 February 1877.[62] The official correspondence indicates that the higher authorities exerted pressure on district magistrates to assert that the situation was under control and that the latter did likewise to sub-divisional officials, who, in turn, reprimanded local police officers entrusted to collect data on mortality for their "inexactness." In March 1877, taking advantage of the confusion of claims and counter-claims, the government suspended all relief work. Later, Henry William Bellew, former sanitary commissioner of Punjab, working from official medical reports filed between 1862 and 1881, estimated that in 1877, the total number of deaths amounted to 21,858 in Noakhali and 19,177 in Bakarganj,[63] while the magistrate's report indicated the number of deaths due to the storm-wave to be 45,443 and from cholera (between 31 October 1876 and 31 March 1877) to be 47,783—giving a total of 93,226.[64]

The Colonial Relief Measures and Indian Landlords

These reports provide access to the mentality of the colonial officials. On an individual level, colonial officials were probably more sympathetic to the suffering peasantry than rapacious Indian landlords and moneylenders. However, colonial officials shared a faith that market forces would ultimately restore the balance following the cyclone. They thus decided to provide only minimal relief to the peasantry through the *zamindar* (landlords) and *sahukar* (moneylenders), powerful Indian allies of the British,

[60] Letter from E. J. Barton, Magistrate of Backergunge, to F. B. Peacock, Commissioner of Dacca, 16 March 1877, BL, OIOC, IOR/P/894/4/118–19, p. 3.

[61] Letter from E. J. Barton, Magistrate of Backergunge, to F. B. Peacock, Commissioner of Dacca, 16 March 1877, BL, OIOC, IOR/P/894/4/118–19, p. 3.

[62] Letter from E. S. Moseley, Joint Magistrate in Charge, Burrisal, to F. B. Peacock, Commissioner of Dacca, 22 February 1877, BL, OIOC, IOR/P/894/4/87, p. 89.

[63] H. W. Bellew, *The History of Cholera in India from 1862–1881* (London: Trübner & Co., 1885), 199.

[64] Letter from R. Porch, Magistrate of Noakholly, to Commissioner of Chittagong Division, 9 April 1877, BL, OIOC, IOR/P/894/4/126, p. 134.

whom they often protected during peasant jacqueries (uprisings).[65] Richard Temple, in his 'Memorandum by the Lieutenant-Governor for Guidance of the Commissioner,' ordered government officials to:

> keep the *zemindars, talookdars* and other landholders personally engaged on their several estates in encouraging and succoring their ryots. It is not sufficient that they should merely send assistance; they must go in person. Do not listen to any applications for remission of land revenue; no such thing was even asked for, much less granted, during the famine of 1874.[66]

Lower-level Indian officials proposed various strategies to involve absentee landholders in providing short-term relief to peasants.[67] For example, on 4 November 1876, Ohiduddin Mahomed, a Bengali clerk on relief duty in Gazipore, advised that the government should issue *perwanahs* (official orders) to "the police and the landholders of respective places to induce the big mahajuns [retailers and moneylenders] to send the articles of food here for sale"[68] in order to bring down the prices of essential commodities. Again, during the cholera epidemic, *perwanahs* were sent to *zemindars* instructing them to remove the rotting corpses of cholera victims from water tanks.

However, landholders (*zamindar* and *talukdars* or leaseholders) and moneylenders resisted official pressure to provide relief to the peasants. Very few were involved in relief efforts in the aftermath of the 1876 cyclone.[69] Officials could ill afford to alienate their powerful allies, who

[65] For an analysis of the role of the government in protecting landlords, see Ranajit Guha, *Elementary Aspects of Peasant Insurgency in Colonial India* (Delhi: Oxford University Press, 1983).

[66] 'Memorandum by the Lieutenant-Governor for Guidance of the Commissioner,' 15 November 1876, BL, OICO, IOR/P/894/4/57, p. 112.

[67] Jack, *Final Report of the Survey and Settlement Operations*, 48. Jack refers to this absentee-landlordism as a Bikrampur invention. Bikrampur, close to Dacca, was the perfect hub for the demand and supply for this undertaking: on the one hand, there were the estate owners who owned forest lands but lacked vigour to bring it under cultivation, and on the other hand, there was a steady supply of men in search of occupation in Dacca district who needed land.

[68] Letter from Ohiduddin Mahomed, Collectorate Clerk on Relief Duty, to E. J. Barton, Magistrate of Backergunge, dated 4 November 1876, BL, OIOC, IOR/P/894/4/49–50, p. 2.

[69] R. H. Hollingbery, *The Zemindary Settlement of Bengal*, vol. 1 (Calcutta: Bengal Secretariat Press, 1879). 320. In an excerpt from the Annual Administrative Reports of 1876–1877, the collector wrote: "In Backergunje, with the exception of Nawab Ahsanoollah, Baboo Mohini Mohun Doss, Rajendro Chunder Roy of Bowal, and Doorga Mohun Doss, a

were fully aware that the government was dependent upon them to retain power on a local level. Indeed, in Dakshin Shabazpore district, it was asserted that "the ryots in this subdivision were extremely well off and many of them can afford to tide over the short time that will elapse before the winter rice crop is harvested."[70] Thus, rather than assist the common people or the small-time tenants, who did not possess guaranteed securities, officials asserted the need "to assist the resident respectable landholders in the distressed tracts with loans in cash, instead of attempting any other large relief measures at present."[71] Despite acknowledging the dire situation of the peasant ***karshadars*** (cultivators),[72] who were "otherwise utterly unable to get plough-cattle and agricultural implements," the Magistrate of Backergunge directed his subordinate officials to advance monetary loans to *zamindars* and the rent receivers at a high rate of interest based on the security of their estates.[73] For example, J. C. Panioty, a co-leaseholder of estates in Backergunge, was granted a loan of 3000 rupees to settle new tenants on the estates and to purchase plough cattle and other agricultural implements.[74]

Dukhinshabazpore zemindar, all of whom largely assisted their ryots after the cyclone, not a single zemindar is mentioned as having taken the smallest interests in his tenantry, or done anything material to assist them over the time of trouble they have been recently passing through. The example thus set by the zemindars has been followed by the talookdars and howladars in the storm affected tracts. Few have done anything to assist the ryots, many of whom would, but for our assistance, have fared badly indeed. Speaking generally of the conduct of the zemindars, Mr. Burton repeats what he said last year, that anything like an enlightened desire to improve their estates or improve the condition of their tenantry is wanting."

[70] Letter from E. J. Barton, Magistrate of Backergunge, to F. B. Peacock, Commissioner of the Dacca Division, dated February 6 1877, BL, OIOC, IOR/P/894/4/84–85, p. 2. It is important to note that the magistrate E. J. Barton in the official letter dated 6 February 1877 specifically outlined: "The cyclone-hurricane swept over the whole district and did not leave a single village or corner untouched." The calamity befell on a population of 831,557 and with little exception destroyed human beings, cattle, crops, and property of all kinds.

[71] Letter from H. N. Harris, District Superintendent of Police, to the Magistrate of Backergunge, dated 8 November 1876, BL, OIOC, IOR/P/894/4/54, p. 111.

[72] Jack, *Final Report of the Survey and Settlement Operations*, 49. Jack described them as "a very depressed body with no rights of any sort and regarded by the tenure holder as little better than farm labourers."

[73] Letter from E. J. Barton, Collector of Backergunge, to the Commissioner of Dacca Division, dated February 19 1877, BL, OIOC, IOR/P/894/4/82, p. 69. The letter also noted that these loans were to be given to the *zamindars* and the leaseholders at a high rate of interest based on the security of their estates and, then with the money left to provide assistance to the *ryots* to help the *ryots*.

[74] Letter from E. J. Barton, Collector of Backergunge, to the Commissioner of Dacca Division, dated 26 March 1877, BL, OIOC, IOR/P/894/4/121, p. 133.

Debt Burden and Agricultural Labour Migration by Small Peasants

In Bakarganj, the successful extraction of surplus value from agricultural labour depended on the availability of a plentiful pool of workers for the estates. This, in turn, depended on demographic growth and in-migration. Traditionally, the district failed to attract as many immigrants as elsewhere in eastern Bengal, but state-aided drives to clear forests and make the wastelands habitable resulted in both the seasonal migration of agricultural labourers from surrounding districts and immigration from further afield to meet the labour deficit. For example, the agricultural estates of central Bakarganj attracted seasonal migrants from the adjoining districts of Faridpur, Jessore, and Tipperah during the rice harvest season from October to December,[75] while Dakshin Shabazpore and the adjoining *chars* (river islands) on the estuary drew *badla* (migrants) from nearby Noakhali to weed, turn the soil, and harvest *aus* (monsoon) and *aman* (winter) paddy.[76]

The 1876 cyclone, which hit during the harvest season, and its aftermath caused great loss of life among these migrant workers and local small peasants. For example, Bhola district experienced a population decline of nearly 11 per cent.[77] In addition, in the worst-affected areas, damage to crops and the loss of cattle provoked migration of surviving peasants to less affected or unaffected regions in search of work to keep their families alive and to avoid paying rents and taxes. Thus, *ryots* in Ganeshpuree, a *khas mehal* (crown land or not part of zamindary estate), which experienced high human and animal mortality and almost total destruction of crops, left the district en masse,[78] as did small peasants from the

[75] Sugata Bose, *Agrarian Bengal: Economy, Social Structure and Politics, 1919–1947* (Cambridge: Cambridge University Press, 1986), 28.

[76] Jack, *Final Report of the Survey and Settlement Operations*, 73. Describing the composition of the hired labourers, Jack wrote, "of the labour employed in the south about one-third comes from the north of the district and another third from Faridpur, while the remainder is procured from Noakhali, Dacca, Jessore, and even Pabna and Mymensingh."

[77] Jack, *Final Report of the Survey and Settlement Operations*, 47.

[78] Letter from Baboo Ananda Chandra Sen, Deputy Collector, and Baboo Surjya Kumar Sen, Deputy Collector of Backergunge, to E. J. Barton, Collector of Backergunge, 3 January 1877, Government and Resumed Estates in the Island of Dukhin Shabazpore, BL, OIOC, IOR/P/894/4/63–64, p. 7.

cyclone-devastated government estate of Bara dhali.[79] Later, the land settlement report of 1915 indicated that the storm-wave of 1876 "converted the south of the district temporarily into a wilderness,"[80] although official reports at the time of the disaster deliberately downplayed its role in forcing *ryots* to migrate and exaggerated both the efficacy of attempts to retain them and agricultural productivity in affected areas.[81] In some cases, however, officials preferred the distressed population to migrate and fend for themselves, removing the need to provide relief measures.[82]

The governmental documents interpreted out-migration as a deliberate measure by peasants to evade paying rent, but a closer look indicates otherwise. The average tenant in Bakarganj, as J. C. Jack pointed out, was forced to take out short-term loans at high rates of interest during the lean months before the principal *aman* harvest. Even more exploitative were the *abwabs* (illegal taxes), known variously as *tahuri*, *mamuli*, *rashami*, and *sahana*, that local landholders imposed on the tenants in addition to the rent and which were forbidden under the Permanent Settlement of 1793.[83] There were, in addition, other special taxes imposed purportedly for purposes such as building or maintaining *khalbandi* (embankments) or *pol kharach* (bridges). Through these levies, the landlords placed a crushing financial burden on ordinary cultivators.[84] In addition, middlemen positioned below the actual tenure holders imposed *beggar* (forced labour) on peasants to clear jungles and plough fields and demanded gifts in kind.[85] As a result, as Jack indicated with reference to the Dhankhali

[79] Letter from Baboo Ananda Chandra Sen, Deputy Collector, and Baboo Surjya Kumar Sen, Deputy Collector of Backergunge, to E. J. Barton, Collector of Backergunge, 3 January 1877, Government and Resumed Estates in the Island of Dukhin Shabazpore, BL, OIOC, IOR/P/894/4/63–64, p. 9.

[80] Jack. *Final Report of the Survey and Settlement Operations*, 23.

[81] Letter from L. Cameron, Civil Surgeon, Backergunge, to the Magistrate of Backergunge, dated 14 March 1877, BL, OIOC, IOR/P/894/4/118–19, p. 5. The civil surgeon Dr L. Cameron reported in the affected sub-divisions, "trading classes and cultivators who could afford were leaving for places not affected by the storm wave." In the side margin, the magistrate indicated this phenomenon to be "an exaggeration."

[82] Letter from E. J. Barton, Collector of Backergunge, to the Commissioner of Dacca Division, dated 1 June 1877, BL, OIOC, IOR/P/894/4/163–64, p. 1.

[83] Jack, *Final Report of the Survey and Settlement Operations*, 79–80. Under Permanent Settlement all such levies were consolidated in the rent. In Bakarganj, landholders perpetually disregarded this clause.

[84] Jack, *Final Report of the Survey and Settlement Operations*, 79–80.

[85] Jack, *Final Report of the Survey and Settlement Operations*, 80.

estates, following the disaster, cultivators were not only left without any form of assistance but were also forced to pay frequent *abwabs*, and if they were unable to pay, they were evicted.[86]

Conclusion

This study of the environmentally vulnerable, highly fertile southern shoreline of the Bay of Bengal region demonstrates how, in times of natural disaster, a combination of colonial capitalism and the land tenure system in Bakarganj led already impoverished peasant labour who survived such events into disaster-induced migration. It would be wrong to project this study of an environmentally vulnerable disaster-prone micro-region as a complete picture of colonial capitalism in the South Asian Indian Ocean littoral. Nonetheless, such micro-studies reveal the complex impact of both the operation of colonial capital and the globalization of commodity production on the subsistence peasant producers. Capital operated through a network of intermediary tenure-holding social classes and moneylending operations that represented class oppression of an economically defenceless working poor.

Disaster and relief operations here signified not the emergence of a developmentalist welfare state but rather a predatory capitalist state that, in times of natural catastrophe, provided minimal relief to the rural poor. The development of communications, the integration of global commodity production, and the creation of large grain markets failed to provide any safety net for peasants. This micro-history provides a portal to the operations of fragments of a global capitalist network. Capitalism and free trade did not bring an end to the power of landholding social classes in this corner of colonial Bengal. Instead, they strengthened the subinfeudation of land and utilized state machinery to sustain the process of the extraction of surplus from a small peasant economy.

[86] Jack, *Final Report of the Survey and Settlement Operations*, 217.

CHAPTER 11

Famine and Slavery in Africa's Red Sea World, 1887–1914

Steven Serels

In the late 1880s, the food economy in much of the Southern Red Sea Region (SRSR) collapsed, and the countries on the African littoral were pushed into a deadly, multi-year famine. The SRSR comprises parts of contemporary Sudan, Eritrea, Ethiopia, Djibouti, Somalia, Yemen, and Saudi Arabia. Historically, the SRSR was home to a unified, though heterogenous, socio-economic system built upon a robust trade in both locally produced and internationally imported grain. Though much of this region is arid or semi-arid, under normal conditions the abundance of grain within this regional food economy supported non-agricultural niche activities, such as pearl diving, animal rearing, and cloth weaving, as well as complex states with unique court cultures. The grain trade that fuelled economic differentiation in the SRSR was made possible by the Indian Ocean monsoon system, which facilitated open-water sailing within and

Researching and writing this paper was made possible by grants from the Gerda Henkel Stiftung, the Alexander von Humboldt Foundation, the Volkswagen Stiftung, the Andrew W. Mellon Foundation, the Weatherhead Initiative on Global History, and the Social Science and Humanities Research Council of Canada.

S. Serels (✉)
Harvard University, Cambridge, MA, United States

G. Campbell (ed.), *Bondage and the Environment in the Indian Ocean World*, Palgrave Series in Indian Ocean World Studies,
https://doi.org/10.1007/978-3-319-70028-1_11

between the Red Sea, the Gulf of Aden, and the grain ports of western India.[1] Maritime shipping routes and overland caravan roads allowed grain and other commodities to be shipped over large distances. These trading facilities also supported the trade in slaves within the SRSR. Generally, women and children were kidnapped in South Sudan and western Ethiopia and marched overland on the caravan roads to inland African markets or to the ports on the African littoral. From these ports, slaves were shipped across the sea on dhows that often also carried grain to be sold in Arabia. The close link between the grain and slave trades ensured that the 1888–1892 famine on the African littoral had long-term consequences for the SRSR slave system.

The 1888–1892 famine precipitated the breakdown of the existing social order. Scholars have shown that, in the early stages, famines are commonly recognized by victims as threats to communal integrity and that, therefore, initial responses tend to focus on collective self-preservation.[2] This was certainly the case during the 1888–1892 famine. As the famine was beginning and hardship was sinking in, whole communities began migrating together under the leadership of traditional elites.[3] In late 1888, hundreds of thousands of destitute and starving men, women, and children walked from the African interior to the European-controlled ports of Sawakin and Massawa. Others made their way to regional capitals and market towns. Starving refugees flooded the streets of these cities, where they were often forced to rely on charity. Unfortunately, conditions continued to deteriorate and communal ties broke under the strain. In desperation, people abandoned their kin and adopted individual survival strategies. They started wandering the countryside on their own and eating taboo foods, such as garbage, the rotting

[1] William Patzert, "Wind-Induced Reversal in Red Sea Circulation," *Deep Sea Research and Oceanographic Abstracts* 21, no. 2 (1974): 109–21; John Meloy, *Imperial Power and Maritime Trade: Mecca and Cairo in the Later Middle Ages* (Chicago: Middle East Documentation Center, 2010), 53–55.

[2] See Amrita Rangasami, "'Failure of Entitlements' Theory of Famine: A Response," part 1, *Economic and Political Weekly* 20, no. 41 (12 October 1985): 1747–52. For a study of communal self-preservation strategies during a famine in the region, see Alex de Waal, *Famine that Kills: Darfur Sudan,* rev. ed. (Oxford: Oxford University Press, 2005).

[3] Intelligence Department, Egyptian Army, *Staff Diary and Intelligence Report, Suakin,* no. 98 (24 December 1888–6 January 1889), Sudan Archive, Durham University (hereafter SAD). For example, on 25 December 1889, 560 Hadandawa pastoralists arrived at Sawakin together with their *shaykhs.*

offal of dead animals, and undigested grain picked out of animal faeces. There were reports of widespread cannibalism.[4]

These survival strategies proved ineffective, and people died in large numbers. Some died from starvation, but it is likely that more died from diseases directly tied to the famine. The mass migration of starving refugees facilitated the spread of deadly diseases, such as cholera and smallpox.[5] There are no precise figures for the human toll of this famine because population statistics for this period are incomplete or, more commonly, nonexistent. Contemporary British observers estimated that the population of Sudan shrank in the 1880s and 1890s from 8.52 million to 1.87 million people as a result of war, famine, and disease.[6] However, these figures are little more than conjecture based on observations of a depopulated country where large swathes of previously cultivated land had reverted to wilderness and where numerous villages had been completely abandoned.[7] Indeed, statistics for other regions are equally unreliable. Nonetheless, the decline in population can similarly be inferred. As in Sudan, much arable land in Eritrea and Ethiopia fell out of cultivation in the late 1880s and early 1890s as a result of the rapid depopulation of the countryside.[8] First-hand accounts of conditions led Richard Pankhurst to estimate that in Ethiopia and Eritrea, the agricultural population decreased by one-third and the pastoral population by two-thirds during the famine.[9]

[4] For descriptions of the famine, see Richard Pankhurst, *The Great Ethiopian Famine of 1888–1892: A New Assessment* (Addis Ababa: Haile Selassie I University, 1964); Steven Serels, "Famines of War: The Red Sea Grain Market and Famine in Eastern Sudan 1889–1891," *Northeast African Studies* 12, no. 1 (2012): 73–94; Rudolph von Slatin, *Fire and Sword in the Sudan: A Personal Narrative of Fighting and Serving the Dervishes, 1879–1895*, trans. F. R. Wingate (London: Edward Arnold, 1896), 452–57; F. R. Wingate, *Ten Years' Captivity in the Mahdi's Camp 1882–1892* (London: Sampson, Low, Marston and Co, 1892) 284–91; Ferdinando Martini, *Nell'Africa italiana*, 8th ed. (Milan: Fratelli Treves, 1925), 29–31.

[5] Administration Report of the Somali Coast for the Year 1891–92, IOR/V/10/1995, British Library, London (hereafter BL).

[6] *British Parliamentary Papers: Command Papers. Reports by His Majesty's Agent and Consul-General on the Finances, Administration and Condition of Egypt and the Soudan in 1903* (Cd 1951, 1904), 79.

[7] See William Garstin, "Report on the Soudan," in *British Parliamentary Papers: Command Papers. Dispatch from Her Majesty's Agent and Consul General at Cairo* (C 9332, 1899), 5; "Report by Mr. Garstin on the Province of Dongola," n.d. [April 1897], FO407/143/12, National Archives, Kew (hereafter NAUK).

[8] *L'Economia Eritrea; nel Cinquatennio dell'occupazione di Assab (1882–1932)* (Florence: Istituto Agricolo Coloniale Italiano, 1932), 7–8.

[9] Pankhurst, *The Great Ethiopian Famine*, 39.

Though profoundly devastating, the impacts of the famine were not evenly distributed. Narrow segments of society with close ties to states were able to weather the famine relatively well. Political elites did not starve to death, and neither did their most valued dependants. However, ensuring the food security of the privileged few was possible only through further harming the unprivileged masses. For example, the leadership of the Mahdist state in Sudan did not personally suffer from the deprivations of the famine because they could use force to appropriate the limited supply of grain produced in the fertile Jazira plain. Similarly, Al-Khalifa Abdallahi ibn Muhammad, the head of the Sudanese state, also ensured that his fellow Baqqara tribesmen remained well supplied with food by allowing them to seize the yields of farms that they passed as they migrated towards the capital during the famine. The Mahdist leadership and their Baqqara allies ate because other people starved.[10] The famine also impacted free individuals differently from slaves. Scholars have previously noted that free men and women historically offered themselves or their dependants as slaves during famines in order to secure food.[11] However, the ways in which slaves experienced famine have previously received little scholarly attention.[12] Slaves, as dependants of their owners, lacked independent resources that they could mobilize for their own self-maintenance. Slave owners may have decided to divert their own limited resources away from their slaves in order to conserve them for themselves and their immediate kin. Slaves may have responded to the stressed conditions of these food crises either by renegotiating their relationships with their owners or by fleeing to try to survive on their own. Regardless of which strategy they chose, slaves faced grim prospects and probably died in large numbers.

[10] Richard Pankhurst and Douglas Johnson, "The Great Drought and Famine of 1888–92 in Northeast Africa," in *The Ecology of Survival: Case Studies from Northeast African History*, ed. Douglas Johnson and David Anderson (Boulder, Colorado: Westview Press, 1988), 51.

[11] See Gwyn Campbell, ed., *The Structure of Slavery in Indian Ocean Africa and Asia* (London: Frank Cass, 2004); Paul Lovejoy and Toyin Falola, eds., *Pawnship, Slavery and Colonialism in Africa* (Trenton, NJ: Africa World Press, 2003).

[12] See Ahmad Alawad Sikainga, *Slaves into Workers: Emancipation and Labor in Colonial Sudan* (Austin: University of Texas Press, 1996); Suzanne Miers, *Slavery in the Twentieth Century: The Evolution of a Global Problem* (Walnut Creek, CA: AltaMira Press, 2003); Richard Pankhurst, *Economic History of Ethiopia, 1800–1935* (Addis Ababa: Haile Selassie I University Press, 1968), 73–134. Accounts of famines are absent from nearly all studies of slavery in Africa's Red Sea world in the nineteenth and twentieth centuries.

This chapter examines the impact of the 1888–1892 famine on the SRSR slave system. Though the famine was confined to the countries of the African littoral, its impacts were felt throughout the wider region. Trans-maritime links supported both the regional food economy and the slave trade. During the famine, widespread social dislocation and high levels of mortality caused the slave population to decline drastically in the countries of the African littoral; in some parts, the famine caused the slave system to collapse entirely. When the acute crisis ended, people began to rebuild their devastated communities. Since slavery was a normal part of communal structure, they also sought to rebuild the slave system. The process of rebuilding required redeveloping the complete structure of the slave trade at every level from the capture of free individuals to their ultimate purchase. For free men and women, recovery meant living among unfree domestic servants, concubines, and labourers. The reconstruction of the slave system was, by definition, a truly regional process. Arabia has historically been an important market for slaves from north-east Africa. In the decades following the famine, recovery could be possible only through expanding the maritime slave trade.

The recovery of the SRSR slave system after the 1888–1892 famine implicated all segments of society. Though slaves were often kept by elites as signs of their conspicuous consumption, slaves also played important roles in the overall regional economy. The slave trade was a source of profit for the merchants who sold slaves, the caravan leaders who led the slave caravans, the camel owners who rented their camels to the caravans, the pastoralists who served as guides, the pastoral political leaders who charged rents to passing caravans, and the ship captains who transported slaves across the sea to Arabia. Slaves worked on the ships that moved other forms of cargo, including grain, and slaves also worked the land in key grain producing regions, especially in Sudan. Moreover, they were structurally important for political rule. In the twentieth century, the Ethiopian court funded annual slave raiding expeditions as a way of both generating income for the court and shoring up its network of dependence. Even European colonial officials were implicated in this trade. This revival often happened with the aid and encouragement of the imperial powers that conquered much of the region in the aftermath of the famine. With the notable exception of Italian officials in Eritrea, European colonial officials, like their Ethiopian counterparts, were complicit in both the maintenance of slavery and the slave trade well into the twentieth century.

The Famine and Its Immediate Aftermath

The 1888–1892 famine was actually a set of distinct, though interrelated and simultaneous, famines in Sudan, Ethiopia, Eritrea, Djibouti, and Somalia precipitated by changes in local political and ecological conditions.[13] In the years immediately preceding this complex food crisis, the political map of the region was redrawn as new states emerged and European powers renewed their interest in local affairs. In September 1882, the British Army invaded Egypt, which ruled Sudan and much of the African littoral of the Red Sea and the Gulf of Aden. At the same time, Muhammad Ahmad ibn 'Abd Allah, a Sudanese religious leader, proclaimed himself to be *al-Mahdi* (that is, the prophesied Islamic eschatological leader who would herald the end of days) and announced a jihad against the Turko-Egyptian rulers of Sudan. As the Mahdist Rebellion in Sudan gained strength in the mid-1880s, British officials ordered the withdrawal of Sudan's Turko-Egyptian government, created new Anglo-Egyptian administrations on the Egyptian-Sudanese Nilotic frontier and on the Sudanese Red Sea coast, assumed control over the Somali ports of Berbera and Zaila, and allowed Italian officials to take over the port of Massawa.[14] French officials, unwilling to see Britain and its allies dominate the important Red Sea maritime route and the lucrative trade with Ethiopia, responded by entering into a new treaty with the Sultan of Tadjoura that solidified French claims to Red Sea ports.[15] The expanding Mahdist Sudanese and Ethiopian states, as well as the growing following of Muhammad 'Abd Allah al-Hasan in Somalia, did not respond passively to European encroachment.[16] As a result, the African littoral of the SRSR

[13] See Pankhurst and Johnson, "The Great Drought and Famine of 1888–92 in Northeast Africa," in *The Ecology of Survival: Case Studies from North East African History*, ed. Douglas H. Johnson and David M. Anderson (London: Lester Crook Academic and Westview Press, 1988), 47–73. Pankhurst and Johnson were the first to suggest that the famine in the Ethiopian plateau was distinct from the famine in the central Sudan.

[14] For a history of the Mahdist Rebellion, see P. M. Holt, *The Mahdist State in the Sudan, 1881–1898: A Study of its Origins, Development and Overthrow*, 2nd ed. (Oxford: Clarendon Press, 1970).

[15] For a history of Djibouti, see Philippe Oberlé, *Histoire de Djibouti: des origines à la République* (Paris: Présence Africaine, 1985).

[16] For a history of Muhammad Abd Allah al-Hasan's movement in Somalia, see Abdi Sheik-Abdi, *Divine Madness: Mohammed Abdulle Hassan (1856–1920)* (London: Zed Books Ltd, 1993). For a history of modern Ethiopia, see Bahru Zewde, *A History of Modern Ethiopia, 1855–1974* (Addis Ababa: Addis Ababa University Press, 1991).

was thrown into an extended, multi-party armed confrontation during which patterns of trade and agricultural production were frequently disrupted.

In Sudan, the famine was caused by policies pursued by British and Mahdist officials. Before the British-ordered withdrawal from the Sudan in 1885, Mahdist forces had been unable to gain a foothold in the important grain-producing region of Northern Nilotic Sudan. Since the local population had actively fought against the invading Mahdist forces, they feared that Mahdist reprisals would follow the withdrawal.[17] As a result, many fled with their slaves when the Egyptian Army pulled out. Official statistics record 12,825 refugees arriving in Upper Egypt from the Sudan in the summer of 1885.[18] However, the actual decline in the agricultural population of Northern Nilotic Sudan was much greater than this number would indicate. Official Egyptian Army statistics do not include those refugees who chose not to register with the government, which was probably most of them. In addition, there are no records of the number of those who did not cross into Egypt as refugees, of the number who died in battle as Mahdist forces conquered Northern Nilotic Sudan, nor of the number of slaves who abandoned their masters to join the advancing Mahdist slave army.[19] The sharp decline in available agricultural labour had major economic repercussions. Irrigation in Northern Nilotic Sudan was dependent on *sāqiyas* (waterwheels), so the decline in population led to an immediate drop in agricultural yields. At the same time, the expanding Mahdist forces on the frontier, which were commanded by 'Abd al-Rahman al-Nujumi, increased local demand for provisions, which proved insufficient to support booth al-Nujumi's men and the local population. Grain could not be imported from elsewhere in Sudan because surplus stocks were being used to feed the Mahdist leadership's allies in western and central Sudan.[20] The result was widespread famine.

The disruption of grain production in Sudan was further compounded by an outbreak of rinderpest, a viral disease that kills up to 90 per cent of infected cattle in virgin populations. Rinderpest was introduced into Africa in 1887 when Italian officials unknowingly imported infected cattle in

[17] Egerton to Granville, 26 August 1884, FO407/62/337, NAUK.

[18] Baring to Granville, 9 June 1885, FO407/65/343, NAUK.

[19] For a history of military slavery in the Sudan, see Douglas H. Johnson, "Sudanese Military Slavery from the Eighteenth to the Twentieth Century," in *Slavery and Other Forms of Unfree Labour*, ed. Leonie Archer (London: Routledge, 1988), 142–55.

[20] See Pankhurst and Johnson, "The Great Drought and Famine," 47–73.

order to feed their large garrison in Massawa. The disease rapidly spread via trade routes into the interior, decimating herds in Eritrea, Ethiopia,[21] and Somalia,[22] as well as in eastern and southern Sudan.[23] The rapid decline in the cattle population throughout the African littoral of the SRSR disrupted regional patterns of agricultural production and trade. In the grain-producing regions of the Eritrean and Ethiopia plateau, cultivation was dependent on cattle-drawn ploughs so that the rapid decline in the cattle population led to an equally rapid reduction in cultivation.[24] Though some cultivators tried to replace their lost animal labour by yoking ploughs to other animals or by using hoes and spades, many could not cultivate sufficient grain to meet even minimum household needs.[25] Decreased yields meant that little grain reached the markets, which, in turn, negatively affected those who depended on trade. Before the outbreak, pastoralists in Eastern Sudan, the lowlands of Eritrea and Ethiopia, and in northern Somalia had diets consisted primarily of grain that they purchased with money earned from economically exploiting their herds.[26] The epizootic destroyed most pastoral wealth and, as a result, pastoralists could not acquire sufficient food.[27] Pastoralists and agriculturalists starved in large numbers as a result.

The famine led to an intensification of fighting. In Ethiopia and Somalia, the famine and the outbreak of rinderpest led to increased raiding. During the famine, Emperor Menelik II, with the assistance of local *rases* (local government rulers), launched a series of large-scale cattle raids into southern and western Ethiopia and also into Somali territory. One of these raids was launched from Harar Province during the height of the famine. At the time, much land in the province had fallen into disuse because the starving farmers lacked the strength to cultivate.[28] In January

[21] Pankhurst and Johnson, "The Great Drought and Famine," 63.

[22] *Administration Report of the Somali Coast for the Year 1891–92*, IOR/V/10/1995, BL.

[23] John Rowe and Kjell Hødnebø, "Rinderpest in the Sudan 1888–1890: The Mystery of the Missing Panzootic," *Sudanic Africa* 5 (1994): 149–79; B. A. Lewis, *The Murle: Red Chiefs and Black Commoners* (Oxford: Clarendon Press, 1972).

[24] See Pankhurst and Johnson, "The Great Drought and Famine of 1888–92 in Northeast Africa," 47–73.

[25] Pankhurst, *The Great Ethiopian Famine*, 29.

[26] See Serels, "Famines of War," 77–81; Jonathan Miran, *Red Sea Citizens: Cosmopolitan Society and Cultural Change in Massawa* (Bloomington, IN: Indiana University Press, 2009), 91; Abdi Ismail Samatar, *The State and Rural Transformation in Northern Somalia, 1884–1986* (Madison, WI: University of Wisconsin Press, 1989), 27.

[27] Serels, "Famines of War," 84–86.

[28] *Administration Report of the Somali Coast for the Year 1891–92*, IOR/V/10/1995, BL.

1891, Ras Makonnen, the provincial governor, led approximately 4000 men into the Ogaden region to raid.[29] He armed his men with over 2000 rifles and 46,000 rounds of ammunition imported from Europe.[30] The volume of arms imports into Ethiopia increased the following year,[31] and, in subsequent years, Ethiopian officials imported increasing numbers of arms through the Côte Française des Somalis (Djibouti) and the southern coast of Eritrea.[32] Ethiopian officials often sold slaves taken in raids to pay for imported arms, which meant that large numbers of slaves were exported to Arabian markets.[33]

Mahdist leaders also responded to the food crisis by launching large-scale, offensive military actions to seize grain yields from Sudanese cultivators in the fertile regions of the Jazira (between the White and Blue Niles) and near Qadarif and Qallabat (on the Ethiopian and Eritrean frontier).[34] When these stocks were exhausted, these leaders organized a raid into southern Sudan. However, the Dinka and Shilluk pastoralists indigenous to the region successfully defended themselves from the invading force.[35] Mahdist officials also led a military invasion of Egypt in the summer of 1889. This too was unsuccessful; during the long trek through the Egyptian desert and the subsequent Battle of Tushki on 3 August 1889, 6500 Sudanese men and women died, 6000 were taken prisoner by the Egyptian army, and 2500 escaped the battlefield.[36] The retreating Mahdist force looted Northern Nilotic Sudan, which precipitated a second wave of mass refugee migration into Upper Egypt: in under two years, official Egyptian statistics recorded the arrival of an additional 13,000 Sudanese refugees.[37] The population loss in Northern Nilotic Sudan led to a further

[29] *Extract from a Report by the Italian Consul General at Aden*, 16 April 1891, IOR/R/20/A/1174, BL.

[30] Tornielli to Salisbury, 12 February 1890 IOR/R/20/A/1173, BL.

[31] Stace to Hardinge, 2 September 1891, IOR/R/20/A/1174, BL.

[32] Rodd to Salisbury, 14 June 1897, FO403/255, NAUK.

[33] Eustace to the Commander-in-Chief of the East Indies, 1 September 1905, IOR/R/20/A/1300, BL.

[34] Von Slatin, *Fire and Sword in the Sudan*, 456.

[35] Pankhurst and Johnson, "The Great Drought and Famine of 1888–92 in Northeast Africa," 63.

[36] Clarke to Salisbury, 26 August 1889, FO407/89/108, NAUK.

[37] See Intelligence Department, Egyptian Army, *Staff Diary and Intelligence Report, Frontier Field Force*, no. 254 (29 March–11 April 1891), SAD. The Egyptian Military Intelligence Report covering 29 March–11 April 1891 records the last arriving refugees.

drop in agricultural productivity and ensured that food insecurity remained endemic.

The famine fundamentally altered patterns of slave owning on the African littoral of the SRSR. In the decades before the famine, changing political and economic conditions had decreased the price of slaves and encouraged wider segments of society to own them. The famine rapidly reversed this trend because it made it impossible for small-scale slave owners to maintain themselves, their kin, and their slaves. For example, innovations introduced under Turko-Egyptian rule (*ca.* 1820–1885), including the application of Islamic religious law to land tenure and the levying of a high tax on land, had encouraged cultivators in Northern Nilotic Sudan to increase the intensity and extent of cultivation by replacing family labour with imported slave labour.[38] During the famine, these cultivators lost control of their slaves. Some of these slaves were absorbed into the Mahdist army, others made their way into the Egyptian Army, and others simply ran away.[39] The magnitude of the decline in the agricultural slave population can be inferred from the limited available statistics. In 1885, there were 6451 *sāqiyas* between the second and fourth Nile cataracts, according to the official tax ledger.[40] To be fully operational, each *sāqiya* required the labour of eight to ten men, and in the late nineteenth century, this labour was primarily provided by slaves.[41] Therefore, it can be assumed that there were between 51,600 and 64,500 male agricultural slaves in the region before the Turko-Egyptian withdrawal. A house-by-house census completed by the conquering Egyptian Army in 1896 counted 1992 "Sudanese" men. This census had only three categories, "Native," "Arab," and "Sudanese," which were how British officers referred to free agriculturalists, free pastoralists, and slaves, respectively.

[38] See Jay Spaulding, "Land Tenure and Social Class in the Northern Turkish Sudan," *International Journal of African Historical Studies* 15, no. 1 (1982): 1–20.

[39] *Petition Submitted by Some Refugees of Dongola Province, at Assuan*, n.d. [August 1885], FO407/66/113, NAUK; Intelligence Department, Egyptian Army, *Staff Diary and Intelligence Report, Frontier Field Force,* no. 229 (SAD, 11–17 May 1890); Intelligence Department, Egyptian Army, *Staff Diary and Intelligence Report, Frontier Field Force,* no. 233 (8–14 June 1890), SAD; Intelligence Department, Egyptian Army, *Staff Diary and Intelligence Report, Frontier Field Force,* no. 244 (28 September–11 October 1890), SAD.

[40] House of Commons. United Kingdom Government. *British Parliamentary Papers: Command Papers. Reports on the Province of Dongola* (C 8427, 1897), 2.

[41] W. Nichols, "The Sakia in Dongola Province," *Sudan Notes and Records* 1, no. 1 (January 1918): 23–24; Spaulding, "Land Tenure and Social Class in the Northern Turkish Sudan," 1–20.

From these figures, it can be deduced that the male agricultural slave population had shrunk by 96 per cent by 1896, but that did not mean that all forms of slavery in Sudan disappeared. Slave soldiers continued to be an important part of the Mahdist military force, and the Mahdist government continued to divert resources to raiding its southern marches to kidnap more men into slave soldiery.[42] Slavery had become, primarily, the prerogative of the state.

Rebuilding the Slave System and the Consolidation of State Power

Though the acute food crisis ended in 1892, the African littoral of the SRSR was subsequently hit by a number of natural and man-made disasters that hindered true recovery. Locust swarms visited the region during the 1892–1893 and 1895–1896 cultivation years, damaging the limited yield.[43] Drought afflicted Eastern Sudan, the Red Sea coast, and the highlands of Eritrea and Ethiopia in 1893 and 1894.[44] The rapid depopulation of the countryside in Eritrea and Ethiopia increased the rangeland of wild animals, and animal attacks became common.[45] As a result of these trying conditions, markets in the highlands of Eritrea and Ethiopia continued to be insufficiently supplied with provisions throughout the 1890s. In addition, grain yields in this region remained low because of a lack of cattle to plough the fields, and trade remained depressed because there were insufficient pack animals to bring harvests to market.[46] Continued fighting in

[42] Intelligence Department, Egyptian Army, *Staff Diary and Intelligence Report, Eastern Sudan*, no. 4 (15–28 April 1891), SAD. On 26 April 1891, an Egyptian government dhow caught two dhows unloading contraband cargo on the Sudanese coast. Arrested crew members subsequently confessed that there was an extensive trade in slaves between the Sudan and Arabia.

[43] Intelligence Department, Egyptian Army, *Intelligence Report, Egypt*, no. 8 (December 1892), 3, SAD; *British Parliamentary Papers: Command Papers. Report for the Year 1896 on the Trade of Suakin*, Commercial no. 1859 (C 8277, 1897), 2.

[44] "Report and Map of Tokar Sub-District by Kaimkam Hickman Bey, Commanding Tokar," in Intelligence Department, Egyptian Army, *Intelligence Report, Egypt*, no. 21 (December 1893), 5, SAD; Barnham to Cromer, 14 February 1894, FO407/126/77, NAUK.

[45] Pankhurst, *The Great Ethiopian Famine*, 39.

[46] Gino Bartolimmei Gioli, "La produzione frumentaria in Eritrea di fronte alle relazioni doganali fra metropoli e colonia," in *Atti della R. Academia dei Geografili*, series V, vol. 1, no. 1 (1904): 86–88.

Northern Nilotic Sudan prevented the return of refugees and ensured that there was an inadequate supply of labour to properly work the land. Yields along the Nile failed to meet even local demand. The Mahdist state responded by monopolizing agricultural resources in central and Eastern Sudan and by becoming the largest factor in the internal grain trade. These policies diverted harvests from producers to the state and state-connected elites and, in so doing, ensured that food insecurity remained endemic in the Sudanese countryside.[47] These disasters created the opening for British, French, Italian, and Ethiopian officials to carve the region up among themselves. Though the three European powers had created footholds in the region in the mid-1880s, it was only during the famine and its immediate aftermath that they began to lay claim to territories in the interior.

The establishment of British rule in Somaliland and the Anglo-Egyptian Sudan, Italian rule in Eritrea, and French rule in the Côte Française des Somalis incorporated these colonies into a trans-imperial order that was officially committed to abolishing the slave trade and emancipating slaves. The Red Sea slave trade had already been outlawed by a number of international treaties orchestrated by the British government in the 1870s, but these were recognized as ineffective from the outset for two reasons. First, they depended on the active participation of the Egyptian rulers of the African littoral and the Ottoman rulers of the Arabian littoral, both of whom were unreliable in their commitment to abolition.[48] Second, enforcement at sea depended on the British Navy, whose boats could be easily outrun by the indigenous dhows that could better navigate the barrier islands, reefs, and winds.[49] The end of Egyptian rule and the division of the coast between the three European powers should have made enforcement more rigorous. However, these international commitments were frequently ignored as local ecological and economic imperatives shaped the policies on slavery pursued by new colonial states. These imperatives were, in turn, the legacy of the 1888–1892 famine.

Following the British-led conquest of Mahdist Sudan (1896–1898), the British Navy was given new orders. Ships were withdrawn from the

[47] Babikr Badri, *The Memoirs of Babikr Bedri*, trans. by George Scott (London: Oxford University Press, 1969), 203.

[48] Miers, *Slavery in the Twentieth Century*, 14–19.

[49] Secretary to the Admiralty to Lister, 8 June 1881, FO84/1597, NAUK; Jones to Hay, 20 October 1885, FO407/67/165, NAUK. British naval officers repeatedly requested permission to acquire and operate a fleet of dhows, but this request went unfulfilled.

Red Sea and moved to the Gulf of Aden in order to support the ongoing campaign against Muhammad 'Abd Allah al-Hasan, whom the British called the "Mad Mullah". As a result, the Sudanese coast was left without a regular coastguard.[50] British naval officers cruising along the Somali coast were instructed to prevent the trade in slaves and arms and to enforce local customs regulations.[51] However, British naval officers routinely ignored orders to police the slave trade, and slave running from British Somaliland coast went virtually unchecked as a result.[52] They understood that their main mission was to end the importation of arms into Somaliland. The famine had precipitated a cycle of raiding and counter-raiding in the interior that was further intensifying because of the steady flow of rifles into the region. Though the British Navy augmented its fleet with a number of armed dhows, its operations were generally a failure. The movement of these dhows was well known and therefore easily evaded.[53] Somaliland was flooded with arms and ammunition, which both intensified the violence and, counterintuitively, led to a decline in the export of slaves from the region. In addition to coming by sea, munitions were also imported into Somaliland from Ethiopia via overland trade routes.[54] Moreover, British officials themselves were supplying arms to the Somalis. These officials were unwilling to launch a sustained military campaign or to invest in a permanent administration in the interior and therefore settled on a policy of arming their indigenous allies.[55] Rather than lead to peace under British protection, this policy further destabilized the region, and prolonged instability changed the nature of raiding between Somali communities. Belligerent parties stopped kidnapping slaves and began to

[50] R. Wingate, "Memorandum by the Governor-General, 1904," in *Reports on the Finances, Administration and Conditions of the Sudan, 1904*, vol. 2 (1904), 35, SAD.

[51] H. E. S. Cordeaux, *Somaliland Protectorate Revenue Cutters. Standing Orders*, 9 March 1903, IOR/L/PS/10/32, BL.

[52] E. B. C. Dickson, Captain, Senior Officer Red Sea Sloops, note, 2 March 1931, IOR/L/PS/12/4094, BL. In 1931, Rear Admiral Deville, the naval officer in command of the Red Sea fleet, stated that his ships had stopped searching dhows for slaves. As Captain Dickson subsequently recounted, Deville "appeared to consider it quite a joke we should do so. Whilst admitting that the capture of slaves might be barbarous, he considered their lot in Arabia so comfortable that they were deserving of no special pity."

[53] Admiralty to Foreign Office, 14 December 1903, IOR/R/20/A/1300, BL.

[54] British Somaliland Protectorate, *Intelligence Report No. 8*, August 1908, CO535/12, NAUK.

[55] Commissioner of the Somaliland Protectorate to Secretary of State for the Colonies, 21 February 1910, CO535/18, NAUK.

focus almost exclusively on killing their opponents. Though the maritime trade in slaves from the coast of British Somaliland dried up, the constant fighting led the population of the interior to dramatically contract.[56]

In the Anglo-Egyptian Sudan, British-led rule brought peace to much of the country, but it also led to the revival of the slave trade and the redevelopment of the slave plantation system along the Sudanese Nile. Rather than oppose these developments, British officials actively encouraged them. These officials believed that the revival of slavery and the slave trade were necessary to end the persistent food insecurity that had helped topple the Mahdist state and that they feared might some day topple theirs. In the early years of British-led rule, grain prices remained high because key Nilotic markets continued to be insufficiently supplied with grain.[57] British officials feared that endemic food insecurity would weaken their hold on Sudan. In addition, they were convinced that northern Sudanese cultivators were unwilling to work their own land and that agriculture was therefore completely dependent on slave labour.[58] Therefore British officials implemented a set of policies and procedures that encouraged Sudanese cultivators to rapidly reconstruct the slave-based agricultural system that had collapsed under Mahdist rule. The number of slaves imported into the region in the early years of British-led rule can only be inferred from the colonial record because officials deliberately obscured their activities.[59] However, a comparison over time of the number of operational *sāqiyas* can give a rough idea of the number of male slaves.[60] Between 1898 and 1912, the number of operational *sāqiyas* increased from 1520 to

[56] Archer to the Secretary of State for the Colonies, 19 August 1913, CO535/31, NAUK. British officials estimated that one-third of the population of Somaliland died during a particularly bloody period of fighting in 1911 and 1912.

[57] *Note—the Price of Grain in Omdurman and Eastern Sudan*, SAD269/2/47–48.

[58] Talbot to Wingate, 10 February 1899, SAD269/2/35–36.

[59] Slatin to Wingate, 2 December 1912, SAD183/3/16–17. Rudolph von Slatin, the Senior Inspector, once even threatened to cut off a finger from the hand of a junior official for having called a "slave" a "slave" and not a "servant" or simply "Sudanese" in an official document.

[60] C. H. Townsend, "Annual Report, Berber Province, 1910," in *Reports on the Finances, Administration and Conditions of the Sudan. 1910* (1910), 193, SAD; H. W. Jackson, "Annual Report, Dongola Province, 1908," in *Reports on the Finances, Administration and Conditions of the Sudan, 1908* (1908), 503, SAD; E. A. Stanton, "Annual Report, Khartoum Province, 1908," in *Reports on the Finances, Administration and Conditions of the Sudan, 1908* (1908), 554, SAD.

12,853.[61] In order to function properly, each *sāqiya* needed the labour of eight to ten male slaves. This suggests that as many as 100,000 male slaves were imported into the region during the first decade and a half of British-led rule. Some of these men were kidnapped runaway slaves who had escaped from their masters during the British-led conquest,[62] while others were captured by private slave raiding militias from the traditional slave-raiding areas in southern Kordofan, western Ethiopia, Darfur, and Bahr al-Ghazal.[63] These militias and the privateers that led them revived and reconstructed some of the features of the nineteenth-century *zariba* system of slave raiding that La Rue describes in his contribution to this volume (Chap. 8).

French officials in the Côte Française des Somalis were also complicit in the slave trade because they saw it as the way of capturing a part of the Ethiopian trade. In the wake of the 1888–1892 famine, slaves became the most important Ethiopian export. Although Menelik II had declared his opposition to the slave trade as early as 1874, the expanding Shawan Ethiopian state depended on the slave system to establish and consolidate power. In the years after the famine, the state-sponsored cattle raids into Somalia turned into large-scale slave and cattle raids in Ethiopia's southern and western marches. These raids continued throughout the first quarter of the twentieth century.[64] Local elites in regions that submitted to the state were frequently required to pay annual tribute in the form of slaves, gold, and cattle.[65] Slave raiding was so prevalent that the local population

[61] *British Parliamentary Papers: Command Papers. Reports by His Majesty's Agent and Consul-General on the Finances, Administration, and Conditions of Egypt and the Soudan, 1899* (Cd 95, 1900), 6; *Memorandum by Mr. Dawkins on the Conditions of the Dongola Province*, n.d. [March 1897] FO407/142/80, NAUK; E. B. Wilkinson, "Annual Report, Department of Agriculture and Forests, 1912," in *Reports on the Finances, Administration and Conditions of the Sudan, 1912*, vol. 2 (1912), 155, SAD.

[62] Quoted in *Reports by His Majesty's Agent and Consul-General on the Finances, Administration, and Conditions of Egypt and the Soudan*, 62.

[63] Cromer to Wingate, 1 January 1906, SAD278/1/2; Wingate to Kitchener, 16 January 1911, SAD300/1/63; Wingate to Kitchener, 24 January 1911, SAD300/1/77; Wingate to Hamilton, 9 December 1908, SAD284/5/8–9.

[64] Rodd to Salisbury, 3 June 1897, FO403/255, NAUK; Cromer to Salisbury, 20 January 1898, FO 403/274, NAUK.

[65] Aboussamad H. Ahmad, "Trading in Slaves in Bela-Shangul and Gumuz, Ethiopia: Border Enclaves in History, 1897–1938," *Journal of African History* 40, no. 3 (November 1999): 433.

in some regions shrank by as much as 90 per cent.[66] Fear of raiding and counter-raiding induced local populations to cluster in larger settlements, a practice that further contributed to the decline in the extent of cultivated land and hence low crop yields.[67] Slaves from the south and west frequently ended up in the households of state-connected elites or passed through the hands of a number of merchants en route to the coast, where they were purchased by Arabian merchants and shipped across the Red Sea.[68] In contrast to their Sudanese counterparts, few small-scale Ethiopian cultivators owned slaves, and slave labour was rarely used when cultivation revived after the famine.[69] The economic importance of the Ethiopian slave trade led French officials in the Côte Française des Somalis to adopt a series of policies designed to divert the maritime trade in Ethiopian slaves, and with it the trade in all other goods, away from ports in neighbouring Eritrea and British Somaliland and towards their territorial ports. French officials, recognizing that the interconnected trade in arms, ammunition, and slaves at the coast was key to lubricating trade generally between the Red Sea and Ethiopia, tacitly condoned this traffic. Officially, French officials were required to free any slave that crossed into French territory, but this was rarely done. Until the late 1920s, French control did not extend beyond the port of Djibouti, and local elites in the rest of the territory actively participated in the slave trade.[70] Traders, aware of the limits of French authority, simply diverted the slave trade away from Djibouti to other natural harbours and ports along the coast.[71]

Unlike their British, French, and Ethiopian counterparts, Italian officials implemented policies in Eritrea designed to end both slavery and the slave trade. They were less concerned about profiting from trade or about reconstructing local agricultural systems than about developing their Red

[66] Lord Noel-Buxton, "Slavery in Abyssinia," *International Affairs* 11, no. 4 (July 1932): 517; Pankhurst, *Economic History of Ethiopia*, 111.

[67] Rennell Rodd to Salisbury, 14 May 1897, FO 403/255, NAUK. This process was noticed as early as 1897.

[68] Philip Zaphiro, *Memorandum on the Slave Traffic between Abyssinia and the Coast of Arabia* (November 1929), IOR/R/20/1/1560, BL.

[69] See James McCann, *From Poverty to Famine in Northeast Ethiopia: A Rural History 1900–1935* (Philadelphia: University of Pennsylvania Press, 1987), 80–81. Cultivation in the grain producing regions in northern Ethiopia continued to be limited by insufficient animal labor resulting from repeated outbreaks of rinderpest.

[70] Philippe Oberlé and Pierre Hugot, *Histoire de Djibouti: des origines à la République* (Paris: Présence Africaine, 1985), 103–05.

[71] J. B. Eustace, Senior Naval Officer, Aden Division, to the Commander-in-Chief, East Indies, 1 September 1905, IOR/R/20/A/1300, BL.

Sea colony into a site for white settlement. Local officials were guided by metropolitan politicians who recognized Eritrea as the solution to the joint problems of underdevelopment and overpopulation in southern Italy. As a result, they sought to induce poor Italian peasants to emigrate to the Eritrean plateau, where local Italian officials seized and reserved 300,000 hectares of vacant farmland for potential emigrants.[72] However, they failed to understand that the region had only recently become depopulated as a result of famine, disease, and war. Since Italian officials did not see slavery as a means of stabilizing their own power, they did not need to go back on their international commitments to combat the slave trade, and, at the beginning of the twentieth century, they ended slave raiding[73] and publicly liberated all slaves in Eritrea.[74] In addition, despite repeated warnings from the Italian Commercial Agent in northern Ethiopia that such policies would divert trade away from Eritrea towards the Anglo-Egyptian Sudan and the Côte Française des Somalis,[75] officials granted freedom and protection to refugee slaves from neighbouring countries. A total of 393 were thus liberated between 1903 and 1913.[76] Initially, officials required these runaway slaves to work as bonded servants for state-connected elites or indigenous soldiers.[77] However, Italian officials ended this policy in 1904, and from then onwards liberated runaway slaves were permitted to remain independent.[78] Nonetheless, these efforts were limited in scope. Italian officials were not willing to commit the manpower and other resources necessary to truly police their colony. The thinness of the Italian colonial administration on the ground limited its surveillance powers and created easy opportunities for illicit trade, including the trade in slaves, to continue.

[72] Istituto Agricolo Coloniale Italiano. *L'Economia Eritrea: nel cinquatennio dell'occupazione di Assab (1882–1932)* (Florence: Istituto Agricolo Coloniale Italiano, 1932), 7–8. This plan was formulated by Leopoldo Franchetti in 1889 at the behest of Prime Minister Francesco Crispi. On 1 July 1890, the Italian government formally adopted this policy.

[73] Alexander Naty, "Environment, Society and the State in Western Eritrea," *Africa: Journal of the International African Institute* 72, no. 4 (2002): 574.

[74] Residente del Sahel to Governor of Eritrea, 26 June 1905, PACCO454, Archivio Eritrea of the Ministero degli Affari Esteri, Rome (hereafter AEMAE).

[75] L'Agente Italiano in Tigré to Governor of Eritrea, 31 August 1913, PACCO580, AEMAE.

[76] *Elenco degli schiavi liberati dalla autoriza della colonia dal 1905 al 1913*, PACCO193, AEMAE.

[77] R. Commissario Civile, Nota, 18 September 1904, PACCO1063, AEMAE.

[78] *Norme per la liberazione degli schiavi provenanti da oltre confine*, n.d. [1904] PACCO64, AEMAE.

Conclusion

By the second decade of the twentieth century, the reconstruction of the SRSR slave system was well under way. Slave raiding had resumed in southern Sudan, as well as in western and southern Ethiopia. Regular slave caravans to northern Sudan and Red Sea ports had been resumed, slave plantations in Northern Nilotic Sudan had been re-established, and African slaves were again being transported across the sea in large numbers to be sold in Arabian markets. Participating in the reconstruction of the slave system was the way in which many segments of local society recovered from the devastation of the 1888–1892 famine. It was also the way in which expanding imperial states sought to establish and consolidate their power on the African littoral of the SRSR. French, British, and Ethiopian officials simultaneously implemented state-building programmes that tied political power to the rapid revival of the slave trade. Though Italian officials more actively respected their international commitments to abolition and emancipation, their impact was limited. As a result, European colonial officials, Ethiopian state leaders, merchants, soldiers, private militia members, pastoralists, cultivators, and urban elites all worked together to recover from the famine by reconstructing the SRSR slave system.

Bibliography

Primary Sources

Archives de l'Académie Académie Malgache, Antananarivo, Madagascar

Raombana, "Histoires".

Archives Départementales de La Réunion, Réunion, France (ADR)

"An 12e, Isle de La Réunion, Quartier Ste Marie, Déclaration de M. Joseph Lory," L 169.

"An 13e [1804], Isle de La Réunion, Quartier Ste Marie, Déclaration de Routier, habitant," L 168.

"An [1805]. Isle de La Réunion, Quartier de Ste Suzanne, Déclaration de Cyrille Routier," L183.2.

Cessions d'engagés, impôts, 12M69.

Furcy, baptismal record, 2 MIEC11 [J38].

Furcy to Louis Gilbert Boucher, Port Louis, Ile Maurice, 15 May 1826, 1Jp2007-1.

"Grand Recensement, R. Recensement des Sieurs Routier Père, et Fils, habitants de la paroisse Sainte Marie pour l'année 1780," 71C.

G. Campbell (ed.), *Bondage and the Environment in the Indian Ocean World*, Palgrave Series in Indian Ocean World Studies,
https://doi.org/10.1007/978-3-319-70028-1

"Isle Bonaparte, Année 1808, Quartier Saint Denis, Paroisse, idem, Recensement de la Dame Ursule Desblottières Ve Routier," L 152/2.
"Isle Bonaparte, Année 1810, Quartier Saint Denis, Paroisse, idem, Recensement de Joseph Lory," L 156/2.
"Isle Bourbon, Année 1813, Quartier [blank], Paroisse, Saint Denis, Recensement de Monsieur Cyrille Routier," L 163–164.
"Isle Bourbon, Année 1814, Quartier de Saint Denis, Déclaration de M. Joseph Lory," L 165.
"Isle Bourbon, Année 1814, Quartier de Saint Denis, Recensement de Monsieur Cyrille Routier, habitant de Sainte Suzanne," L 166/2.
Registre Paroissial de Saint-Denis, 1786, 2 MIEC11 [J38].
"Recensement de dame Vve Routier et des héritiers de feu Sr. [Routier] pour l'année 1787," 74C.
"Saint Denis, Recensement de la Citoyenne Veuve Routier, pour l'année 4me," L 142/1.
Statistiques du travail, 10M9.
Transferts d'engagés, taxes, 12M69.

Archives Nationales d'Outre-Mer, Aix-en-Provence, France (ANOM)

"Commission d'enquête de 1877," c 379 d 3213.
"Coup de vent du 5 mars 1832," Rapport au ministre du 13 mars 1832, FM SG/Reu c 124 d 931.
"Coup de vent du 8 et 9 mars 1845," Rapport au ministre du 31 mars 1845, FM SG/Reu c 124 d 934.
"Coup de vent du 25 janvier 1847," FM SG/Reu c 124 d 935.
"Cyclone du 12–13 mars 1868," FM SG/Reu c 124 d 945.
"Cyclone du 7 et 8 janvier 1873," FM SG/Reu c 124 d 947.
"Cyclones des 11 et 26 février 1860, inondation du 11 février," FM SG/Reu c 124 d 941.
"Enquête réalisée sur un domaine 'ravine des figues," FM SG/Reu c 379 d 3211.
"Eruption volcanique du 19 mars 1860," Rapport de Hugoulin, pharmacien de la Marine, au ministre des colonies, FM SG/Reu c 124 d 940.
"Etablissement du baril," C 379 d 3210.
Etat nominative des passagers embarqués pour les colonies, pendant les années 1769–1780, Port Lorient, COL F/5B/51.
"Inondation du 4 janvier 1844," Rapport du gouverneur du 6 janvier 1845, FM SG/Reu c 124 d 932.

"Ouragan du premier mars 1850," FM SG/Reu 124 d 937.
"Ouragan du 16 au 17 janvier 1858," Le gouverneur au ministre, 3 février 1858, FM SG/Reu c 124 d 939.
"Ouragan du 27 février 1861," FM SG/Reu c 124 d 942.
"Ouragans de 1863," FM SG/Reu c 124 d 943.
Personnel colonial ancien: Routier, Charles, major des milices à Bourbon 1746/1779, COL E 358 bis.
"Protestations anglaises pour les traitements des coolies. Mai 1866," c 319 d 3317.
"Rapport des syndics des immigrés du 15 juin 1858, après visite de 107 établissements, situation sur le premier trimestre 1858," FM SG/Reu, c 384 d 3348; c382 d 3321.
"Raz de marée de juillet 1853," FM SG/Reu c 124 d 938.
Recensement, Ile Bourbon, 1776. Col. G/1/479.
"Réunion, tableau de l'immigration africaine à la Réunion de 1848 à 1869," FM SG/Reu C 454, d 5042–5074.
"Sinistre de 1872 cyclone du 17 février," FM SG/Reu c.124 d. 946.

Archivio Eritrea of the Ministero degli Affari Esteri, Rome, Italy

L'Agente Italiano in Tigre' to Governor of Eritrea, 31 August 1913, PACCO580.
Elenco degli Schiavi Liberati dalla Autoriza della Colonia dal 1905 al 1913, PACCO193.
Norme per la Liberazione degli Schiavi Provenanti da Oltre Confine, n.d. [1904], PACCO64.
R. Commissario Civile. Nota., 18 September 1904, PACCO1063.
Residente del Sahel to Governor of Eritrea, 26 June 1905, PACCO454.

Archivo General de Indias, Filipinas, Seville, Spain

"No. 7 Governor General a Senor Secretario de Estado, 4 June 1896."

Archivo Histórico Nacional de España, Madrid, Spain

"Juan Cencilli a Senor conde de Aranda," 16 April 1774, Estado 2845, caja 2.

British Library, London, UK (BL)

Administration Report of the Somali Coast for the Year 1891–92, IOR/V/10/1995.

Admiralty to Foreign Office, 14 December 1903, IOR/R/20/A/1300.

Anon. 1790. "Mémoire historique et politique sur l'Isle de Madagascar," Add. 18126.

Coghlan, W.M. (in charge Muscat, Zanzibar Commission), letter dated 1st November 1860, No. 14, OIOC, IOR/R/15/1/171/63-65.

Cordeaux, H.E.S. 1903. *Somaliland Protectorate Revenue Cutters. Standing Orders*, 9 March, IOR/L/PS/10/32.

Despatch from the Honorable the Court of Directors, 1st March No 1 of 1854, OIOC, IOR/R/15/1/143/367.

Dickson, E.B.C. 1931. Captain, Senior Officer Red Sea Sloops, Note, 2 (March), IOR/L/PS/12/4094.

Dumaine, "Voyage à la côte de l'ouest, autrement dite pays des Séclaves," January 1793, Add. 18128.

Extract from a Report by the Italian Consul General at Aden, 16 April 1891, IOR/R/20/A/1174.

Ethersey to Jones, Bassadore, 24 July 1854, OIOC, IOR/R/15/1/157/227.

Eustace, J.B. 1905. Senior Naval Officer, Aden Division, to the Commander-in-Chief, East Indies, 1 September, IOR/R/20/A/1300.

Geographical Letters from and Despatches to India, Madras and Bombay, 1867–1879 IOR/L/E/3/205.

Goodfellow, W.R. 1869. Memorandum of Number of Slaves Landed and Liberated at Aden, and How Disposed of, Aden Residency, 13 July, OIOC, IOR/L/P&S/18/B84/69.

Government of Bengal, Proceedings A. Financial Department Circular V. Industry and Science Head no. 3 "Meteorology" Calcutta 1876a, IOR/P/894.

Government of Bengal, Proceedings A. Home Department Home Affairs Richard Temple, "Cyclone and Storm-Wave in Backergunge and Noacolly," Minute by the Lieutenant-Governor of Bengal, 21 November 1876b, Mss Eur/F/86/148.

Hennell to Porter, Camp near Bushire, 26 June 1851, OIOC, IOR/R/15/1/127/20.

Hennell to Malet, Bushire, 30 October 1851, OIOC, IOR/R/15/1/127/31.

Merewhether to Secretary to the Bombay Government, Aden, 17 May 1865, OIOC, IOR/L/P&S/9/42/352.

Montrion to Clarke, Aden, 3 May 1854, OIOC, IOR/R/15/1/143/320-321.

Moollah Houssein to Hennell, s.l., 25 Rujub/6 June 1850, IOR/R/15/1/123/19-20.

Moollah Houssein to Hennell, s.l., 17 May 1851, OIOC, IOR/R/15/1/127/18-19.

Forrest, Thomas. 1779. The Memorial of Thomas Forrest to the Court of Directors of the East India Company, 29 June, 21 July, East India Company Court Minutes, vol. 88, F. 188.

Houssein, Moollah to Hennell, s.l., 13 June 1851, OIOC, IOR/R/15/1/127/23.

Hugon, "Aperçu de mon dernier voyage à ancova de l'an 1808," Add. 18137.

Jones to Anderson, Bushire, 26 May 1856, OIOC, IOR/R/15/1/157/193-197.
Jones to Anderson, Bushire, 28 August 1856, OIOC, IOR/R/15/1/157/208.
Jones to Anderson, Bushire, 25 April 1857, OIOC, IOR/R/15/1/168/34-35.
Jones to Jenkins, Amulgavine Roads, 30 April 1859, OIOC, IOR/R/15/1/168/46-47.
Jones to Forbes, Bushire, 2 September 1861, OIOC, IOR/R/15/1/168/109.
Jones to Hay, 20 October 1885, FO407/67/165.
Kemball to Malet, Bushire, 12 November 1852, OIOC, IOR/R/15/1/130/292.
Kemball to Thomson, Bushire, 12 January 1854, OIOC, IOR/R/15/1/143/304.
Mayeur. 1777. "Voyage au pays d'ancove, autrement dit des hovas ou Amboilamba dans l'intérieur des terres, Isle de Madagascar," Add. 18128.
———. 1785. "Voyage au pays d'ancove, par le pays d'ancaye autrement dit des Baizangouzangoux," Add. 18128.
Petition Submitted by Some Refugees of Dongola Province, at Assuan, n.d. [August 1885], FO407/66/113.
Playfair to Coghlan, Aden, 7 April 1860, OIOC, IOR/R/20/A1A/285/73-76.
Playfair to Russell, Zanzibar, 4 January 1864, OIOC, IOR/L/P&S/9/41/240.
Playfair to Haveloch, Zanzibar, 3 February 1864, OIOC, IOR/L/P&S/9/41/245.
Rennell, James. 1762–1763. *Journal of a Voyage to the Sooloo Islands and the Northwest Coast of Borneo, from and to Madras with Descriptions of the Islands, 1762–1763*, Add. 19299.
Report on the Slave Trade in the Persian Gulf extending from January 1 1852 to June 30 1858, compiled by H. Disbrowe, OIOC (Oriental and India Office Collection), IOR/R/15/1/171/23.
Rigby to Anderson, Zanzibar, 13 September 1858, OIOC, IOR/R/15/1/171/101-102.
Secretary to the Admiralty to Lister, 8 June 1881, FO84/1597.
Stace to Hardinge, 2 September 1891, IOR/R/20/A/1174.
Taylor to Hennell, Basra, 16 September 1851, OIOC, IOR/R/15/1/127/56.
Tornielli to Salisbury, 12 February 1890, IOR/R/20/A/1173.
Yacoob, Hajee to Jones, s.l., 2 Moohurrum 1276, OIOC, IOR/R/15/1/171/150.
Zaphiro, Philip. 1929. *Memorandum on the Slave Traffic between Abyssinia and the Coast of Arabia* (November), IOR/R/20/1/1560.

Mahārāshtra State Archives, Mumbai, India (MAHA)

Crawfold to Anderson, Bombay, 5 January 1856, PD/1856/93/335/207-212.
Hennell to Willoughby, Bushire, 24 September 1837, PD/1837/78/854/393-400.
Playfair to Havelock, s.l., 1 May 1864, PD/1864/54/942/12-13.
Rigby to Anderson, Zanzibar, 28 March 1860, PD/1860/159/830/194-195.
Rigby to Anderson, Zanzibar, 30 November 1859, PD/1860/159/12/7.
Souter to Secretary to the Bombay Government, Bombay, 6 March 1867, PD/1867/126/509/55-56.

National Archives, Kew, London, UK (NAUK)

Archer to the Secretary of State for the Colonies, 19 August 1913, CO535/31.

Baring to Granville, 9 June 1885, FO407/65/343.

Barnham to Cromer, 14 February 1894, FO407/126/77.

British Somaliland Protectorate, *Intelligence Report No 8*, August 1908, CO535/12.

Clarke to Salisbury, 26 August 1889, FO407/89/108.

Commissioner of the Somaliland Protectorate to Secretary of State for the Colonies, 21 February 1910, CO535/18.

Crawfold to Walker, H.M.S. *Sidon* at Mauritius, 30 June 1861, ADM123/179.

Cromer to Salisbury, 20 January 1898, FO403/274.

Dawkins. 1897. *Memorandum by Mr. Dawkins on the Conditions of the Dongola Province*, n.d. [March 1897], FO407/142/80.

Egerton to Granville, 26 August 1884, FO407/62/337.

James Hastie, "Diary" 1817, 187, CO167-/34.

Le Sage, "Mission to Madagascar," 1816, CO167/34.

Pelly to Stewart, Zanzibar, 10 July 1862, FO800/234/95.

Playfair to Russell, Zanzibar, 4 January 1864, FO84/1224/181-182.

Rennell Rodd to Salisbury, 14 May 1897, FO403/255.

Rodd to Salisbury, 3 June 1897 and 14 June 1897, FO403/255.

Unfoliated Khosrew Effendi to Col. Barnett, Kafr Mugar, 6 Decembre 1842, PRO/FO141.8.

London Missionary Society Archives, School of Oriental & African Studies, London

Freeman to Ellis, Tamatave, 3 Oct 1833 and Vohitsara, 11 Oct 1833, B4.F4.JC (Madagascar Incoming Letters).

Nationaal Archief, The Hague (NA)

Nationaal Archief, The Hague (hereafter NA), VOC 4073, p. 584, Q. 5.

National Archives, Mauritius

Papiers doyen, 2 October 1806, Archives de la Marine, Île de France, 16105-07.

Peabody and Essex Museum, Salem, USA

F. R. Webb to E. D. Ropes, Zanzibar, 13 April 1869, Philips Library, MH201/Box 3.

Sudan Archive, Durham University, UK (SAD)

Cromer to Wingate, 1 January 1906, 278/1/2.
Intelligence Department, Egyptian Army. *Staff Diary and Intelligence Report, Suakin,* no. 98 (24 December–6 January 1889).
———, *Frontier Field Force,* no. 229 (11–17 May 1890).
———, *Frontier Field Force,* no. 233 (8–14 June 1890).
———, *Frontier Field Force,* no. 244 (28 September–11 October 1890).
———, *Frontier Field Force,* no. 254 (29 March–11 April 1891).
———, *Eastern Sudan,* no. 4 (15–28 April 1891).
Intelligence Report, Egypt, no. 8 (December 1892).
Intelligence Report, Egypt, no. 21 (December 1893).
Jackson, H.W. 1908. "Annual Report, Dongola Province, 1908," in *Reports on the Finances, Administration and Conditions of the Sudan, 1908.*
Note – The Price of Grain in Omdurman and Eastern Sudan, 269/2/47-48.
Slatin to Wingate, 2 Dec 1912, 183/3/16-17.
Stanton, E.A. 1908. "Annual Report, Khartoum Province, 1908," in *Reports on the Finances, Administration and Conditions of the Sudan, 1908.*
Talbot to Wingate, 10 February 1899, 269/2/35-36.
Townsend, C.H. 1910. "Annual Report, Berber Province, 1910," in *Reports on the Finances, Administration and Conditions of the Sudan. 1910.*
Wilkinson, E.B. 1912. "Annual Report, Department of Agriculture and Forests, 1912," in *Reports on the Finances, Administration and Conditions of the Sudan, 1912,* vol. 2.
Wingate, R. 1904. "Memorandum by the Governor-General, 1904," in *Reports on the Finances, Administration and Conditions of the Sudan, 1904,* vol. 2.
Wingate to Hamilton, 9 December 1908, 284/5/8-9.
Wingate to Kitchener, 16 January 1911, 300/1/63 and 24 January 1911, 300/1/77.

Western Cape Regional Archives, Cape Town, South Africa (CA)

Marsch route der rebellen onder commando van Louis, CJ 515, C.
Resolutions of the Council of Policy, 15 July–19 August 1760, CA, C 138.
Stellenbosch Dagregister, 2 July 1748, CA, C 653.
Testimony of Isaac van Boegies, 4 August 1760, CA, CJ 373.
Testimony of Manika van Bengal, 9 January 1749, CA, 1/STB 3/8.

Zanzibar National Archives, Zanzibar, Tanzania (ZZBA)

Hamerton to Chief Secretary to the Bombay Government, 25 October 1849, AA12/29/85.

Hamerton to Secretary to the Bombay Government, Zanzibar, 17 September 1855, AA12/29/136-37.
Rigby to Bombay Government, Zanzibar, 14 May 1861, AA12/2/117.
Rigby to Bombay Government, Zanzibar, 5 September 1861, AA12/2/12.

Secondary Sources

Abu-Lughod, J.L. 1989. *Before European Hegemony The World System A.D. 1250–1350*. New York: Oxford University Press.

Adshead, Samuel A.M. 2004. *T'ang China: The Rise of the East in World History*. Basingstoke: Palgrave Macmillan.

Ahmad, Aboussamad H. 1999. Trading in Slaves in Bela-Shangul and Gumuz, Ethiopia: Border Enclaves in History, 1897–1938. *Journal of African History* 40 (3): 433–446.

Ahuja, Ravi. 2003. 'The Bridge Builders': Some Notes on Railways Pilgrimage and 'British Civilizing Mission' in Colonial India. In *Colonialism and Civilizing Mission*, ed. Harold Fischer-Tine and Michael Mann, 89–110. London: Anthem Press.

Aitchison, Charles U. 1909. *A Collection of Treaties, Engagements and Sanads: Relating to India and Neighbouring Countries*. Vol. 7. Calcutta: Office of the Superintendent of Government Printing, India.

al-Sayyid Marsot, Afaf Lutfi. 1982. Muhammad Ali's Internal Politics. In *L'Égypte au XIXe siècle*, ed. Robert Mantran, 153–173. Paris: Éditions du Centre National de la Recherche Scientifique.

al-Tunisi, Muhammad ibn Umar. 1845. *Voyage au Darfour*. Trans. Dr Perron. Paris: Benjamin Duprat.

———. 1851. *Voyage au Ouaday*. Paris: Benjamin Duprat.

Allen, Richard B. 1999. *Slaves, Freedmen, and Indentured Laborers in Colonial Mauritius*. Cambridge: Cambridge University Press.

———. 2001. Licentious and Unbridled Proceedings: The Illegal Slave Trade to Mauritius and the Seychelles During the Early Nineteenth Century. *Journal of African History* 42 (1): 91–116.

———. 2004. The Mascarene Slave Trade and Labour Migration in the Indian Ocean During the Eighteenth and Nineteenth Centuries. In *The Structure of Slavery in Indian Ocean Africa and Asia*, ed. Gwyn Campbell, 33–50. London: Frank Cass.

———. 2006. A Traffic Repugnant to Humanity: Children, the Mascarene Slave Trade and British Abolitionism. *Slavery & Abolition* 27 (2): 219–236.

———. 2008. The Constant Demand of the French: The Mascarene Slave Trade and the Worlds of the Indian Ocean and Atlantic During the Eighteenth and Nineteenth Centuries. *Journal of African History* 49 (1): 43–72.

———. 2010. Satisfying the 'Want for Labouring People': European Slave Trading in the Indian Ocean, 1500–1850. *Journal of World History* 21 (1): 45–73.

———. 2014. *European Slave Trading in the Indian Ocean 1500–1850*. Athens, OH: Ohio University Press.

Allibert, Claude. 2011. Le *Kitāb-i bahriyye* de Pīrī Re'īs et l'océan Indien dans le contexte vohémarien: analyse des versions de 1521 et 1526. *Études Océan Indien* 46–47: 197–220.

Alpers, Edward A. 1975. *Ivory and Slaves in East Central Africa: Changing Patterns of International Trade to the Later 19th Century*. Berkeley: University of California Press.

———. 2013. Pawnship and Slavery in Nineteenth-Century East Africa. In *Bonded Labour and Debt in the Indian Ocean World*, 31–43. London: Routledge.

Alvarez, Francisco. 1881. *Narrative of the Portuguese Embassy to Abyssinia During the Years 1520–1527 by Father Francisco Alvrez*. Trans. and ed. Lord Stanley of Alderley. London: Hakluyt Society.

Anderson, Philip. 1856. *The English in Western India: Being the Early History of the Factory at Surat, of Bombay, and the Subordinate Factories on the Western Coast, from the Earliest Period until the Commencement of the Eighteenth Century. Drawn from Authentic Works and Original Documents*. London: Smith, Elder and Co.

Anderson, Clare. 2005a. The Bel Ombre Rebellion: Indian Convicts in Mauritius, 1815–53. In *Abolition and its Aftermath in Indian Ocean Africa and Asia*, ed. Gwyn Campbell, 50–65. London: Routledge.

Anderson, Katharine. 2005b. *Predicting the Weather. Victorians and the Science of Meteorology*. Chicago: University of Chicago Press.

Anderson, David M. 2016. The Beginning of Time? Evidence for Catastrophic Drought in Baringo in the Early Nineteenth Century. *Journal of Eastern African Studies* 10 (1): 45–66.

Anon. 1892. Early Notices of Madagascar from the Old Voyagers, Part III: Hamilton's Description of Madagascar, 1744. *Antananarivo Annual and Madagascar Magazine* 16: 497–499.

———. 2015. Red Skies Over Tambora. *Tempo*, April 12, 48–58.

Armstrong, James C. 1983–1984. Madagascar and the Slave Trade in the Seventeenth Century. *Omaly sy Anio* 17–20: 211–233.

Armstrong, James, and Nigel Worden. 1989. The Slaves, 1652–1834. In *The Shaping of South African Society, 1652–1840*, ed. Richard Elphick and Hermann Giliomee, 109–183. Middletown, CT: Wesleyan University Press.

Arnold, David. 1986. Cholera and Colonialism in British India. *Past & Present* 113 (1): 118–151.

Atwell, William S. 2001. Volcanism and Short-Term Climatic Change in East Asian and World History, c.1200–1699. *Journal of World History* 12 (1): 31–40.

Austen, Ralph A. 1979. The Trans-Saharan Slave Trade: A Tentative Census. In *The Uncommon Market: Essays in the Economic History of the Atlantic Slave Trade*, ed. H.A. Gemery and J.S. Hogendorn, 23–75. New York: Academic Press.

———. 1989. The 19th Century Islamic Slave Trade from East Africa (Swahili and Red Sea Coasts): A Tentative Census. In *Economics of the Indian Ocean Slave Trade*, ed. William Gervase Clarence-Smith, 21–44. London: Cass.

Avine, Grégoire. 1961. *Voyages aux isles de France d'Anjouan de Madagascar, de Mosambique, de Zanzibar et de la côte Coromandel*, *Mauritius Archives Publications* 5. Paris: G. Durassié.

Axelson, Eric. 1969. *Portuguese in South-East Africa 1600–1700*. Johannesburg: Witwatersrand University Press.

Aynekulu, E., et al. 2011. Dieback Affects Forest Structure in a Dry Afromontane Forest in Northern Ethiopia. *Journal of Arid Environments* 75 (5): 499–503.

Badri, Babikr. 1969. *The Memoirs of Babikr Bedri*. Trans. George Scott. London: Oxford University Press.

Baer, Gabriel. 1969. *Studies in the Social History of Modern Egypt*. Chicago: University of Chicago Press.

Balce, Guillermo R., and Emmanuel G. Ramos. 1987. Geologic Hazards in the Philippines: A Definition and Overview. In *Geologic Hazards and Disaster Preparedness Systems*, ed. Ann Varley, 1–21. Manila: Philippines Institute of Volcanology and Seismology.

Balfour, Edward. 1885. *The Cyclopædia of India and of Eastern and Southern Asia: Commercial, Industrial and Scientific, Products of the Mineral, Vegetable, and Animal Kingdoms, Useful Arts and Manufactures*. Vol. 3. London: Bernard Quaritch.

Banaji, Dady R. 1933. *Slavery in British India*. Bombay: D.B Taraporevala Sons.

Bankoff, Greg. 2002. *Cultures of Disaster*. London: Routledge Curzon.

Bankoff, Greg, and Joseph Christensen, eds. 2016. *Natural Hazards and Peoples in the Indian Ocean World*. New York: Palgrave Macmillan.

Bankoff, Greg, George Frerks, and Dorothea Hilhorst, eds. 2004. *Mapping Vulnerability Disasters, Development and People*. London: Earthscan.

Baré, Jean-François. 1983. Remarques sur le vocabulaire monarchique sakalava du nord. In *Les souverains de Madagascar: l'histoire royale et ses résurgences contemporaines*, ed. Françoise Raison-Jourde, 154–173. Paris: Karthala.

Barendse, René J. 2002. *The Arabian Seas: The Indian Ocean World of the Seventeenth Century*. Armonk, NY: M. E. Sharpe.

———. 2009. *Arabian Seas, 1700–1763*. Vol. 1. Leiden: Brill.

Barnes, Robert. 1996. *Sea Hunters of Indonesia: Fishers and Weavers of Lamalera*. New York: Oxford University Press.

Barr, Juliana. 2012. Captivity, Gender and Social Control in the Texas-Louisiana Borderlands. In *Major Problems in the History of North American Borderlands*, ed. Pekka Hamalainen and Benjamin H. Johnson, 68–82. Boston: Wadsworth.

Barrantes, Vicente. 1878. *Guerras piraticas de Filipinas contra Mindanaos y Joloanos*. Madrid: Imprenta de Manuel H. Herrandez.

Barreto, Manuel. 1883. Moçambique e Madagascar: informação do estado e conquista dos rios de Cuama vulga e verdadeiramente chamados rios de Ouro. *Boletim da Sociedade de Geographia de Lisboa* 4 (1): 33–59.

Barrett, Timothy H. 1996. *Taoism Under the Tang: Religion and Empire during the Golden Age of Chinese History*. London: Wellsweep.

Barth, Heinrich. 1965. *Travels and Discoveries in North and Central Africa*. Vol. 2 [orig. pub. 1890]. London: Frank Cass.

Basu, Babu Pyari Mohan. 1896. *Survey and Settlement of the Dakhin Shahbazpur Estates in the District of Bakergunge, 1889–1895*. Calcutta: Bengal Secretariat Press.

Bates, Crispin. 1985. Regional Dependence and Rural Development in Central India: The Pivotal Role of Migrant Labour. *Modern Asian Studies* 19 (3): 573–592.

Batou, Jean. 1991. Muhammad-'Ali's Egypt: A Command Economy in the 19th Century? In *Between Development and Underdevelopment: The Precocious Attempts at Industrialization of the Periphery (1800–1870)*, ed. Jean Batou, 181–218. Genevaz: Droz.

Beachey, Raymond W. 1976. *The Slave Trade of Eastern Africa*. New York: Barnes and Noble.

Bellew, Henry Walter. 1885. *The History of Cholera in India From 1862–1881*. London: Trübner.

Benn, Charles D. 2002. *Daily Life in Traditional China: The Tang Dynasty*. Westport, CT: Greenwood.

Berg, Gerald M. 1981. Riziculture and the Founding of Monarchy in Imerina. *Journal of African History* 22 (3): 289–308.

Beveridge, Henry. 1876. *The District of Bakarganj: Its History and Statistics*. London: Trubner.

Birla, Ritu. 2009. *Stages of Capital: Law, Culture and Market Governance in India*. Durham, NC: Duke University Press.

Bjorkelo, Anders. 1989. *Prelude to the Mahdiyya*. Cambridge: Cambridge University Press.

Black, C.C.D. 1877. Backergunj Cyclone of November 1876. In *Letters from India Geographical Department*, 145–146. Calcutta: Bengal Secretariat Press.

Blaikie, Piers, et al. 1994. *At Risk: Natural Hazards, People's Vulnerability and Disasters*. London: Routledge.

Blake. 1838. *Commander Blake to Rear Admiral F. Maitland, 13 August 1838*, Admiralty 125/133-Sulu Piracy; The Illanoons. *Singapore Free Press and Mercantile Advertiser*, December 13.

Bloch, Maurice. 1971. *Placing the Dead: Tombs, Ancestral Villages and Kinship Organisation in Madagascar*. London: Seminar Press.

Blue, A.D. 1970. Chinese Emigration and the Deck Passenger Trade. *Journal of the Hong Kong Branch of the Royal Asiatic Society* 10: 79–93.

Boer, Jelle Zeilinga de, and Donald Theodore Sanders. 2002. *Volcanoes in Human History: The Far-Reaching Effects of Major Eruptions.* Princeton: Princeton University Press.

Boomgaard, Peter. 2001. Crisis Mortality in Seventeenth Century Indonesia. In *Asian Population History*, ed. Ts'ui-jung Liu et al., 193–197. Oxford: Oxford University Press.

———. 2004. Human Capital, Slavery and Low Rates of Economic and Population Growth in Indonesia, 1600–1910. In *The Structure of Slavery in Indian Ocean Africa and Asia*, ed. Gwyn Campbell, 83–96. London: Frank Cass.

Boothby, Richard. 1745. A Brief Discovery or Description of the Most Famous Island of Madagascar, or St. Laurence, in Asia, Near unto the East-Indies. In *in A Collection of Voyages and Travels*, ed. Thomas Osborne, vol. 2, 626–663. London: Thomas Osborne.

Borgaonkar, H.P., et al. 2009. Assessment of Tree-Ring Analysis of High-Elevation *Cedrus deodara* D. Don from Western Himalaya (India) in Relation to Climate and Glacier Fluctuations. *Dendrochronologia* 27 (1): 59–69.

Bose, Sugata. 1986. *Agrarian Bengal: Economy, Social Structure and Politics, 1919–1947.* Cambridge: Cambridge University Press.

———. 1993. *The New Cambridge History of India, Peasant Labour and Colonial Capital: Rural Bengal Since 1770.* Vol. 3, Part 2. Cambridge: Cambridge University Press.

Bourdillon, J.A. 1883. *Report on the Census of Bengal, 1881.* Vol. 1. Calcutta: Bengal Secretariat Press.

Bourquin, Alexandre. 2005. *Histoire des petits-blancs à la Réunion.* Paris: Karthala.

Bowring, John. 1840. *Report on Egypt and Candia.* British Parliamentary Papers, Reports from Commissioners, XI.

Breman, Jan, I. Guérin, and A. Prakash, eds. 2009. *India's Unfree Workorce. Old and New Practices of Labour Bondage.* New Delhi: Oxford University Press.

Brenner, Robert. 1993. *Merchants and Revolution: Commercial Change, Political Conflict, and London's Overseas Traders, 1550–1653.* Princeton, NJ: Princeton University Press.

Bridet, H.G. 1861. *Étude sur les ouragans de l'hémisphère austral: manœuvres à faire pour s'en éloigner et se soustraire aux avaries qu'ils peuvent occasionner, éd.* Saint-Denis la Réunion: Rambosson.

British Parliamentary Papers: Command Papers. Garstin, William, Report on the Soudan. In *Dispatch from Her Majesty's Agent and Consul General at Cairo* (C 9332, 1899).

———. *Report for the Year 1896 on the Trade of Suakin*, Commercial no. 1859 (C8277, 1897).

———. *Reports on the Province of Dongola,* (C8427, 1897).

———. *Reports by His Majesty's Agent and Consul-General on the Finances, Administration, and Conditions of Egypt and the Soudan, 1899* (Cd95, 1900).

———. *Reports by His Majesty's Agent and Consul-General on the Finances, Administration and Condition of Egypt and the Soudan in 1903* (Cd1951, 1904).

British Parliamentary Papers: Slave Trade. Vols. 24–25, 39–41, 46, 48–52. Shannon: Irish University Press.

Brockey, Liam Matthew. 2008. *Portuguese Colonial Cities in the Early Modern World.* Farnham: Ashgate.

Brook, George A., et al. 1999. A High-Resolution Proxy Record of Rainfall and ENSO since AD 1550 from Layering in Stalagmites from Anjohibe Cave, Madagascar. *The Holocene* 9 (6): 695–706.

Bruce-Chwatt, Leonard. 1980. *Essential Malariology.* London: Heinemann.

Bryson, Reid A., and Christine Padoch. 1981. On the Climates of History. In *Climate and History*, ed. Robert I. Rotberg and Theodore K. Rabb, 3–18. Princeton: Princeton University Press.

Buckeridge, Nicholas. 1973. *Journal and Letter Book of Nicholas Buckeridge 1651–1654.* Ed. John R. Jenson. Minneapolis: University of Minnesota Press.

[Bureau of Insular Affairs, War Department]. 1903. *Description of the Philippines Compiled by the Bureau of Insular Affairs, War Department.* Manila: Government Printing Office.

Burke, Peter. 1992. *History and Social Theory.* Oxford: Polity Press.

Burney, David A. 1997. Late Holocene Vegetational Change in Central Madagascar. *Quaternary Research* 28 (3): 130–143.

Burton, Richard F. 1872. *Zanzibar: City, Island, and Coast.* Vol. 1. London: Tinsley Brothers.

Buxton, Thomas Fowell. 1967. *The African Slave Trade and Its Remedy* [orig. pub. 1840]. London: Cass.

Cairns, Margaret. 1980. The Smuts Family Murders: 14.7.1760. *Cabo* 2 (3): 13–16.

Callet, [François] 1974. *Histoire des rois: Tantaran ny Andriana.* Trans. George S. Chapus and Emmanuel Ratsimba. Tananarive: Éditions de la Librairie de Madagascar.

Campbell, Perisa. 1971. *Chinese Coolie Emigration to Countries within the British Empire: To Countries within the British Empire.* London: Routledge.

Campbell, Gwyn. 1981. Madagascar and the Slave Trade, 1810–1895. *Journal of African History* 22 (2): 203–227.

———. 1988. Slavery and Fanompoana: The Structure of Forced Labour in Imerina (Madagascar), 1790–1861. *Journal of African History* 2 (3): 463–486.

———. 1991. The State and Pre-colonial Demographic History: The Case of Nineteenth Century Madagascar. *Journal of African History* 31 (3): 415–445.

———. 2004. Introduction: Slavery and Other Forms of Unfree Labour in the Indian Ocean World. In *The Structure of Slavery in Indian Ocean Africa and Asia*, ed. Gwyn Campbell, vii–xxxii. London: Frank Cass.
———. 2005a. *An Economic History of Imperial Madagascar, 1750–1895: The Rise and Fall of an Island Empire*. Cambridge: Cambridge University Press.
———, ed. 2005b. *Abolition and its Aftermath in Indian Ocean Africa and Asia*. London: Routledge.
———. 2008. Slave Trades and the Indian Ocean World. In *India in Africa, Africa in India: Indian Ocean Cosmopolitanisms*, ed. John C. Hawley, 17–51. Bloomington: Indiana University Press.
———. 2013a. Debt and Slavery in Imperial Madagascar, 1790–1861. In *Bonded Labour and Debt in the Indian Ocean World*, ed. Gwyn Campbell and Alessandro Stanziani, 45–55. London: Routledge.
———. 2013b. Servitude and the Changing Face of Demand for Labor in the Indian Ocean World, c.1800–1900. In *Slavery and the Slave Trades in the Indian Ocean World: Global Connections and Disconnections*, ed. Bob Harms, Bernard Freamon, and David W. Blight, 23–44. New Haven: Yale University Press.
———. 2016. Africa and the Early Indian Ocean World Exchange System in the Context of Human-Environment Interaction. In *Africa and the Early Indian Ocean World Exchange to circa 1300*, ed. Gwyn Campbell, 1–24. London: Palgrave Macmillan.
Campbell, Gwyn, and Alessandro Stanziani. 2013. *Bonded Labour and Debt in the Indian Ocean World*. London: Routledge.
Cannadine, David. 2001. *Ornamentalism: How the British Saw their Empire*. Oxford: Oxford University Press.
Cannon, Terry. 1994. Vulnerability Analysis and the Explanation of 'Natural Disasters'. In *Disasters, Development and Environment*, ed. Ann Varley, 13–30. Chichester: John Wiley and Sons.
Cao, Shuji, Yushang Li, and Bin Yang. 2012. Mt. Tambora, Climatic Changes, and China's Decline in the Nineteenth Century. *Journal of World History* 23 (3): 587–607.
Carey, Mark. 2012. Climate and History: A Critical Review of Historical Climatology and Climate Change Historiography. *WIREs Climate Change* 3 (3): 233–249.
Carter, Marina. 1995. *Servants, Sidars and Settlers. Indians in Mauritius, 1834–1874*. Delhi: Oxford University Press.
Carter, Marina, and Hubert Gerbeau. 1988. Covert Slaves and Coveted Coolies in the Early 19th Century Mascareignes. *Slavery & Abolition* 9 (3): 194–208.
Castañeda, Isla S., Josef P. Werne, and Thomas C. Johnson. 2007. Wet and Arid Phases in the Southeast African Tropics since the Last Glacial Maximum. *Geology* 35 (9): 823–826.

Chapus, G.S., and G. Mondain. 1951–52. Un chapitre inconnu: des rapports de Maurice et de Madagascar. *Bulletin de l'Académie Malgache* 30: 3–8.
Chaudhuri, K.N. 1985. *Trade and Civilisation in the Indian Ocean: An Economic History from the Rise of Islam to 1750.* Cambridge: Cambridge University Press.
———. 1992. *Asia Before Europe: Economy and Civilisation of the Indian Ocean from the Rise of Islam to 1750.* Cambridge: Cambridge University Press.
Cheke, Anthony S., and Julian P. Hume. 2010. *Lost Land of the Dodo: The Ecological History of Mauritius, Réunion and Rodrigues.* London: A & C Black.
Clancey, Greg. 2004. *Symposium on Natural Disaster in Asian History, Culture and Memory.* Unpublished draft, National University of Singapore, December.
Clarence-Smith, William Gervase. 2005. Islam and the Abolition of the Slave Trade in the Indian Ocean. In *Abolition and its Aftermath in Indian Ocean Africa and Asia*, ed. Gwyn Campbell, 137–149. London: Routledge.
———. 2013. Debt and Coercion of Labour in Islamic Legal Tradition. In *Bonded Labour and Debt in the Indian Ocean World*, ed. Gwyn Campbell and Alessandro Stanziani, 21–30. London: Pickering & Chatto.
Clot-Bey, Antoine Barthélemy. 1840. *De la peste observée en Egypte: recherches et considerations sur cette maladie.* Paris: Fortin, Masson. Cohn, Bernard S. 2000. Representing Authority of Tradition in Victorian India. In *Invention of Tradition*, ed. E.J. Hobsbawm and Terrence Ranger, 165–210. Cambridge: Cambridge University Press.
Colomb, Philip H. 1870. *Slave-Catching in the Indian Ocean: A Record of Naval Experiences.* New York: Longmans, Green.
Comyn, Tomas de. 1969. *State of the Philippines in 1810, Being a Historical Statistical and Descriptive Account of the Interesting Portion on the Indian Archipelago.* Manila: Filipiniana Book Guild.
Congbin, Fu, and J. Fletcher. 1988. Large Signals of Climatic Variation Over the Ocean in the Asian Monsoon Region. *Advances in Atmospheric Sciences* 5 (4): 389–404.
Copland, Samuel. 1822. *A History of the Island of Madagascar.* London: Burton & Smith.
Cordell, Denis. 1977a. *Dar al-Kuti and the Last Years of the Trans-Saharan Slave Trade.* Madison, WI: University of Wisconsin Press.
———. 1977b. Eastern Libya, Wadai and the Sanusiya: A Tariqa and a Trade Route. *Journal of African History* 18 (1): 21–36.
Cordell, Dennis D., and Joel W. Gregory, eds. 1987. *African Population and Capitalism: Historical Perspectives.* Boulder: Westview Press.
Couland, Daniel. 1973. *Les Zafimaniry, un groupe ethnique de Madagascar à la poursuite de la forêt.* Tananarive: Imprimerie Fanontam-Boky-Malagasy.
Cox, John D. 2002. *Storm Watchers: The Turbulent History of Weather Prediction from Franklin's Kite to El Niño.* Hoboken, NJ: Wiley.

Crosby, Alfred. 1986. *Ecological Imperialism: The Biological Expansion of Europe*. New York: Cambridge University Press.

Cruz, Jerome B. 1987. Volcanism and Associated Hazards in the Philippine Setting. In *Geologic Hazards and Disaster Preparedness Systems*, 22–37. Quezon City: Philippine Institute of Volcanology and Seismology.

Cui, Jianxin, and Hong Chang. 2013. The Possible Climate Impact on the Collapse of An Ancient Urban City in Mu Us Desert, China. *Regional Environmental Change* 13 (2): 353–364.

Cuno, Kenneth M. 2010. African Slaves in Nineteenth-Century Rural Egypt: A Preliminary Assessment. In *Race and Slavery in the Middle East: Histories of Trans-Saharan Africans in Nineteenth-Century Egypt, Sudan, and the Ottoman Mediterranean*, ed. Terence Walz and Kenneth M. Curto, 77–98. Cairo: American University in Cairo Press.

d'Arrigo, Rosanne, and Rob Wilson. 2006. On the Asian Expression of the PDO. *International Journal of Climatology* 26 (12): 1607–1617.

Dampier, William. 1937. *A New Voyage Round the World* [orig. pub. 1697]. London: A & C Black.

Dandamaev, Muhammad A. 2009. *Slavery in Babylonia*. DeKalb, IL: Northern Illinois University Press.

Danvers, Frederick Charles. 1896. *List of Marine Records of the Late East India Company, and of Subsequent Date, Presented in the Record Department of the India Office, London*. London: Eyre and Spottiswoode.

Darbyshire, Iain, Henry Lamb, and Mohammed Umer. 2003. Forest Clearance and Regrowth in Northern Ethiopia during the Last 3000 Years. *The Holocene* 13 (4): 537–546.

Das, Pallavi. 2011. Colonialism and the Environment in India: Railways and Deforestation in 19th Century Punjab. *Journal of Asian and African Studies* 46 (1): 38–53.

Dasgupta, Keya. 2005. Plantation Labour in the Brahmaputra Valley: Regional Enclaves in a Colonial Context. In *Abolition and its Aftermath in Indian Ocean Africa and Asia*, ed. Gwyn Campbell, 169–179. London: Routledge.

Datoo, B.A. 1970. Misconceptions About the Use of Monsoons by Dhows in East African Waters. *East African Geographical Review* 8: 1–10.

Datta, Rajat. 1996. Crises and Survival: Ecology, Subsistence and Coping in Eighteenth-Century Bengal. *Calcutta Historical Journal* 18 (1): 1–34.

Davis, Mike. 2001. *Late Victorian Holocausts: El Niño Famines and the Making of the Third World*. London: Verso.

Dawson, Marc H. 1981. Disease and Population Decline of the Kikuyu of Kenya, 1890–1925. In *African Historical Demography*, ed. Christopher Fyfe and David McMaster, vol. 2, 121–138. Edinburgh: University of Edinburgh.

Decary, Raymond. 1957. Poids et mesures d'autrefois. *Bulletin de l'Académie Malgache* 35: 125–131.

———. 1966. *Coutumes guerrières et organisation militaire chez les anciens Malgaches*. Vol. 2. Paris: Éditions Maritimes et d'Outremer.

Defensor, Benjamin. 1960. What Ails the Weather Bureau? *Manila Chronicle*, 18 July, 1.

Delaye, Karine. 2004. Slavery and Colonial Representations in Indochina from the Second Half of the Nineteenth to the Early Twentieth Centuries. In *The Structure of Slavery in Indian Ocean Africa and Asia*, ed. Gwyn Campbell, 129–142. London: Frank Cass.

Delivré, Alain. 1974. *L'histoire des rois d'Imerina: interprétation d'une tradition orale*. Paris: Klincksieck.

Deschamps, Hubert. 1972. *Histoire de Madagascar*. Paris: Berger-Levrault.

———. 1976. Tradition and Change in Madagascar, 1790–1870. In *Cambridge History of Africa*, ed. John E. Flint, vol. 5, 393–417. Cambridge: Cambridge University Press.

Deutsch, Jan-George. 2007. Notes on the Rise of Slavery and Social Change in Unyamwezi, c. 1860–1900. In *Slavery in the Great Lake Region of East Africa*, ed. Henri Médard and Shane Doyle, 76–110. Oxford: James Currey.

Devereux, William Cope. 1869. *A Cruise in the 'Gorgon',: or, Eighteen Months on H.M.S. 'Gorgon', Engaged in the Suppression of the Slave Trade on the East Coast of Africa. Including a Trip Up the Zambesi with Dr. Livingstone*. London: Bell and Daldy.

Dewar, Robert E., and Henry T. Wright. 1993. The Culture History of Madagascar. *Journal of World Prehistory* 7 (4): 417–466.

Dez, Jacques. 1962. Développement économique et tradition à Madagascar. *Cahiers de l'institut de science économique appliquée* 4 (129): 79–108.

———. 1965. Le fokonolona malgache: institution désuète ou cellule de développement? *Cahiers de l'ISEA: économie et sociologie rurales* 8 (160): 189–252.

———. 1970. Éléments pour une étude de l'économie agro-sylvo-pastorale de l'Imerina ancienne. *Terre Malgache* 8: 9–60.

Dixon, Gary. 2011. *Slavery in the West: The Untold Story of the Slavery of Native Americans in the West*. Bloomington, IN: Xlibris Corporation.

Douin, Georges. 1938. *Histoire du règne du Khédive Ismail, tome III L'Empire Africain*. Cairo: Société Royale de Géographie d'Égypte.

Dove, Heinrich Wilhelm. 1886. *The Law of Storms Considered in Connection with the Ordinary Movement of the Atmosphere*. London: Longman.

Drompp, Michael R. 2005. Imperial State Formation in Inner Asia: The Early Tuikic Empires (6th to 9th Centuries). *Acta Orientalia Academiae Scientiarum Hung* 58 (1): 101–111.

Drury, Robert. 1890. *Madagascar; or, Robert Drury's Journal, During Fifteen Years' Captivity on that Island*. Ed. Pasfield Oliver. London: Fisher Unwin.

Dubois, Laurent. 2004. *A Colony of Citizens: Revolution and Slave Emancipation in the French Caribbean, 1787–1804*. Chapel Hill: University of North Carolina Press.

Dull, Jack L., ed. 1972. *Han Social Structure*. Seattle: University of Washington Press.

Durand, John D. 1960. The Population Statistics of China, A. D. 1953. *Population Studies* 13 (3): 209–256.

Dutt, Romesh Chunder. 1874. *The Peasantry of Bengal*. Calcutta: Thacker, Spink & Co.

Eberhard, Wolfram. 1957. *Review of Les aspects économiques du bouddhisme dans la société chinoise du Ve au Xe siècle*, by Jacques Gernet. *Pacific Affairs* 30 (3): 272–274.

Ecormier, Jacques. 1992. *Cyclones tropicaux du Sud-ouest de l'Océan Indien, le cas de l'île de la Réunion*. Saint-Denis: La Réunion.

Ekblom, Anneli. 2008. Forest-Savanna Dynamics in the Coastal Lowland of Southern Mozambique since c. AD 1400. *The Holocene* 18 (8): 1247–1257.

Ekblom, Anneli, et al. 2016. Migration and Interaction between Madagascar and Eastern Africa, 500 BCE–1000 CE: The Archaeological Perspective. In *Early Exchange between Africa and the Wider Indian Ocean World*, ed. Gwyn Campbell, 195–230. New York: Palgrave Macmillan.

el Zein, Abdul Hamid M. 1974. *The Sacred Meadows. A Structural Analysis of Religious Symbolism in an East African Town*. Evanston, IL: Northwestern University Press.

Eliott, John. 1877. *Report of the Vizagapatnam and Backergunj Cyclones of October 1876*. Calcutta: Bengal Secretariat Press.

Ellis, William. 1838. *History of Madagascar*. Vol. 1. London: Fisher.

Ellis, Stephen. 1979. Un texte du XVIIème siècle sur Madagascar. *Omaly sy Anio* 9: 151–166.

Ennin. 1992. *Ru Tang Qiu Fa Xunli Xingji Jiaozhu*, [*Ennin's Diary: The Record of a Pilgrimage to China in Search of the Law*], Annotated by Ono Katsutoshi. Shijiazhuang: Huashan Literature and Art Publishing House.

Eno, Omar A. 2005. The Abolition of Slavery and the Aftermath Stigma: The Case of the Bantu/Jareer People on the Benadir Coast of Southern Somalia. In *Abolition and its Aftermath in Indian Ocean Africa and Asia*, ed. Gwyn Campbell, 83–89. London: Routledge.

Ève, Prosper. 1992. *Île à peur: la peur redoutée ou récupérée à La Réunion des origines à nos jours*. Saint André, La Réunion: Graphica.

———. 2000. *Naître et mourir à Bourbon à l'époque de l'esclavage*. Paris: L'Harmattan.

———. 2012. *Le corps des esclaves de l'île Bourbon*. Paris: PUPS—PU Paris Sorbonne.

Ewald, Janet. 1990. *Soldiers, Traders, and Slavers*. Madison, WI: University of Wisconsin Press.

Fagan, Brian. 1999. *Floods, Famines and Emperors: El Niño and the Fate of Civilizations*. New York: Basic Books.

Fahmy, Khaled. 1997. *All the Pasha's Men: Mehmed Ali, his Army and the Making of Modern Egypt.* Cambridge: Cambridge University Press.

Fan, Ka-wai. 2010. Climatic Change and Dynastic Cycles in Chinese History: A Review Essay. *Climate Change* 101: 565–573.

Ferguson, Niall. 2006. The Next War of the World. *Foreign Affairs* 85 (5): 61–74.

Feuerwerker, Albert. 1958. *China's Early Industrialization: Sheng Hsuan-Huai (1844–1916) and Mandarin Enterprise.* Cambridge, MA: Harvard University Press.

Filliot, J.M. 1974. *La traite des esclaves vers les Mascareignes.* Paris: ORSTOM.

Finley, Moses. 1980. *Ancient Slavery and Modern Ideology.* London: Chatto and Windus.

Flannery, John. 2013. *The Mission of the Portuguese Augustinians to Persia and Beyond (1602–1747).* Leiden: Brill.

Forrest, Thomas. 1779. *A Voyage to New Guinea and the Moluccas from Balambangan: Including an Account of Maguindanao, Sooloo and Other Islands.* London: C. Scott.

Foster, William, ed. 1899. *The Embassy of Sir Thomas Roe to the Court of the Great Mogul, 1615–1619.* Vol. 1. London: Hakluyt Society.

———. 1912a. *The English Factories in India 1637–1641: A Calendar of Documents in the India Office, British Museum, and Public Record Office.* Oxford: Clarendon Press.

———. 1912b. An English Settlement in Madagascar in 1645–6. *English Historical Review* 27: 239–250.

Frake, Charles. 1998. Abu Sayyaf Displays of Violence and the Proliferation of Contested Identities among Philippine Muslims. *American Anthropologist* 100 (1): 41–54.

Freeman, J.J. 1828. Extracts from a Letter from the Rev. J.J. Freeman, dated Tananarivou, Madagascar, October 23, 1827; Addressed to the Late Secretary [Burder?]. *Evangelical Magazine and Missionary Chronicle,* April, 169.

Fresnel, Fulgence. 1850. Mémoire sur le Waday. *Bulletin de la Société de Géographie de Paris* 14: 153–192.

Fuma, Sudel. 1992. *L'esclavage à la Réunion, 1794–1848.* Paris: L'Harmattan.

Galas, Jacques, J.-P. Locci, René Grosso, and Sylvestre Clap. 1993. *Histoire de Vaucluse.* Avignon: A. Barthélemy.

Gandhi, M.K. 1938. *Hind Swaraj or Indian Home Rule.* Ahmedabad: Navjivan Publishing House.

Garnier, Emmanuel, and Jérôme Desarthe. 2013. Cyclones and Societies in the Mascarene Islands 17th–20th Centuries. *American Journal of Climate Change* 2 (1): 1–13.

Gasse, Françoise. 2000. Hydrological Changes in the African Tropics since the Last Glacial Maximum. *Quaternary Science Reviews* 19: 189–211.

Gasse, F., and E. van Campo. 1998. A 40,000-Yr Pollen and Diatom Record from Lake Tritrivakely, Madagascar, in the Southern Tropics. *Quaternary Research* 49 (3): 299–311.

Gastrell, J.E. 1868. *Geographical and Statistical Report of the Districts of Jessore, Fureedpore and Backergunge*. Calcutta: Office of Superintendent of Government Printing.

Gerard, Gabriel. 2006. *Un grand créole, Joseph Hubert, 1747–1825*. Saint-Denis la Réunion: Éditions Azalée.

Gerbeau, Hubert. 1980. Quelques aspects de la traite illégale des esclaves à l'île Bourbon au XIXe siècle. In *Mouvements de populations dans l'Océan Indien*, 273–308. Paris: Champion.

———. 1986. Engagés and Coolies on Réunion Island, Slavery's Masks and Freedom's Constraints. In *Colonialism and Migration: Indentured Labour Before and After Slavery*, ed. Piet C. Emmer, 209–236. Dordrecht, The Netherlands: Martinus Nijhoff.

Gernet, Jacques. 1964. Review of Denis C. Twitchett's *Financial Administration Under the T'ang Dynasty* (Cambridge: University Press, 1963). *Journal of the Economic and Social History of the Orient* 7 (3): 324–325.

———. 1995. *Buddhism in Chinese Society: An Economic History from the Fifth to the Tenth Centuries*. Trans. Franciscus Verellen. New York: Columbia University Press.

Ghinassi, Massimiliano, et al. 2012. Shoreline Fluctuations of Lake Hayk (Northern Ethiopia) During the Last 3500 Years: Geomorphological, Sedimentary, and Isotope Records. *Palaeogeography, Palaeoclimatology, Palaeoecology* 365–366: 209–226.

Gilbert, Erik. 2004. *Dhows and the Colonial Economy of Zanzibar, 1860–1970*. Oxford: James Currey.

Gioli, Gino Bartolimmei. 1904. La produzione frumentaria in Eritrea di fronte alle relazioni doganali fra metropoli e colonia. *Atti della R. Academia dei Geografili*, Series V, 1 (1): 86–88.

Gong, Zhang. 1986. Tangdai Siyuan Nubi Jieung Lueshuo [Brief Survey on the Monastic Slave Class in the Tang Dynasty]. *Shehui Kexue Zhanxian* [*Social Science Front*] 3.

Goody, Jack. 1980. Slavery in Time and Space. In *Asian and African Systems of Slavery*, ed. James L. Watson, 16–42. Berkeley: University of California Press.

Goswami, Manu. 2004. *Producing India: From Colonial Economy to National Space*. Chicago: University of Chicago Press.

Grandidier, Guillaume. 1928. *Histoire physique, naturelle et politique de Madagascar*. Paris: Hachette.

Grandidier, Alfred. n.d. Souvenirs de voyages, 1865–1870. In *Documents anciens sur Madagascar*. Tananarive: Association Malgache d'Archéologie.

Grandidier, Guillaume, and Raymond Decary. 1958. *Histoire physique, naturelle et politique de Madagascar*. Vol. 5, no. 3, fasc. 1. Tananarive: Imprimerie Officielle.

Grandidier, Alfred, and Guillaume Grandidier. 1908. *Histoire physique, naturelle et politique de Madagascar: ethnographie de Madagascar*. Paris: Imprimerie Nationale.

Grassby, Richard. 2001. *Kinship and Capitalism: Marriage, Family, and Business in the English Speaking World, 1580–1740*. New York: Cambridge University Press.

Gray, Rivchard. 1961. *A History of the Southern Sudan, 1839–1889*. London: Oxford University Press.

Gribben, John, and H.H. Lamb. 1978. Climatic Change in Historical Times. In *Climatic Change*, ed. John Gribben, 68–82. Cambridge: Cambridge University Press.

Grove, Richard. 2005. El Niño Chronology and the History of Socio-Economic and Agrarian Crisis in South and Southeast Asia 1250–1900. In *Land Use—Historical Perspectives: Focus on Indo-Gangetic Plains*, ed. Y.P. Abrol, Satpal Sangwan, and Mithilesh K. Tiwari, 133–172. New Delhi: Allied Publishers.

———. 2007. The Great El Niño of 1789–93 and its Global Consequences: Reconstructing an Extreme Climate Event in World Environmental History. *Medieval History Journal* 10 (1–2): 75–98.

Guang, Sima. 1976. *Zi Zhi Tong Jian* [*Comprehensive Mirror to Aid in Government*]. Vol. 248. Kowloon: Chung Hwa Book Co.

Gudmundson, Candace. 2003. El Niño and Climate Prediction. In *El Niño: Overview and Bibliography*, ed. A.M. Babkina, 5–28. New York: Nova Science.

Guelke, L. 1989. Freehold Farmers and Frontier Settlers, 1657–1780. In *The Shaping of South African Society, 1652–1840*, ed. R. Elphick and H. Giliomee, 2nd ed., 69–73. Cape Town: Maskew Miller Longman.

Guha, Ranajit. 1983. *Elementary Aspects of Peasant Insurgency in Colonial India*. Delhi: Oxford University Press.

Guha, Sumit. 2001. The Population History of South Asia from the Seventeenth to the Twentieth Centuries: An Exploration. In *Asian Population History*, ed. Cuirong Liu et al., 63–78. Oxford: Oxford University Press.

Habib, Irfan. 1985. Studying a Colonial Economy Without Perceiving Colonialism. *Modern Asian Studies* 19 (3): 355–381.

Hamilton, Alexander. 1744. *A New Account of the East Indies*. Vol. 1. London: C. Hitch and A. Millar.

Hansen, Valerie. 2003. New Work on the Sogdians, the Most Important Traders on the Silk Road, A.D. 500–1000. *T'oung Pao, Second Series* 89 (1/3): 149–161.

Hardyman, J.T. 1964. The Madagascar Slave-Trade to the Americas (1632–1830). In *Association Historique Internationale de l'Ocean Indien Congrès, Océan Indien et Méditerrannée: Travaux du Sixiéme Colloque International d'Histoire Maritime et du Deuxième Congrès de l'Association Historique Internationale de l'Océan Indien, session de Lourenço Marques: 13–18 August 1962*, 501–521. Lisbon: SEVPEN.

Harris, Edward M. 2002. Did Solon Abolish Debt-Bondage? *Classical Quarterly* 52 (2): 415–430.

Harvey, David. 1989. *The Condition of Modernity: An Enquiry into the Origins of Cultural Change*. Oxford: Blackwell.

He, Ziquan, ed. 1986. *Wushi Nian Lai Han Tang Fojiao Siyuan Jingji Yanjiu* [*Studies for the Past Fifty Years on Buddhist Monastery Economy in Han and Tang Dynasties*]. Beijing: Beijing Normal University Press.

Heersink, Christian. 1998. Environmental Sdaptations in Southern Sulawesi. In *Environmental Challenges in Southeast Asia*, ed. Victor T. King, 95–120. London: Gunon.

Helal, Emad Ahmed. 2010. Muhammad Ali's First Army: The Experiment in Building an Entirely Slave Army. In *Race and Slavery in the Middle East: Histories of Trans-Saharan Africans in Nineteenth-Century Egypt, Sudan, and the Ottoman Mediterranean*, ed. Terence Walz and Kenneth M. Curto, 17–42. Cairo: American University in Cairo Press.

Hill, Richard Leslie. 1959. *Egypt in the Sudan, 1820–1881*. London: Oxford University Press.

———. 1970. *On the Frontiers of Islam: Two Manuscripts Concerning the Sudan Under Turco-Egyptian Rule, 1822–1845*. Oxford: Clarendon Press.

Hill, Richard, and Peter Hogg. 1995. *A Black Corps d'Elite*. East Lansing, MI: Michigan State University Press.

Hilsenberg, Charles Theodore, and Wenceslaus Bojer. 1833. A Sketch of the Province of Emerina, in the Island of Madagascar, and of the Huwa, its Inhabitants; Written During a Year's Residence. In *Botanical Miscellany*, ed. William Jackson Hooker, vol. 3, 246–277. London: John Murray.

Hirata, Lucie C. 1979. Free, Indentured, Enslaved: Chinese Prostitutes in Nineteenth-Century America. *Signs* 5 (1): 3–29.

Ho, Hai Quang. 2004. *Histoire économique de l'île de la Réunion, 1849–1881: engagisme, croissance et crise*. Paris: Lavoisier.

Hobsbawm, Eric. 1968. *Industry and Empire: The Making of Modern English Society, Vol. 2, 1750 to the Present*. New York: Partheon Books.

Hodgkin, J. Eliot. 1897. *Manuscripts of J. Eliot Hodgkin, Esq., F.S.A., of Richmond, Surrey*. London: Eyre and Spottiswoode.

Holder, Preston. 1974. *The Hoe and the Horse and the Plains: A Study of Cultural Development Among North American Indians*. Lincoln: University of Nebraska Press.

Hollingbery, R.H. 1879. *The Zemindary Settlement of Bengal*. Vol. 1. Calcutta: Bengal Secretariat Press.

Holt, Peter Malcolm. 1970. *The Mahdist State in the Sudan, 1881–1898: A Study of Its Origins, Development and Overthrow*. 2nd ed. Oxford: Clarendon Press.

Holt, Peter Malcolm, and M.W. Daly. 1979. *The History of the Sudan: From the Coming of Islam to the Present Day*. Boulder, CO: Westview Press.

Hopkins, Keith. 1978. *Conquerors and Slaves: Sociological Studies in Roman History*. Cambridge: Cambridge University Press.
Hopper, Matthew S. 2013. Debt and Slavery among Arabian Gulf Pearl Divers. In *Bonded Labour and Debt in the Indian Ocean World*, 103–118. London: Routledge.
Hunt, Robert. 1905. Île d'Asada. In *Collections des ouvrages anciens concernant Madagascar*, ed. Alfred Grandidier and Guillaume Grandidier, vol. 3, 259–266. Paris: Comité de Madagascar.
Hunt, J. 1967. Some Particulars Relating to Sulo in the Archipelago of Felicia. In *Notices of the Indian Archipelago and Adjacent Countries*, ed. J. H. Moor [orig. pub. 1820], 31–60. London: Cass.
Hutchins, David Ernest. 1889. *Cycles of Drought and Good Seasons in South Africa*. Wynberg: "Times" Office.
Ileto, Reynaldo Clemena. 1971. *Maguindanao 1860–1888: The Career of Dato Uto of Buayan*. Ithaca: Cornell University.
Iqbal, Iftekhar. 2010. *The Bengal Delta: Ecology, State and Social Change, 1840–1943*. Basingstoke: Palgrave Macmillan.
Istituto Agricolo Coloniale Italiano. 1932. *L'Economia Eritrea: nel cinquatennio dell'occupazione di Assab (1882–1932)*. Florence: Istituto Agricolo Coloniale Italiano.
Jack, James Charles. 1915. *Final Report of the Survey and Settlement Operations in the Bakarganj District 1900 to 1908*. Calcutta: Bengal Secretariat Book Depot.
Jaschok, Maria, and Suzanne Miers, eds. 1994. *Women and Chinese Patriarchy: Submission, Servitude and Escape*. London: Zed Books.
Javellana, Rene B. 1991. *Wood and Stone for God's Greater Glory: Jesuit Art and Architecture in the Philippines*. Manila: Ateneo de Manila Press.
———. 1997. *Fortress of Empire: Spanish Colonial Fortifications of the Philippines 1565–1898*. Manila: Bookmark.
Johnson, Douglas H. 1988. Sudanese Military Slavery from the Eighteenth to the Twentieth Century. In *Slavery and Other Forms of Unfree Labour*, ed. Leonie Archer, 142–155. London: Routledge.
Jordan, Elizabeth Grzymala. 2005. 'Unrelenting Toil': Expanding Archaeological Interpretations of the Female Slave Experience. *Slavery & Abolition* 26 (2): 217–232.
———. 2007. It All Comes Out in the Wash: Engendering Archaeological Interpretations of Slavery. In *Women and Slavery, Vol. 1, Africa, the Indian Ocean World, and the Medieval North Atlantic*, ed. G. Campbell, S. Miers, and J.C. Miller, 335–357. Athens: Ohio University Press.
Junker, Wilhelm. 1890. *Travels in Africa during the Years 1875[–1886]*. Vol. 1. Trans. Augustus Henry Keane. London: Chapman and Halle.
Keller, Conrad. 1901. *Madagascar, Mauritius and the Other East African Islands*. London: S. Sonnenschein.

Kelly, John B. 1968. *Britain and the Persian Gulf 1795–1880*. Oxford: Clarendon Press.

Kemball, Arnold B. 1856. Paper Relative to the Measures Adopted by the British Government, Between the Years 1820 and 1844, for Effecting the Suppression of the Slave Trade in the Persian Gulf. In *Arabian Gulf Intelligence*, ed. Robert H. Thomas, 635–687. Bombay: Bombay Education Society Press.

Kent, Raymond K. 1968. The Sakalava, Origins of the First Malagasy Empire. *Revue française d'histoire d'outre-mer* 55 (199): 145–189.

———. 1970. *Early Kingdoms in Madagascar 1500–1700*. New York: Holt, Rinehart and Winston.

———. 1992. Madagascar and the Islands of the Indian Ocean. In *Africa from the Sixteenth to the Eighteenth Century*, ed. Bethwell A. Ogot, 849–894. Paris: UNESCO.

Kerr, Ian. 2001. Reworking a Popular Religious Practice: The Effects of Railways in 19th and 20th Century South Asia. In *Railways in Modern India*, ed. Ian Kerr, 304–327. Delhi: Oxford University Press.

Kim, Bok-rae. 2004. *Nobi*: A Korean System of Slavery. In *The Structure of Slavery in Indian Ocean Africa and Asia*, ed. Gwyn Campbell, 155–168. London: Frank Cass.

———. 2013. Debt Slaves in Old Korea. In *Bonded Labour and Debt in the Indian Ocean World*, ed. Gwyn Campbell and Alessandro Stanziani, 165–172. London: Routledge.

Klein, Ira. 1973. Death in India. *Journal of Asian Studies* 32 (4): 639–659.

———. 1984. When the Rains Failed: Famine Relief and Mortality in British India. *Indian Economic and Social History Review* 21 (2): 185–214.

———. 1988. Plague Policy and Popular Unrest in British India. *Modern Asian Studies* 22 (4): 723–755.

———. 1989. Population Growth and Mortality, Part 1 – the Climatric of Death. *Indian Economic and Social History Review* 26 (4): 387–404.

Klein, Martin A., ed. 1993. *Breaking the Chains: Slavery, Bondage and Emancipation in Modern Africa and Asia*. Madison, WI: University of Wisconsin Press.

———. 1994. Simulating the African Slave Trade. *Canadian Journal of African Studies* 28 (2): 296–299.

———. 2005. The Emancipation of Slaves in the Indian Ocean. In *Abolition and its Aftermath in Indian Ocean Africa and Asia*, ed. Gwyn Campbell, 198–218. London: Routledge.

Kolb, Peter. 1968. *The Present State of the Cape of Good Hope*. Vol. 2. New York: Johnson Reprint Corporation.

Kopytoff, Igor, and Suzanne Miers. 1977. African 'Slavery' as an Institution of Marginality. In *Slavery in Africa: Historical and Anthropological Perspectives*, ed. Igor Kopytoff and Suzanne Miers, 3–81. Madison: University of Wisconsin Press.

Krapf, Johann L. 1860. *Travels, Researches and Missionary Labours during an Eighteen Years' Residence in Eastern Africa*. London: Trübner.

Kuhnke, LaVerne. 1990. *Lives at Risk: Public Health in Nineteenth-Century Egypt*. Berkeley: University of California Press.

Kuriyama, Yasuyuki. 2012. *Umi to tomoni aru rekishi: Iemen kaijyou kouryuu-shi no kenkyuu*. Tokyo: Cyuuou University Press.

La Rue, George Michael. 1984a. *Land and Social Stratification in Dar Fur, 1785–1875: The Hakura System*. Boston: Boston University.

———. 1984b. Khabir 'Ali at Home in Kubayh: A Brief Biography of a Dar Fur Caravan Leader. *African Economic History* 13: 56–83.

———. 1986. The Export Trade of Dar Fur, ca. 1785–1875. In *Figuring African Trade*, ed. Gerhard Liesegang, H. Pasch, and A. Jones, 636–668. Berlin: Kölner Beiträge zur Afrikanistik 11.

———. 1987. Land and Social Stratification in Dar Fur, 1785–1875: The Hakura System. In *The State and the Market*, ed. Clive Dewey, 24–44. New Delhi: Manohar Press.

———. 2002. The Capture of a Slave Caravan: The Incident at Asyut (Egypt) in 1880. *African Economic History* 30: 81–106.

———. 2007. African Slave Women in Egypt, from ca. 1820 to the Plague Epidemic of 1834–1835. In *Women and Slavery, Vol. 1, Africa and the Western Indian Ocean Islands*, ed. G. Campbell, S. Miers, and J.C. Miller, 168–189. Athens, OH: Ohio University Press.

———. 2010. 'My Ninth Master Was a European': Enslaved Blacks in European Households in Egypt, 1798–1848. In *Race and Slavery in the Middle East: Histories of Trans-Saharan Africans in Nineteenth-Century Egypt, Sudan, and the Ottoman Mediterranean*, ed. Terence Walz and Kenneth M. Curto, 99–124. Cairo: American University in Cairo Press.

———. 2016. Treating Black Deaths in Egypt: Clot-Bey, African Slaves and the Plague Epidemic of 1834–35. In *Histories of Medicine and Healing in the Indian Ocean World*, ed. Anna Winterbottom and Facil Tesfaye, 27–59. New York: Palgrave Macmillan.

Laarhoven, Ruurdje. 1989. *Triumph of Moro Diplomacy: The Maguindanao Sultanate in the 17th Century*. Quezon City: New Day Publishers.

———. 1990. Lords of the Great River: The Maguindanao Port and Polity during the Seventeenth Century. In *The Southeast Asian Port and Polity: Rise and Demise*, ed. J. Kathirithamby-Wells and John Villiers, 161–186. Singapore: Singapore University Press.

Laborde, Léon de. 1838. Chasse aux Nègres. *Revue française* 9: 319–333.

Lacour, Raoul. 1871. *L'Égypte d'Alexandrie à la seconde cataracte*. Paris: Hachette.

Lamb, Hubert H. 1995. *Climate, History and the Modern World*. London: Routledge.

———. 2011. *Climate: Present, Past and Future*. Vol. 2. London: Routledge.

Lamb, Henry F., et al. 2007. Oxygen and Carbon Isotope Composition of Authigenic Carbonate from an Ethiopian Lake: A Climate Record of the Last 2000 Years. *The Holocene* 17 (4): 517–526.

Landsberg, Helmut E. 1981. Past Climates from Unexploited Written Sources. In *Climate and History*, ed. Robert I. Rotberg and Theodore K. Rabb, 631–642. Princeton: Princeton University Press.

Lasker, Bruno. 1950. *Human Bondage in Southeast Asia*. Chapel Hill: University of North Carolina Press.

Legesse, Dagnachew, et al. 2002. Environmental Changes in a Tropical Lake (Lake Abiyata, Ethiopia) during Recent Centuries. *Palaeography, Palaeoclimatology, Palaeoecology* 187 (3–4): 233–258.

Leibbrandt, H.C.V. 1896. *Precis of the Archives of the Cape of Good Hope: Journal, 1699–1732*. Cape Town: W. A. Richardson.

———. 1897. *Precis of the Archives of the Cape of Good Hope, January 1659–May 1662. Riebeeck's Journal, &c.* Part 3. Cape Town: W. A. Richards.

Leiser, Gary, and Michael Dols. 1987. Evliyā Chelebi's Description of Medicine in Seventeenth-Century Egypt: Part I: Introduction. *Sudhoffs Archiv* 71 (2): 197–216.

Lerner, Gerda. 1983. Women and Slavery. *Slavery & Abolition* 4 (3): 173–198.

Levy, H.S. 1955. Review: *The Background of the Rebellion of An Lu-shan* by Edwin G. Pulleyblank. *Journal of the American Oriental Society* 75 (3): 188–192.

Lewis, Bazett Annesley. 1972. *The Murle: Red Chiefs and Black Commoners*. Oxford: Clarendon Press.

Lewis, Mark. 2009. *China between Empires: The Northern and Southern Dynasties*, History of Imperial China. Cambridge, MA: Belknap Press of Harvard University Press.

Li, Fang, ed. 1982. *Wen Yuan Ying Hua [Finest Blossoms in the Garden of Literature]*. Vol. 429. Beijing: Zhongguo Book Company.

Limbert, Mandana E. 2013. 'If You Catch Me Again at It, Put Me to Death': Slave Trading, Paper Trails, and British Bureaucracy in the Indian Ocean. In *Indian Ocean Slavery in the Age of Abolition*, ed. Robert Harms, Bernard L. Freamon, and David W. Blight, 120–140. New Haven: Yale University Press.

Locher, Fabien. 2008. *Le savant et la tempête: étudier l'atmosphère et prévoir le temps au XIXe siècle*. Rennes: Presses Universitaires de Rennes.

Lorimer, John G. 1986. *Gazetteer of the Persian Gulf, Oman and Central Arabia*. Vol. 1. London: Archive Editions.

Lovejoy, Paul E. 2013. Pawnship and Seizure for Debt in the Process of Enslavement in West Africa. In *Debt and Slavery in the Mediterranean and Atlantic Worlds*, ed. Gwyn Campbell and Alessandro Stanziani, 63–75. London: Pickering & Chatto.

Lovejoy, Paul, and Stephen Baier. 1975. The Desert-Side Economy of the Central Sudan. *International Journal of African Historical Studies* 8 (4): 551–581.

Lovejoy, Paul, and Toyin Falola, eds. 2003. *Pawnship, Slavery and Colonialism in Africa*. Trenton, NJ: Africa World Press.

Low, Charles R. 1985. *History of the Indian Navy 1613–1863*. Vol. 2. New Delhi: Manas.

Lyne, Robert N. 1969. *Zanzibar in Contemporary Times: A Short History of the Southern East in the Nineteenth Century*. New York: Negro University Press.

MacDonald, Glen M., and Roslyn A. Case. 2005. Variations in the Pacific Decadal Oscillation over the Past Millennium. *Geophysical Research Letters* 32 (8): L08703.

Mackerras, Colin. 2000. Uygur-Tang Relations, 744–840. *Central Asian Survey* 19 (2): 223–234.

Malzac, R.P. 1912. *Histoire du Royaume Hova depuis ses origins jusqu'à sa fin*. Tananarive: Imprimerie Catholique.

Manning, Patrick. 1990a. *Slavery and African Life: Occidental, Oriental and African Slave Trades*. Cambridge: Cambridge University Press.

———. 1990b. The Slave Trade: The Formal Demography of a Global System. *Social Science History* 14 (2): 255–279.

Mariano, Luis. 1904. Relation du voyage de découverte fait à l'île Saint-Laurent dans les années 1613–1614, par le capitaine Paulo Rodrigues da Costa et les pères jésuites Pedro Freire et Luis Mariano à bord d la caravelle Nossa Senhora da Esperança. In *Collections des ouvrages anciens concernant Madagascar*, ed. Alfred Grandidier and Guillaume Grandidier, vol. 2, 1–64. Paris: Comité de Madagascar.

Marsot, Afaf Lufti Al-Sayyid. 1990. *Egypt in the Reign of Muhammad Ali*. Cambridge: Cambridge University Press.

Martini, Ferdinando. 1925. *Nell'Africa italiana*. 8th ed. Milan: Fratelli Treves.

Maso, Miguel Saderra. 1904. *Volcanos and Seismic Centers of the Philippine Archipelago*. Washington: Government Printing Office.

———. 1915. *Historia del observatorio de Manila fundado y dirigio por los padres de la mision de la Compañia de Jesus de Filipinas 1865–1915*. Manila: E.C. McCullough.

Mason, John Edwin. 1990. Hendrik Albertus and his Ex-Slave Mey: A Drama in Three Acts. *Journal of African History* 31 (3): 423–445.

———. 1991. The Slaves and their Protectors: Reforming Resistance in a Slave Society: The Cape Colony, 1826–1834. *Journal of Southern African Studies* 17 (1): 103–128.

Mayoka, Mireille. 1998. *Cyclones à La Réunion*. La Réunion: Centre des Cyclones Tropicaux de la Réunion, Météo France.

McBride, Theresa. 1992. Public Authority and Private Lives: Divorce after the French Revolution. *French Historical Studies* 17 (2): 747–768.

McCann, James. 1987. *From Poverty to Famine in Northeast Ethiopia: A Rural History 1900–1935*. Philadelphia: University of Pennsylvania Press.

McCarthy, Justin A. 1976. Nineteenth-Century Egyptian Population. *Middle Eastern Studies* 12 (3): 1–39.

McCarthy, James. 2006. *Selim Aga: A Slave's Odyssey*. Edinburgh: Luath Press.

McCoan, J.C. 1882. *Egypt As It Is*. New York: H. Holt.

McGlynn, Gayle, et al. 2010. Palaeolimnological Evidence of Environmental Change over the Last 400 Years in the Rwenzori Mountains of Uganda. *Hydrobiolgia* 648 (1): 109–122.

McKeown, Adam. 2004. *Global Chinese Migration, 1850–1940*. Paper presented at ISSCO V, Helsignor, Denmark, May.

McNeill, William H. 1976. *Plagues and Peoples*. New York: Anchor Books.

Médard, Henri. 2013. La plus ancienne et la plus récente des traites: panorama de la traite et de l'esclavage en Afrique orientale et dans l'Océan Indien. In *Traites et esclavage en Afrique orientale et dans l'Océan Indien*, ed. Henri Médard et al., 65–118. Paris: Karthala.

Mednick, Melvin. 1957. Some Problems of Moro History and Political Organisation. *Philippine Sociological Review* 5: 39–52.

Meloy, John. 2010. *Imperial Power and Maritime Trade: Mecca and Cairo in the Later Middle Ages*. Chicago: Middle East Documentation Center.

Mengin, Félix. 1839. *Histoire sommaire de l'Égypte*. Paris: Librarie de Firmin Didot.

Mentzel, O.F. 1921. *Description of the Cape of Good Hope*. Vol. 1. Ed. H. Mandelbrote. Cape Town: Van Riebeeck Society.

Michaud, Joseph Fr, and Baptistin Poujoulat. 1835. *Correspondence d'Orient, 1830–1831*. Vol. 7. Paris: Ducollet.

Miers, Suzanne. 2003. *Slavery in the Twentieth Century: The Evolution of a Global Problem*. Walnut Creek, CA: AltaMira Press.

———. 2005. Slavery and the Slave Trade in Saudi Arabia and the Arab States on the Persian Gulf, 1921–63. In *Abolition and its Aftermath in Indian Ocean Africa and Asia*, ed. Gwyn Campbell, 120–136. London: Routledge.

Miers, Suzanne, and Martin A. Klein, eds. 1999. *Slavery and Colonial Rule in Africa*. London: Frank Cass.

Mikhail, Alan. 2010. An Irrigated Empire: The View from Ottoman Fayyum. *International Journal of Middle East Studies* 42 (4): 569–590.

———. 2013. Unleashing the Beast: Animals, Energy, and the Economy of Labor in Ottoman Egypt. *American Historical Review* 118 (2): 317–348.

Miles, Samuel B. 1919. *The Countries and Tribes of the Persian Gulf*. London: Harrison and Sons.

Mimura, Nobuo, et al. 2007. Small Islands. In *Climate Change: Impacts, Adaptation and Vulnerability. Contribution of Working Group II to the Fourth Assessment Report of the Intergovernmental Panel on Climate Change*, ed. M.L. Parry et al., 687–716. Cambridge: Cambridge University Press.

Miran, Jonathan. 2009. *Red Sea Citizens: Cosmopolitan Society and Cultural Change in Massawa*. Bloomington, IN: Indiana University Press.

Mizukami, Chiyuki. 2004. Kaiyou jiyuu no keisei. *Hiroshima Hougaku* 28 (1 and 2): 1–22 and 1–25.

Monro, J. 1878. *Report on the Police of the Lower Bengal Presidency for the Year 1877*. Calcutta: Bengal Secretariat Press.

Mookherrji, Sudhansu Bimal. 1962. *The Indenture System in Mauritius, 1837–1915*. Calcutta: K.L. Mukhopadhyay.

Morton, Fred. 2009. Small Change: Children in the Nineteenth-Century East African Slave Trade. In *Children in Slavery Through the Ages*, ed. Gwyn Campbell, Suzanne Miers, and Joseph C. Miller, 55–70. Athens: Ohio University Press.

Mukherjee, Upamanyu Pablo. 2013. *Natural Disasters and Victorian Empire: Famines, Fevers and the Literary Cultures of South Asia*. London: Palgrave Macmillan.

Murakami, Ei. 2013. Two Bonded Labour Emigration Patterns in Mid-Nineteenth-Century Southern China: The Coolie Trade and Emigration to Southeast Asia. In *Bonded Labour and Debt in the Indian Ocean World*, 153–164. London: Routledge.

Nachtigal, Gustav . 1971. *Sahara and Sudan*, Vol. 4, *Wadai and Darfur*. Trans. Allen G. S. Fisher and Humphrey J. Fisher. London: C. Hurst.

Naty, Alexander. 2002. Environment, Society and the State in Western Eritrea. *Africa: Journal of the International African Institute* 72 (4): 569–597.

Nelson, Thomas. 2004. Slavery in Medieval Japan. *Monumenta Nipponica* 59 (4): 463–492.

Newitt, Malyn. 1984. *The Comoro Islands: Struggle against Dependency in the Indian Ocean*. Boulder, CO: Westview.

———. 1995. *A History of Mozambique*. Bloomington: Indiana University Press.

———. 2005. *A History of Portuguese Overseas Expansion 1400–1668*. London: Routledge.

Newton-King, Susan. 1999. *Masters and Servants on the Cape Eastern Frontier, 1760–1803*. Cambridge: Cambridge University Press.

Nichols, W. 1918. The Sakia in Dongola Province. *Sudan Notes and Records* 1 (1): 21–24.

Nicholson, Sharon E. 1981. The Historical Climatology of Africa. In *Climate and History: Studies in Past Climates and their Impact on Man*, ed. T.M.L. Wigley et al., 249–270. Cambridge: Cambridge University Press.

———. 2013. Temperature Variability over Africa during the Last 2000 Years. *The Holocene* 23 (8): 1085–1094.

Noel-Buxton, Noel Edward. 1932. Slavery in Abyssinia. *International Affairs* 11 (4): 512–526.

Northrup, David. 1995. *Indentured Labor in the Age of Imperialism, 1834–1922*. Cambridge: Cambridge University Press.

O'Fahey, R.S. 1973. Slavery and the Slave Trade in Dar Fur. *Journal of African History* 14 (3): 29–43.

Oberlé, Philippe. 1985. *Histoire de Djibouti: des origines à la République*. Paris: Présence Africaine.

Oberlé, Philippe, and Pierre Hugot. 1985. *Histoire de Djibouti: des origines à la République*. Paris: Présence Africaine.

Oliver, Samuel Pasfield. 1886. *Madagascar: An Historical and Descriptive Account of the Island and its Former Dependencies*, 2 Vols. London: Macmillan.

———. 1891. Sir Robert Townsend Farquhar and the Malagasy Slave Trade. *Antananarivo Annual and Madagascar Magazine* 15: 319–321.

Oliver-Smith, Anthony. 1994. Peru's Five-Hundred Year Earthquake. In *Disasters, Development and Environment*, ed. Ann Varley, 3–48. Chichester: Wiley.

Oppenheimer, Clive. 2011. *Eruptions that Shook the World*. Cambridge: Cambridge University Press.

Owen, Roger. 1969. *Cotton and the Egyptian Economy, 1820–1914: A Study in Trade and Development*. Oxford: Clarendon Press.

Ozoux, Léon. 1926. *Vieux principes d'économie rurale malgache*. Tananarive: Imprimerie Industrielle.

Pajot, Élie. 1887. *Simples renseignements sur l'île Bourbon*. Paris: Challanel Ainé, Librairie Coloniale.

Pankhurst, Richard. 1964. *The Great Ethiopian Famine of 1888–1892: A New Assessment*. Addis Ababa: Haile Selassie I University.

———. 1968. *Economic History of Ethiopia, 1800–1935*. Addis Ababa: Haile Selassie I University Press.

———. 1982. The Medical Activities in Eighteenth Century Ethiopia of James Bruce the Explorer. *Medizinhistorisches Journal* 17 (3): 256–276.

Pankhurst, Richard, and Douglas Johnson. 1988. The Great Drought and Famine of 1888–92 in Northeast Africa. In *The Ecology of Survival: Case Studies from Northeast African History*, ed. Douglas Johnson and David Anderson, 47–72. Boulder, CO: Westview Press.

Panzac, Daniel. 1985. *La peste dans l'empire Ottoman, 1700–1850*. Leuven: Editions Peeters.

———. 1987. The Population of Egypt in the Nineteenth Century. *Asian and African Studies* 21 (1): 11–32.

Parat. 1978. Le mémoire de Parat sur Madagascar (1714), ed. J.C. Hebert. *Bulletin de l'Académie Malgache* 56 (1–2): 50–75.

Parthasarathi, Prasannan. 2014. *Why Europe Grew Rich and Asia Did Not: Global Economic Divergence, 1600–1850*. Cambridge: Cambridge University Press.

Patnaik, Utsa, and Manjari Dingwaney, eds. 1985. *Chains of Servitude: Bondage and Slavery in India*. Hyderabad: Sangam Books.

Patzert, William. 1974. Wind-Induced Reversal in Red Sea Circulation. *Deep Sea Research and Oceanographic Abstracts* 21 (2): 109–121.

Payton, Walter. 1905. A Journall of All Principall Matters Passed in the Twelfth Voyage to the East India, Observed by mee Walter Payton, in the Good Ship the Expedition: The Captaine whereof was M. Christopher Newport, Being Set Out, Anno 1612. In *Hakluytus Posthumus or Purchas his Pilgrimes Containing a History of the World in Sea Voyages and Lande Travells by Englishmen and Others*, ed. Samuel Purchas, vol. 4, 180–213. Glasgow: James MacLehose and Sons.

Peabody, Sue. 2009. Furcy, la question raciale et le 'sol libre de France': une micro-histoire. *Annales: histoire, sciences sociales* 64 (6): 1305–1334.

———. 2017. *Madeleine's Children: Family, Freedom, Secrets, and Lies in France's Indian Ocean Colonies*. New York: Oxford University Press.

Penn, Nigel. 1999. *Rogues, Rebels and Runaways: Eighteenth-Century Cape Characters*. Cape Town: David Philip.

———. 2005. *The Forgotten Frontier: Colonist and Khoisan on the Cape's Northern Frontier in the Eighteenth Century*. Athens, OH: Ohio University Press.

Petherick, John, and Katherine Harriet Petherick. 1869. *Travels in Central Africa and Explorations of the Western Nile Tributaries*. Vol. 1. London: Tinsley Brothers.

Piddington, Henry. 1848. *The Sailor's Horn-Book for the Law of Storms*. London: Smith/Elder.

Pincus, Steven C.A. 1996. *Protestantism and Patriotism: Ideologies and the Making of English Foreign Policy, 1650–1668*. Cambridge: Cambridge University Press.

Pinker, Steven. 2011. *The Better Angels of our Nature: Why Violence has Declined*. London: Penguin.

Platt, Virginia Bever. 1969. The East India Company and the Madagascar Slave Trade. *William and Mary Quarterly* 26 (4): 548–577.

Pulleyblank, Edwin G. 1955. *The Background of the Rebellion of An Lu-shan*, London Oriental Series. Vol. 4. London: Oxford University Press.

———. 1958. The Origins and Nature of Chattel Slavery in China. *Journal of the Economic and Social History of the Orient* 1 (2): 185–220.

Punongbayan, Raymundo S. 1987. Disaster Preparedness Systems for Natural Hazards in the Philippines: An Assessment. In *Geologic Hazards and Disaster Preparedness Systems*, 77–101. Quezon City: Philippine Institute of Volcanology and Seismology.

Purchas, Samuel. 1905. *Hakluytus Posthumus or Purchas his Pilgrimes Containing a History of the World in Sea Voyages and Lande Travells by Englishmen and Others*. Vol. 4. Glasgow: James MacLehose and Sons.

QuanSheng, Ge, et al. 2013. General Characteristics of Climate Changes during the Past 2000 Years in China. *Science China Earth Sciences* 56: 321–329.

Raison, Jean-Pierre. 1977. Perception et réalisation de l'espace dans la société merina. *Annales: économies, sociétés, civilisations* 32 (3): 412–432.

Rangarajan, Mahesh. 1996. *Fencing the Forest: Conservation and Ecological Change in India's Central Provinces 1860–1914*. Delhi: Oxford University Press.

Rangasami, Amrita. 1985. 'Failure of Entitlements' Theory of Famine: A Response, Part 1. *Economic and Political Weekly* 20 (41): 1747–1752.

Raven-Hart, Rowland. 1971. *Cape of Good Hope 1652–1702: The First Fifty Years of Dutch Colonization as Seen by Callers*. Vol. 2. Cape Town: Balkema.

Reid, William. 1838. *An Attempt to Develop the Law of Storms*. London: John Weale.

Reid, Anthony, ed. 1983. *Slavery, Bondage and Dependency in Southeast Asia*. St Lucia: University of Queensland Press.

Reimer, Michael J. 1994. Ottoman Alexandria: The Paradox of Decline and the Reconfiguration of Power in Eighteenth-Century Arab Provinces. *Journal of the Economic and Social History of the Orient* 37 (2): 107–146.

Reis, Pîrî. 2013. *The Book of Bahriye*. Istanbul: Boyut.

Repetti, W.C. n.d. *The Manila Observatory*. Washington: Georgetown University.

Ricks, Thomas M. 1988. Slaves and Slave Traders in the Persian Gulf, 18th and 19th Centuries: An Assessment. *Slavery & Abolition* 9 (3): 60–70.

Ricquebourg, Lucien-Jacques-Camille. 1983. *Dictionnaire généalogique des familles de l'île Bourbon (La Réunion) 1665–1810*. Vol. 3. Mayenne: Imprimérie de la Manutention.

Ross, Robert. 1977. Smallpox at the Cape of Good Hope in the Eighteenth Century. *African Historical Demography* 1: 416–428. (Edinburgh: University of Edinburgh).

———. 1983. *Cape of Torments: Slavery and Resistance in South Africa*. London: Routledge.

Rowe, John, and Kjell Hødnebø. 1994. Rinderpest in the Sudan 1888–1890: The Mystery of the Missing Panzootic. *Sudanic Africa* 5: 149–179.

Roy, Tirthankar. 2010. 'The Law of Storms': European and Indigenous Responses to Natural Disasters in Colonial India, c. 1800–1850. *Australian Economic History Review* 50 (1): 6–22.

Russell, J.M., and T.C. Johnson. 2007. Little Ice Age Drought in Equatorial Africa: Intertropical Convergence Zone Migrations and El Niño–Southern Oscillation Variability. *Geology* 35 (1): 21–24.

Sainsbury, W. Noel, ed. 1892. *East Indies and Persia, 1630–34, Calendar of State Papers, Colonial Series*. London: Eyre and Spottiswoode.

Saldanha, Jerome A., ed. 1986. *The Persian Gulf Précis*. Vol. 3. Gerrards Cross, Archive Editions.

Saleeby, Najeeb M. 1905. *Studies in Moro History, Law and Religion*. Manila: Bureau of Public Printing.

Salman, Michael. 2005. The Meaning of Slavery: The Genealogy of 'an Insult to the American Government and to the Filipino People'. In *Abolition and its Aftermath in Indian Ocean Africa and Asia*, ed. Gwyn Campbell, 180–197. London: Routledge.

Samatar, Abdi Ismail. 1989. *The State and Rural Transformation in Northern Somalia, 1884–1986*. Madison, WI: University of Wisconsin Press.

Sandin, Benedict. 1967. *The Sea Dayaks of Borneo before White Rajah Rule*. London: Macmillan.

Santo, Domingos do Espírito. 1630. *Breve relaçam das christandades que os religiosos de N. Padre Sancto Agostinho tem a sua conta nas partes do Oriente, & do fruyto que nellas se faz, tiradas principalmente das cartas que nestes annos de lâ se escrevem; em que se contem cousas muy notaveis*. Lisbon: por Antonio Alvarez.

Santoro, Michael M., et al. 2015. An Aggregated Climate Teleconnection Index Linked to Historical Egyptian Famines of the Last Thousand Years. *The Holocene* 25 (5): 8972–8979.

Sarma, A., and K. Sen. 2012. Henry Piddington (1797–1858): A Bicentennial Tribute. *Weather* 52 (6): 187–193.

Saul, S.B. 1960. *Studies in British Overseas Trade, 1870–1914*. Liverpool: Liverpool University Press.

Scarr, Deryck. 1998. *Slaving and Slavery in the Indian Ocean*. New York: Saint Martin's Press.

Schiffman, Daniel A. 2010. Rabbinical Perspectives on Money in Seventeenth-Century Ottoman Egypt. *European Journal of the History of Economic Thought* 17 (2): 163–197.

Schoeman, Karel. 2007. *Early Slavery at the Cape of Good Hope, 1652–1717*. Pretoria: Protea.

Schottenhammer, Angela. 2004. Slaves and Forms of Slavery in Late Imperial China (Seventeenth to Early Twentieth Centuries). In *The Structure of Slavery in Indian Ocean Africa and Asia*, ed. Gwyn Campbell, 143–154. London: Frank Cass.

Schuler, Monica. 1986. The Recruitment of African Indentured Labourers for European Colonies in the 19th Century. In *Colonialism and Migration: Indentured Labour Before and After Slavery*, ed. Peter Emmer, 125–161. Dordrecht: Martinus Nijhoff.

Schweinfurth, Georg August. 1874. *The Heart of Africa*. Vol. 2. New York: Harper and Bros.

Scully, Pamela. 1990. *The Bouquet of Freedom: Social and Economic Relations in the Stellenbosch District, c.1870–1900*. Cape Town: Centre for African Studies, University of Cape Town.

———. 1997. *Liberating the Family? Gender and British Slave Emancipation in the Rural Western Cape, South Africa, 1823–53*. Portsmouth, NH: Heinemann.

Sekoto, Gerard. 1995. *My Life and Work*. Johannesburg: Viva Books.

Sen, Amartya. 1982. *Poverty and Famines: An Essay on Entitlement and Deprivation*. Oxford: Clarendon Press.

Serels, Steven. 2012. Famines of War: The Red Sea Grain Market and Famine in Eastern Sudan 1889–1891. *Northeast African Studies* 12 (1): 73–94.

Shaoliang, Zhou, ed. 2000. *Quan Tang Wen Xin Bian* [*Complete Works of Tang New Edition*]. Vol. 753. Changchun: Jilin Science Literature Press.
Sharf, Robert H. 2002. *Coming to Terms with Chinese Buddhism: A Reading of the Treasure Store Treatise*. Honolulu: Kuroda Institute, University of Hawai'i Press.
Sheik-Abdi, Abdi. 1993. *Divine Madness: Mohammed Abdulle Hassan (1856–1920)*. London: Zed Books.
Sheriff, Abdul. 1987. *Slaves, Spices & Ivory in Zanzibar*. London: James Currey.
———. 2005. The Slave Trade and its Fallout in the Persian Gulf. In *Abolition and its Aftermath in Indian Ocean Africa and Asia*, ed. Gwyn Campbell, 103–119. London: Routledge.
Shu, Jiu Tang. 1963. [*Old Book of Tang*]. Ed. Liu Xu et al. Shanghai: Zhonghua Book Company.
Sikainga, Ahmad Alawad. 1996. *Slaves into Workers: Emancipation and Labor in Colonial Sudan*. Austin: University of Texas Press.
von Slatin, Rudolph. 1896. *Fire and Sword in the Sudan: A Personal Narrative of Fighting and Serving the Dervishes, 1879–1895*. Trans. F.R. Wingate. London: Edward Arnold.
Sociats, Arturo Garin y. 1881. Memoria sobre el Archipielago de Jolo. *Boletin de la Sociedad Geografica de Madrid* 10: 110–133.
Soon, Willie, et al. 2003. Reconstructing Climatic and Environmental Changes of the Past 1000 Years: A Reappraisal. *Energy & Environment* 14 (2–3): 233–296.
Sorokin, Pitirim A. 1976. *Manual Society and Calamity the Effects of War, Revolution, Famine, Pestilence, upon Human Mind, Behavior, Social Organisation and Cultural Life*. New York: Simon and Schuster.
Southon, Michael. 1995. *The Navel of the Perahu: Meaning and Values in the Maritime Trading Economy of a Butonese Village*. Canberra: Department of Anthropology, Australian National University.
Souza, Teotonio R. De. 1999. *Goa through the Ages: An Economic History*. Vol. 2. Goa: Goa University.
Spaulding, Jay. 1982. Land Tenure and Social Class in the Northern Turkish Sudan. *International Journal of African Historical Studies* 15 (1): 1–20.
Spinage, C.A. 2012. *African Ecology: Benchmarks and Historical Perspectives*. Berlin: Springer.
Stanziani, Alessandro. 2013a. Beyond Colonialism. Servants, Wage Earners and Indentured Migrants in Rural France and Reunion Island, c. 1750–1900. *Labour History* 54 (1): 64–87.
———. 2013b. Local Bondage in Global Economies: Servants, Wage Earners and Indentured Migrants in 19th Century France, Great Britain and the Mascarene Islands. *Modern Asian Studies* 47 (4): 1218–1251.
Stigland, C.H. 1966. *The Land of Zinj: Being an Account of British East Africa, its Ancient History and Present Inhabitants*. London: Frank Cass.

Stommel, Henry, and Elizabeth Stommel. 1983. *Volcano Weather: The Story of 1816, the Year Without a Summer*. Newport: Seven Seas Press.

Stothers, Richard B. 1984. The Great Tambora Eruption in 1815 and its Aftermath. *Science* 224 (4654): 1191–1198.

Strachan, Michael, and Boies Penrose, eds. 1971. *The East India Company Journals of Captain William Keeling and Master Thomas Bonner, 1615–1617*. Minneapolis: University of Minnesota Press.

Sulivan, George L. 2003. *Dhow Chasing in Zanzibar Waters* [orig. pub. 1873]. Zanzibar: Gallery Publications.

Sussman, George D. 1982. *Selling Mother's Milk: The Wet-Nursing Business in France, 1715–1914*. Urbana: University of Illinois Press.

Suzuki, Hideaki. 2011. Indo-you nishi-kaiiki ni okeru 'dorei-sen gari': 19-seiki dorei koueki haizetsu katudou no ichidanme'. *Afurika Kenkyuu* 79: 13–25.

———. 2013a. Indo-you nishi-kaiiki sekai no kanousei: kaiiki-shi kara sekai-shi he. *Rekishigaku-kenkyuu* 911: 178–185.

———. 2013b. Arabia-hantou Omān-wan, Persia-wan enganbu to Mangrōbu: Indo-you nishi-kaiiki ni okeru koukan ga sasaeru Arabu shakai no jyuu kuukan. In *Mangrōbu: Arabu no nariwai seitaikei*, ed. Ryou Nakamura and Hiroshi Nawata, vol. 3, 151–168. Kyoto: Rinkawa Shobou.

Theal, George McCall. 1901. *Records of South-Eastern Africa*. Vol. 5. London: William Clowes.

———. 1904. *Records of the Cape Colony*. Vol. 20. London: William Clowes.

———. 1916. *History of Africa South of the Zambezi*. Vol. 1. London: George Allen & Unwin.

———. 2010. *History and Ethnography of Africa south of the Zambesi*. Vol. 3. New York: Cambridge University Press.

Thom, Hendrik Bernardus, ed. 1954. *Journal of Jan van Riebeeck*. Vol. 2. Cape Town: Van Riebeeck Society.

Tinker, Hugh. 1974. *A New System of Slavery: The Export of Indian Labour Overseas, 1830–1920*. London: Oxford University Press.

Tomlinson, B.R. 1982. The Political Economy of the Raj: The Decline of Colonialism. *Journal of Economic History* 42 (1): 133–137.

Torry, William. 1979. Anthropological Studies in Hazardous Environments: Past Trends and New Horizons. *Current Anthropology* 20 (30): 517–540.

Truter, Paul. 2005. The Robben Island Rebellion of 1751: A Study of Convict Experience at the Cape of Good Hope. *Kronos: Journal of Cape History* 31 (2005): 34–49.

Tsoar, Haim. 1995. Desertification in Northern Sinai in the Eighteenth Century. *Climatic Change* 29 (1995): 429–438.

Tubiana, Marie-José. 1960. Un document inédit sur les sultans du Wadday. *Cahiers d'études africaines* 1 (2): 49–112.

Tyson, P.D., W. Karien, K. Holmgren, and G.A. Heiss. 2000. The Little Ice Age and Medieval Warming in South Africa. *South African Journal of Science* 96 (3): 121–126.

Ulrich, Nicole. 2010. Abolition from Below: The 1808 Revolt in the Cape Colony. In *"Humanitarian Intervention" and Changing Labour Relations: Long-Term Consequences of the British Act on the Abolition of the Slave Trade, 1807*, ed. M. van der Linden, 193–222. Leiden: Brill.

Valentyn, François, et al. 1971. *Description of the Cape of Good Hope*. Vol. 1. Cape Town: Van Riebeeck Society.

Vaux, Charles Grant de. 1801. *The History of Mauritius: Or the Isle of France, and the Neighbouring Islands; from their First Discovery to the Present Time*. London: W. Bulmer and Company.

Velthoen, Esther. 1997. Wanderers, Robbers and Bad Folk: The Politics of Violence, Protection and Trade in Eastern Sulawesi 1750–1850. In *The Last Stand of Autonomous States, 1750–1870: Responses to Modernity in the Diverse Worlds of Southeast Asia and Korea*, ed. Anthony Reid, 367–388. London: Macmillan.

Vergunst, Nicolaas. 2000. *Hoerikwaggo: Images of Table Mountain*. Cape Town: South African National Gallery.

Vernal, Fiona. 2011. Discourse Networks in South African Slave Society. *African Historical Review* 43 (2): 1–36.

Vernet, Thomas. 2003. Le commerce des esclaves sur la côte swahili, 1500–1750. *Azania* 38 (1): 69–97.

———. 2006. Les réseaux de traite de l'Afrique orientale: côte swahili, Comores et nord-ouest de Madagascar (vers 1500–1750). *Cahiers des anneaux de la mémoire* 9: 67–107.

———. 2009. Slave Trade and Slavery on the Swahili Coast (1500–1750). In *Slavery, Islam and Diaspora*, ed. Paul Lovejoy, Behnaz A. Mirzai, and Ismael M. Montana, 37–76. Trenton, NJ: Africa World Press.

Verschuren, Dirk. 2004. Decadal and Century-Scale Climate Variability in Tropical Africa during the Past 2000 Years. In *Past Climate Variability through Europe and Africa*, ed. R.W. Battarbee et al., 139–158. Dordrecht: Kluwer.

Vidal, Jose Montero y. 1888. *Historia General de Filipinas desde el descubrimiento de diches Islas hasta nuestras dias*. Vol. 2. Madrid: Imprenta de M. Tello.

Virah-Samwy, M., K.J. Willis, and L. Gillson. 2010. Evidence for Drought and Forest Declines during the Recent Mega Faunal Extinctions in Madagascar. *Journal of Biogeography* 37 (3): 506–519.

Waal, Alex de. 2005. *Famine that Kills: Darfur Sudan*. Rev. ed. Oxford: Oxford University Press.

Wallace, Ernest, and E. Adamson Hoebel. 1952. *The Comanches: Lords of the South Plains*. Norman: University of Oklahoma Press.

Wang Pu, ed. 1963. *Tang Hui Yao* [*Institutional History of Tang*]. Vol. 84. Taipei: World Book.

Wang, Xunming, et al. 2010. Climate Desertification, and the Rise and Collapse of China's Historical Dynasties. *Human Ecology* 38: 157–172.

Wanquet, Claude. 1981. *Histoire d'une révolution: La Réunion 1789–1803*. Vol. 1. Marseille: Jeanne Lafitte.

Ware, Rudolph T., III. 2011. Slavery in Islamic Africa. In *The Cambridge World History of Slavery: Volume 3, AD 1420–AD 1804*, ed. David Eltis and Stanley L. Engerman, 47–80. Cambridge: Cambridge University Press.

Warren, James Francis. 1981. *The Sulu Zone 1768–1898: The Dynamics of External Trade, Slavery and Ethnicity in the Transformation of a Southeast Asian Maritime State*. Singapore: Singapore University Press.

———. 1994. Chinese Prostitution in Singapore: Recruitment and Brothel Organisation. In *Women and Chinese Patriarchy. Submission, Servitude and Escape*, ed. Maria Jaschok and Suzanne Miers, 77–105. London: Zed Books.

———. 1998. *The Sulu Zone, the World Capitalist Economy and the Historical Imagination*. Amsterdam: VU University Press/CASA.

———. 2001. *Iranun and Balangingi: Globalisation, Maritime Raiding and the Birth of Ethnicity*. Quezon City: New Day Publishers.

Watts, Sheldon. 1999. British Development Policies and Malaria in India 1897–c. 1929. *Past & Present* 165 (1): 141–181.

Weber, Jacques. 2002. L'émigration indienne à la Réunion: 'contraire à la morale' ou 'utile à l'humanité'. In *Esclavage et abolition dans l'Océan Indien, 1723–1869*, ed. Edmond Maestri, 309–328. Paris: L'Harmattan.

Werne, Ferdinand. 1852. *African Wanderings; or, An Expedition from Sennaar to Taka, Basa, and Beni-Amer, with a Particular Glance at the Races of Bellad Sudan*. London: Longman, Brown, Green and Longmans.

White, Sam. 2010. Rethinking Disease in Ottoman History. *International Journal of Middle East Studies* 42 (4): 549–567.

Wilbur, C. Martin. 1943. *Slavery in China during the Former Han Dynasty 206 BC–AD 25*. New York: Russell & Russell.

Wingate, Francis Reginald. 1892. *Ten Years' Captivity in the Mahdi's Camp 1882–1892*. London: Sampson, Low, Marston and Co.

Wink, André. 1996. *Al-Hind: The Making of the Indo-Islamic World*. Leiden: Brill.

Wood, Gillen D'Arcy. 2008. The Volcano Lover: Climate, Colonialism, and the Slave Trade in Raffle's 'History of Java' (1817). *Journal for Early Modern Cultural Studies* 8 (2): 33–55.

Worden, Nigel. 1985. *Slavery in Dutch South Africa*. Cambridge: Cambridge University Press.

———. 1989. Adjusting to Emancipation: Freed Slaves and Farmers in the Mid-Nineteenth Century Western Cape. In *The Angry Divide: Social and Economic History of the Western Cape*, ed. W. James and M. Simons, 31–39. Cape Town: David Philip.

———. 2005. Indian Ocean Slavery and its Demise in the Cape Colony. In *Abolition and its Aftermath in Indian Ocean Africa and Asia*, ed. Gwyn Campbell, 29–49. London: Routledge.

———. 2007. Revolt in Cape Colony Slave Society. In *Resisting Bondage in Indian Ocean Africa and Asia*, ed. E. Alpers, G. Campbell, and M. Salman, 10–23. London: Routledge.

———. 2010. 'Armed with Swords and Ostrich Feathers': Militarism and Cultural Revolution in the Cape Slave Uprising of 1808. In *War, Empire and Slavery, 1770–1830*, ed. R. Bessel, N. Guyatt, and J. Rendall, 121–138. London: Palgrave.

———. 2013. Forgotten Revolutionaries: Slave Resistance at the Cape, 1760–1808. In *L'Atlantique révolutionnaire: une perspective ibéro-américaine*, ed. C. Thibaud et al., 421–446. Les Perséides: Bécherel.

Worden, Nigel, and Gerald Groenewald. 2005. *Trials of Slavery: Selected Documents Concerning Slaves from the Criminal Records of the Council of Justice at the Cape of Good Hope, 1705–1794.* Cape Town: Van Riebeeck Society.

Wright, Henry T. 2006. Early State Dynamics as Political Experiment. *Journal of Anthropological Research* 62 (3): 305–319.

Xu, Hai, Yetang Hong, and Bin Hong. 2012. Decreasing Asian Summer Monsoon Intensity after 1860 AD in the Global Warming Epoch. *Climatic Dynamics* 39 (7–8): 2079–2088.

Yajima, Hikoichi. 2006. *Kaiiki kara mita rekishi: Indo-you to Chicyuu-kai wo musubu kouryuu-shi.* Nagoya: Nagoya University Press.

Yang, Jiping. 1996. Tangdai de Nubi, Buqu Yu Tongpu, Jiaren, Jingren [Slaves, Private Retainers, and Servants, Family Persons, Pure Persons]. *Zhongguo Shi Yanjiu [Chinese History Studies]* 3: 53–63.

Yang, Anand A. 1998. *Bazar India: Markets Society and the Colonial State in Bihar.* Berkeley and Los Angeles: University of California Press.

Zewde, Bahru. 1991. *A History of Modern Ethiopia, 1855–1974.* Addis Ababa: Addis Ababa University Press.

Zhang, Zhibin, et al. 2010. Periodic Climate Cooling Enhanced Natural Disasters and Wars in China during AD 10–1900. *Proceedings of the Royal Society B* 277: 3745–3753.

Theses etc.

Banerjee, Sandeep. 2013. *Landscaping India: From Colony to Postcolony.* PhD dissertation, Syracuse University.

Courteau, Anaïs. 2005. *La canne à sucre et l'environnement à La Réunion – Revue bibliographique.* Rapport de stage de Maîtrise, Université de Franche-Comté. http://reunionmayotte.cirad.fr/ index.php/dr/content

Devi, S. Kamala. 2011. *History of Cyclones on the Coromandel Coast with Special Reference to Tamil Nadu, 1800–1900.* PhD dissertation, Bharathidasan University.

Géraud, Jean-François. 2002. *Des habitations sucreries aux usines sucrières, la 'mise en sucre' de l'île Bourbon, 1783–1848*. PhD dissertation, Université de la Réunion.

Grove, Craig Alexander. 2012. *Madagascar's Climate History Unlocked by Giant Corals*. PhD dissertation, Vrije Universiteit.

Kruse, Elaine Marie. 1983. *Divorce in Paris, 1792–1804: Window on a Society in Crisis*. PhD dissertation, University of Iowa.

La Rue, George Michael. 1989. *The Hakura System: Land and Social Stratification in the Social and Economic History of the Sultanate of Dar Fur (Sudan), ca. 1785–1875*. PhD dissertation, Boston University.

Mednick, Melvin. 1965. *Encampment of the Lake: The Social Organisation of a Moslem Philippine (Moro) People*. PhD dissertation, University of Chicago.

Nicholson, Sharon. 1976. *A Climate Chronology for Africa: Synthesis of Geological, Historical and Meteorological Information and Data*. PhD dissertation, University of Wisconsin-Madison.

Warnich, Pieter Gabriel. 1988. *Die Toepassing en invloed van slawewetgewing in die landdrosdistrik Tulbagh-Worcester, 1816–1830*. PhD dissertation, University of Stellenbosch.

Wastell, R.E.P. 1944. *British Imperial Policy in Relation to Madagascar, 1810–96*. PhD dissertation, London University.

Web References

Aga, Selim. 1846. *Incidents Connected with the Life of Selim Aga, a Native of Central Africa: Electronic Edition*, 20–23. http://docsouth.unc.edu/neh/aga/aga.html

Augnet. *Website of the Order of St Augustine*. http://www.augnet.org/?ipageid=685

Bureau of Meteorology. *Guide to Tropical Cyclone Forecasting*. http://www.bom.gov.au/cyclone/about/names.shtml

Centre canadien de prévision d'ouragan. *Comment catégorise-t-on les ouragans*?http://www.ec.gc.ca/ouragans-hurricanes/default.asp?lang=Fr&n=AB062B74-1

Damodaran, Vinita, Rob Allen, and James Hamilton. 2015. *Climate Signals, Environment and Livelihoods in C17th India in a Comparative Context*. http://foodsecurity.exeter.ac.uk/wp-content/uploads/2015/09/17th-century.pdf. Accessed 30 August 2017.

Daocheng. Shishi Yaolan [Manual of Buddhist Practices], *Dazheng Xinxiu Dazang Jing* [*Taisho Tripitaka*], 54, No. 2127. Last modified 15 June 2016. Chinese Buddhist Electronic Text Association. http://www.cbeta.org/result/normT54/2127.003.htm. Accessed 8 August 2017.

Games, Alison. 2003. Oceans, Migrants, and the Character of Empires: English Colonial Schemes in the Seventeenth Century. *History Cooperative*. http://webdoc.sub.gwdg.de/ebook/p/2005/history_cooperative/ www.historycooperative.org/proceedings/seascapes/games.html

Greshko, Michael. 2016. 201 Years Ago, This Volcano Caused a Climate Catastrophe. *National Geographic*, 8 April. http://news.nationalgeographic.com/2016/04/160408-tambora-eruption-volcano-anniversary-indonesia-science/

Kumarajiva. Foyi Jiaojing Lunshu Jieyao [Extracted Commentary of Sutra on the Buddha's Bequeathed Teaching], *Dazheng Xinxiu Dazang Jing* [*Taisho Tripitaka*] 40, no. 1820. Last modified 15 June 2016. Chinese Buddhist Electronic Text Association. http://www.cbeta.org/result/normal/T40/1820_001.htm. Accessed 8 August 2017.

La Rue, George Michael. 2016. Land documents in Dār Fūr sultanate (Sudan, 1785–1875): Between Memory and Archives. *Afriques*. http://afriques.revues.org/1896

Lambert, J. 2005. *Base de données SisFrance 'Océan Indien' et site Internet associé*. BRGM/RP 53711-FR (March), 25. http://www.planseisme.fr/IMG/pdf/rapport_SisFrance_ ocean_indien.pdf

Laughton, J.K. 2004. Keeling, William (1577/8–1620). Rev. G.G. Harris, *Oxford Dictionary of National Biography*. Oxford: Oxford University Press. http://www.oxforddnb.com/view/article/15243

McCullough, P.E. Buckeridge, John (d. 1631), Bishop of Ely. *Oxford Dictionary of National Biography*. http://www.oxforddnb.com/ view/printable/3854

"Ring of Fire". 2009. *U.S. Geological Survey Earthquakes Visual Glossary*. http://earthquake.usgs.gov/learn/glossary/?term=Ring of Fire

Seager, Richard. 2006. *Blueprints for Medieval Hydroclimate*. http://www.ldeo.columbia.edu/res/div/ocp/pub/seager/Seager_etal_QSR.pdf. Accessed 28 August 2017.

Zhipan. n.d. Fozu Tongji [A Chronicle of Buddhas and the Patriarchs]. *Dazheng Xinxiu Dazang Jing (Taisho Tripitaka)* vol. 49, no. 2035 (Chinese Buddhist Electronic Text Association). http://www.cbeta.org/result/normal/T49/2035_004.htm

INDEX[1]

[1]Note: Page number followed by 'n' refers to notes.

G. Campbell (ed.), *Bondage and the Environment in the Indian Ocean World*, Palgrave Series in Indian Ocean World Studies,
https://doi.org/10.1007/978-3-319-70028-1

Q

R

T

Z

The manufacturer's authorised representative in the EU is Springer Nature Customer Service Centre GmbH, Europaplatz 3, 69115 Heidelberg, Germany. If you have any concerns regarding our products, please contact ProductSafety@springernature.com

Printed and bound by CPI Group (UK) Ltd, Croydon, CR0 4YY
25/09/2026
02222890-0006